T0323363

Human Migration

Human Migration

Biocultural Perspectives

Edited by

MARIA DE LOURDES MUÑOZ-MORENO AND
MICHAEL H. CRAWFORD

OXFORD
UNIVERSITY PRESS

Oxford University Press is a department of the University of Oxford. It furthers the University's objective of excellence in research, scholarship, and education by publishing worldwide. Oxford is a registered trade mark of Oxford University Press in the UK and certain other countries.

Published in the United States of America by Oxford University Press
198 Madison Avenue, New York, NY 10016, United States of America.

Library of Congress Control Number: 2021941963
ISBN 978-0-19-094596-1

DOI: 10.1093/oso/9780190945961.001.0001

1 3 5 7 9 8 6 4 2

Printed by Integrated Books International, United States of America

Contents

III. REGIONAL MIGRATION

IV. CULTURE AND MIGRATION

V. DISEASE AND MIGRATION

Preface

The first international conference on migration was held in Lawrence, Kansas, on March 1–2, 2010. The primary emphasis of the conference was the consequences of worldwide cultural diversity and evolution for migration, particularly from a molecular perspective. More than 100 scholars from 12 countries and 24 institutions participated in the conference. A volume based on the proceedings of the conference, *Causes and Consequences of Human Migration: An Evolutionary Perspective,* was published by Cambridge University Press in 2012.

This new volume on migration, *Human Migration: Biocultural Perspective,* published by Oxford University Press, compiles 21 chapters focusing on what research on genomic and cultural diversity can tell us about the process of migration. The chapters were written by participants in an interdisciplinary conference on human migration held at the Center for Research and Advanced Studies of the National Polytechnic Institute (CINVESTAV-IPN) in Mexico City from October 17 to 21, 2017. The conference was organized by Maria de Lourdes Muñoz-Moreno and was funded by the following programs and Institutions: CINVESTAV-IPN, the Instituto Politécnico Nacional, the University of Kansas (Lawrence), Consejo Mexiquense de Ciencia y Tecnología EdoMex, Red Mexicana de Virologia, Red Temática: Envejecimiento, Salud y Desarrollo Social, Instituto Nacional de Antropología e Historia, and DNS Microgenomic S.A. de C.V. Inc. More than 200 scholars from 11 different countries and 27 institutions attended the formal portion of the 4-day conference. In addition, the conference included a strong multidisciplinary plenary approach to human migration with the participation of anthropologists, molecular anthropologists, archaeologists, molecular geneticists, and human biologists.

In this volume, migration is defined as the mass directional movements of large numbers of species and populations from one location to another. Communities that respond to changes in resource availability or habitat quality over the course of time are included. From the earliest human existence to contemporary times, migration (voluntary or involuntary) has been a basic survival strategy. Humans migrate to search for work, land, education, safety, and new opportunities. Migration research recognizes the effects of culture, language, genetics, law, economics, and the environment. The primary aim of this book is to examine the causes and consequences of ancient and contemporary human migration from a multidisciplinary perspective. This includes hominin spread from Africa to the far reaches of the world. Human dispersion through time resulted in the transfer of genetic variation from one population to another (gene flow), genetic admixture, and genetic drift in smaller populations. Sociocultural consequences of migration include language transfers and dispersion of diseases like tuberculosis, dengue fever, chikungunya, Zika, and West Nile virus infection (as well as disease transmission by mosquito vectors like *Aedes aegypti*, *A. albopictus*, and *Culex* spp.). Importantly, leading academic scientists, researchers, and research scholars have contributed to this volume their vast experience and knowledge of the broad range of genomic studies on migration and culture as well as social structures of the past and present. Therefore, this book will aid a better understanding of human migration from ancient times until the present, and of migration as a major contributor to globalization, and it will serve as an excellent tool for researchers, professionals, and students from different study areas.

The organizers of the second conference plan to continue the exploration of the causes and consequences of migration with another international meeting and a third volume, this time emphasizing the causes and consequences of human population movements in Africa. The forthcoming conference, to be held in Morocco, is being organized by Dr. Abdelmajid Hannoum (University of Kansas).

Contributors

Enrique Alcalá-Castañeda
Department of Archaeologic Studies, Instituto Nacional de Antropología e Historia, Mexico City, Mexico.

Sarah Alden
Laboratory of Biological Anthropology, University of Kansas, Lawrence, KS, USA

A. S. Barashkova
Scientific-Research Institute of Regional Economy of the North, North-Eastern Federal University, Yakutsk, Russia.

Kristine G. Beaty
Laboratory of Biological Anthropology, University of Kansas, Lawrence, KS, USA; Laboratory of Anthropology, University of Oklahoma, Norman, OK, USA.

Noel Boaz
Anatomy Laboratory, Emory & Henry College School of Health Sciences, Marion, VA, USA.

Michael H. Crawford
Laboratory of Biological Anthropology, University of Kansas, Lawrence, KS, USA.

Randy E. David
Laboratory of Biological Anthropology, Department of Anthropology, University of Kansas, Lawrence, KS, USA

Bartholomew Dean
Laboratory of Biological Anthropology, Department of Anthropology, University of Kansas, Lawrence, KS, USA

Alvaro Diaz-Badillo
Department of Human Genetics & South Texas Diabetes and Obesity Institute, The University of Texas Rio Grande Valley, McAllen, TX, USA.

A. G. Egorova
Yakut Science Centre of Complex Medical Problems, Yakutsk, Russia.

Constanza de la Fuente
Department of Human Genetics, University of Chicago, Chicago, IL, USA.

R. Gómez
Department of Toxicology, Centro de Investigación y de Estudios Avanzados del IPN, Mexico City, Mexico.

Michael F. Hammer
Interdisciplinary Program in Statistics, University of Arizona, Tucson, AZ, USA.

Abdelmajid Hannoum
Department of Anthropology, University of Kansas, Lawrence, KS, USA.

I. King Jordan
School of Biological Sciences, Georgia Institute of Technology, Atlanta, GA, USA; IHRC-Georgia Tech Applied Bioinformatics Laboratory (ABiL), Atlanta, GA, USA; PanAmerican Bioinformatics Institute, Cali, Valle del Cauca, Colombia.

Tatiana M. Karafet
University of Arizona, Tucson, AZ, USA.

W. Leonard
Anthropology Department, Northwestern University, Evanston, IL, USA.

Franz Manni
CNRS UMR 7206, Département «Homme et Environnement», Musée de l'Homme, National Museum of Natural History, University of Paris-Diderot, Paris, France.

Adrián Martínez Meza
Department of Physical Anthropology, Instituto Nacional de Antropología e Historia, Mexico City, Mexico.

M. A. Meraz-Ríos
Department of Toxicology, Centro de Investigación y de Estudios Avanzados del Instituto Politécnico Nacional, Mexico City, Mexico.

Brent E. Metz
Department of Anthropology, University of Kansas, Lawrence, KS, USA.

Igor Mokrousov
Laboratory of Molecular Epidemiology and Evolutionary Genetics, St. Petersburg Pasteur Institute, St. Petersburg, Russia.

J. Víctor Moreno-Mayar
Lundbeck Foundation GeoGenetics Centre, Globe Institute, University of Copenhagen, Copenhagen, Denmark.

Maria de Lourdes Muñoz-Moreno
Department of Genetics and Molecular Biology, Centro de Investigación y de Estudios Avanzados del Instituto Politécnico Nacional, Mexico City, Mexico.

María Teresa Navarro-Romero
Department of Genetics and Molecular Biology, Centro de Investigación y de Estudios Avanzados del Instituto Politécnico Nacional, Mexico City, Mexico.

John Nerbonne
Germanistische Linguistik, Albert Ludwigs University of Freiburg, Freiburg im Breisgau, Germany.

Lauren Norman
Department of Anthropology, University of Kansas, Lawrence, KS, USA.

Emily T. Norris
School of Biological Sciences, Georgia Institute of Technology, Atlanta, GA, USA; IHRC-Georgia Tech Applied Bioinformatics Laboratory (ABiL), Atlanta, GA, USA; PanAmerican Bioinformatics Institute, Cali, Valle del Cauca, Colombia.

Dennis H. O'Rourke
Department of Anthropology, University of Kansas, Lawrence, KS, USA.

Mirna Isabel Ochoa-Lugo
Department of Genetics and Molecular Biology, Centro de Investigación y de Estudios Avanzados del Instituto Politécnico Nacional, Mexico City, Mexico.

Ludmila P. Osipova
Institute of Cytology and Genetics, Siberian Branch of the Russian Academy of Sciences, Novosibirsk, Russia.

Gerardo Pérez-Ramírez
Departement of Genetics and Molecular Biology, Centro de Investigación y de Estudios Avanzados del Instituto Politécnico Nacional, Mexico City, Mexico.

Christine Phillips-Krawczak
Laboratory of Biological Anthropology, University of Kansas, Lawrence, KS, USA.

Maanasa Raghavan
Department of Human Genetics, University of Chicago, Chicago, IL, USA.

Lavanya Rishishwar
School of Biological Sciences, Georgia Institute of Technology, Atlanta, GA, USA; IHRC-Georgia Tech Applied Bioinformatics Laboratory (ABiL), Atlanta, GA, USA; PanAmerican Bioinformatics Institute, Cali, Valle del Cauca, Colombia.

T. G. Schurr
Department of Anthropology, University of Pennsylvania, Philadelphia, PA, USA.

Mark Stoneking
Department of Evolutionary Genetics, Max Planck Institute for Evolutionary Anthropology, Leipzig, Germany.

S. A. Sukneva
Scientific-Research Institute of Regional Economy of the North, North-Eastern Federal University, Yakutsk, Russia.

Justin Tackney
Department of Anthropology, University of Kansas, Lawrence, KS, USA.

Larissa Tarskaia
Laboratory of Biological Anthropology, University of Kansas, Lawrence, KS, USA.

Eladio Terreros-Espinosa
Templo Mayor Museum, Instituto Nacional de Antropología e Historia, Mexico City, Mexico.

Introduction

What Can Research on Genomic and Cultural Diversity Tell Us About the Process of Migration?

Maria de Lourdes Muñoz-Moreno and Michael H. Crawford

Human migration is a major contributor to globalization that facilitates gene flow and the exchange of culture and ideas. This edited volume compiles research on many aspects of migration, population development, and human genetics. It also provides an interdisciplinary platform for researchers, professionals, and students from different fields to review and discuss the most recent innovations and trends, as well as practical challenges encountered in the fields of migration, demography, and human genetics. Leading academic scientists, researchers, and research scholars contributed to this volume based on their vast experience and knowledge of the broad range of genomic studies on migration, culture, and social structures in the past and present. The volume first presents a theoretical overview of the genomic evidence related to the out-of-Africa dispersal of modern humans (Chapter 1), followed by investigations into human migration using ancient mitochondrial DNA (mtDNA) sequences distributed from the Arctic to South America. The evolutionary consequences of the settlement of the Aleutian Islands are discussed, indicating the loss of genetic variability stimulated by population fission and founder effect. These conclusions are confirmed by data demonstrating the loss of mitochondrial haplogroup B in contemporary Aleuts, although it was present in prehistoric Aleut populations. Crawford et al. (Chapter 2) discuss a key factor, the genetic consequence of kin migration, which caused rapid genetic differentiation of the more distant island populations along the Aleutian insular chain.

Regional Dimensions of Migration

The next section includes chapters on regional migration in Samoyedic-speaking populations from Siberia, early human migrations in Gabon Africa and the Republic of Sakha (formerly, Yakutia), and African migration to Europe during the 21st century. Several chapters discuss human adaptive evolution, the influence of admixture, the high endogamy index of the Yakut (Sakha) and the Russian population of

María de Lourdes Muñoz-Moreno and Michael H. Crawford, *Introduction* In: *Human Migration*. Edited by: Maria de Lourdes Muñoz-Moreno and Michael H. Crawford, Oxford University Press. © Oxford University Press 2021. DOI: 10.1093/oso/9780190945961.003.0001 the Republic of Sakha (Yakutia) (Chapter 9), and the Y-chromosome diversity in Aztlan descendants associated with the history of Central Mexico (Chapter 10). The volume also focuses on how human migration is influenced by cultural practices. Biocultural approaches to migration and urbanization in the Peruvian Amazonia and the Ch'orti' Maya diaspora in search of fertile forests and political security are discussed. Evidence of human migration in the Puyil cave (Puxcatán, Tabasco) and migration of the Zoques to the mountain region of Tabasco (from linguistic and archaeological perspectives) are also considered. Two chapters document the effect of the migration of specific populations on the geographic distribution of diseases like dengue and mycobacterial infections.

Theoretical Overview

Mark Stoneking's chapter examines theories on the origin and earliest migration and expansion of anatomically modern humans, such as how the "out-of-Africa" hypothesis (OOA) is in agreement with the more recent single-origin hypothesis (RSOH), in conjunction with other theories on the origin and the earliest migration of anatomically modern humans (*Homo erectus*). Stoneking (Chapter 1) discusses the theories and presents arguments in support of one of the models that best describes the origins of our species (i.e., the recent African origin model), followed by assimilation of archaic humans as modern humans dispersed across and out of Africa. Although numerous questions remain, the chapter focuses on one significant question: Was there a single dispersal or multiple dispersals of modern humans from Africa? Stoneking weighs the evidence derived from genomic research.

Stoneking's chapter is followed by an overview by Maanasa Raghaven (Chapter 3) on the early peopling of the Americas. Her chapter includes the genomic evidence for human migration from the Bering Strait to South America, thousands of years before Clovis—the earliest widespread cultural manifestation south of the glacial ice. In the Americas, the evidence from human remains documents the path that the Native American ancestors followed to reach South America. Two of the chapters in this volume include ancient DNA evidence from Native Americans. The chapter by Dennis H. O'Rourke and his colleagues (Chapter 4) focuses on migration in the Arctic, and the chapter integrates genomic, archaeological, and paleo-ecological data. The authors discuss the timing and mode of arrival of the first people into the Western Hemisphere. The current understanding of the early peopling of the Americas is summarized. This includes the sources, numbers, timings, and routes of the early migrations into the Americas, mainly in light of recent genome-wide ancient DNA studies using diversification models of the founder population(s) and the past population structure, including ancient DNA from the continents.

The Maya populations have some of the most influential cultures of Mesoamerica. Muñoz et al. (Chapter 5) examined migration based on mtDNA sequencing of the hypervariable region from ancient osseous remains found in different Maya archaeological sites. This research suggests that communication occurred among the Maya populations represented by different archeological sites and that there was migration from north to south and from south to north. Navarro et al. also discussed recent studies of the genetic origins of pre-Hispanic human remains from Puxcatán, Tacotalpa, Tabasco, Mexico (Chapter 6).

Local migration of different populations is highly significant from an evolutionary perspective, and the migrations of the Samoyedic-speaking populations from Siberia were reconstructed from a genetic perspective. Tatiana Karafet et al. (Chapter 7) examined 567,096 autosomal genome-wide SNPs and 147 Y-chromosome data from 15 Siberian and 12 reference populations to assess the affinities of Siberian populations and to address hypotheses on the origin of the Samoyedic peoples.

Linguistic data have been utilized to reconstruct patterns of human migration. In their chapter, Franz Manni and John Nerbonne (Chapter 8) focus on the early migrations of Bantu-speaking populations from Gabon, Africa, based upon linguistic diversity. They suggest that the first Bantu-speaking groups spoke KOTA-KELE (B20) languages. The other varieties concern four different immigration waves (B10; B30; B40; B50-B60-B70—Guthrie nomenclature) that penetrated Gabon later in history.

The chapter by Norris et al. addresses how rapid, adaptive evolution was accelerated by admixture in the Americas. The authors present evidence for the contribution of gene flow between archaic and modern human populations and the admixed populations in the Americas (formed relatively recently via admixture among African, European, and Indigenous American ancestral populations) in driving rapid, adaptive evolution in human populations (Chapter 10).

The chapter on local migration by R. Gómez and her colleagues focuses on Y-chromosome diversity in descendants of populations from Aztlan and its implications for the history of Central Mexico (Chapter 11). The authors examined Y-chromosome variation in 1,614 Mexican mestizos from different geographic regions to delineate the indigenous and colonial history of Mexico. Their findings reveal considerable diversity of paternal lineages within Mexican males, as well as a limited number of shared haplotypes among them.

The chapter by Crawford et al. (Chapter 12) examines a series of local migrations and relocations of the Garifuna (Black Caribs) from St. Vincent to the Bay Islands and to the coast of Honduras. The migrants from St. Vincent were forcibly relocated by the British and intermixed in Central America with Creole and Native American populations. This triracially admixed population resulted in high levels of genetic diversity and a relatively high incidence of hemoglobinopathies (Hb S and C), which gave an evolutionary advantage to these populations in malarial environments when compared to the hemoglobin monomorphic Native American populations. The establishment of this Afro-Native American hybrid population resulted in high fertility and massive geographical expansion of Garifuna along the coast of Central America.

Sociocultural Dimensions of Migration

The comparison of migration studies using sociocultural dimensions is important in detecting the effects of cultural variables resulting from unprecedented flows of contemporary human migration. This volume has five chapters that focus on migration consequences based on sociocultural anthropology. The authors attempt to answer questions like: Why do African youth migrate? Even taking into account the great danger and high risk associated with the migration from North Africa to Europe, why do so many youth migrate? They ignore the risks of migration because of the religious belief that death is a destiny awaiting us all, and its timing is not up to us. This belief mirrors part of their culture and is discussed by Abdelmajid Hannoum in his chapter, "Out of Africa, Again: African Migration to Europe in the Twenty-first Century (Chapter 13)."

The chapter by Randy David and Bartholomew Dean focuses on contemporary Amazonian population mobility and urbanization in Peru in the historically important urban centers, such as Yurimaguas and the Huallaga Valley. They describe the influence of population adaptation to diverse ecosystems, political-economic constraints and opportunities, novel infrastructure, changing lifeways, and violence and social upheaval in shaping the migratory histories of the people in those urban centers. The chapter integrates genetic and cultural dimensions of the migration of populations into an urban setting (Chapter 14).

In his chapter, Brent Metz explains the causes of migration to and from the Ch'orti' Maya area located on the borders of Guatemala, Honduras, and El Salvador. The longitudinal research for this history was based on ethnography, ethnohistory, review of official documents, and a compilation of secondary historical, demographic, and genetic sources. Metz (Chapter 15) shows how immigrants like Ch'orti's, Pipils, Spaniards, and Ladinos have been attracted to the region for its minerals and agricultural potential. He also explains the immigrants' reasons for leaving their residences, which include flight from oppression and economic exploitation, lack of opportunities, epidemics, climate change, crime, and rapidly expanding populations.

Enrique Alcalá-Castañeda (Chapter 16) presents archaeological evidence of human migration over time found in the Puyil cave, Puxcatán, Tabasco, where osseous materials from different time periods have been discovered in conjunction with an assortment of cultural offerings. In addition, skulls with different forms of deformation have been identified, suggesting that some practices probably started earlier than was originally thought.

The last chapter of this section, by Eladio Terreros-Espinosa, describes the migration of the Zoques to the mountain region of Tabasco. The discussion focuses on the linguistic and archaeological evidence. Linguistic evidence suggests that the Proto-Mixe-Zoque speakers from several centuries BC were among the first foreign groups to migrate to Tabasco, and they merged with the local inhabitants. In addition, the author presents analysis of pre-Hispanic pottery recovered in this region and proposes a chronology from the Early Preclassic to the Protoclassic period, continuing into the Late Terminal-Classic through the Late Postclassic period (Chapter 17).

Migration and Disease

The last section of the book discusses the relationship between human migration and disease, which has been increasing recently due to greater in-migration flow toward specific regions, such as North America and Europe. This form of migration has increased the risk that the native populations will be infected by new pathogenic microorganisms, because the countries fail to implement a good surveillance system to detect emergent and re-emergent infections, and because of faulty response of the healthcare systems, poor vector control, and lack of education programs to inform the population about strategies to avoid infections. Therefore, infectious diseases associated with population mobility represent an important cause of morbidity and mortality and the geographic establishment of specific infectious diseases. This section has two chapters. The first, by Alvaro Diaz-Badillo and Maria de Lourdes Muñoz-Moreno (Chapter 18), focuses on the impact of human migration and the spread of arboviral diseases on the border of the United States and Mexico. They establish how virus dispersal on the border is associated with the distribution of the mosquito vectors *Aedes aegypti* and *A. albopictus*.

The second chapter, by Igor Mokrousov, examines the impact of massive migration on the spread of *Mycobacterium tuberculosis* strains. Mokrousov included in his study the *M. tuberculosis* strains resistant to antibiotics and the distribution of the most important strains and diseases throughout the world (Chapter 19).

In summary, the various approaches to human migration included aspects as population development, human genetics, archaeology, anthropology, biology, linguistics, diseases, a broad range of genomic studies, and cultural and social structures in the past and present. They also explain human migration as a major contributor to globalization that facilitates gene flow and exchange in ideas and culture.

PART I
THEORETICAL OVERVIEW

1

Genomic Insights into the Out-of-Africa Dispersal(s) of Modern Humans

Mark Stoneking

Background

A question that occupied a great many of us for many years was how our species, modern humans, originated. While many different ideas have been proposed over the years, beginning in the 1980s, three major hypotheses were hotly debated, all based on fossil and archaeological evidence (for more detail, see Stoneking 2008). The evidence showed that all human evolution took place in Africa until ~ 2 million years ago (mya), when fossils and stone tools attributed to one or more species of *Homo* begin to spread outside Africa. The various groups of pre-modern *Homo*, referred to here as "archaic humans," became widespread both outside and within Africa, and the three different hypotheses about modern human origins (multiregional evolution, recent African origin with replacement, and recent African origin with assimilation) differ in the extent to which archaic humans inside vs. outside Africa would have directly participated in human origins. The multiregional evolution (MRE) hypothesis is based primarily on claims of regional continuity in the fossil record, which holds that there are features in the skeletons of modern humans living in a particular region of the Old World that are characteristic of archaic humans from that same region, and so modern humans in Africa, Europe, Asia, etc., are descended primarily from the archaic humans in Africa, Europe, Asia, etc., respectively. However, modern humans are clearly one interbreeding species, and so they could not have evolved independently in different parts of the Old World. Therefore, to account for the traits that all modern humans share, MRE invokes extensive gene flow and selection. Gene flow (migration) is necessary to spread the traits that all modern humans share across the Old World, while selection is necessary to maintain regional continuity despite persistent gene flow, which should erase (or at least minimize) regional differences. According to MRE, then, regardless of where a particular trait or mutation originally arose, it was spread quickly by gene flow to other populations, and so all of the Old World populations of archaic humans—spread across thousands and thousands of kilometers, from sub-Saharan Africa, to western Europe, to eastern Asia and Australasia—evolved in concert by this complicated interplay of gene flow and regional selection to become modern humans.

Mark Stoneking, *Genomic Insights into the Out-of-Africa Dispersal(s) of Modern Humans* In: *Human Migration.*
Edited by: Maria de Lourdes Muñoz-Moreno and Michael H. Crawford, Oxford University Press. © Oxford University Press 2021.
DOI: 10.1093/oso/9780190945961.003.0002

In contrast to MRE, which holds that archaic humans who were spread across all the Old World by 1–2 mya were involved in the origin of modern humans, the recent African origin (RAO) hypothesis holds that the transformation from archaic to modern humans occurred in a single place, Africa, and that this occurred relatively recently, 250–300 thousand years ago (kya) or so. Although the fossil evidence used to support the RAO is the same as that used to support MRE, obviously the interpretations are different: whereas proponents of MRE pointed to skeletal features that they interpreted as indicating regional continuity in the fossil record, proponents of RAO instead pointed to skeletal features that they interpreted as indicating that "modern humans" first appeared in Africa, and that the earliest modern human fossils outside Africa are more similar to early modern Africans than to archaic non-Africans.

RAO comes in two flavors: RAO with replacement (hereafter referred to as "replacement") and RAO with assimilation (hereafter referred to as "assimilation"). The replacement hypothesis holds that, after modern humans originated in Africa, they spread across and out of Africa, and completely replaced all the archaic non-African humans they may have encountered, without any interbreeding. By contrast, the assimilation hypothesis agrees with the replacement hypothesis that modern humans originated in and spread from Africa, but it differs in proposing that some interbreeding took place between the modern humans coming from Africa and the archaic non-Africans. There are a variety of assimilation hypotheses, varying in the amount of interbreeding invoked and where it did or did not occur, but they all agree on the most important distinction from the replacement hypothesis—that at least some archaic non-Africans contributed some ancestry to some modern non-African populations.

It is important to emphasize that the three hypotheses—MRE, RAO with replacement, and RAO with assimilation—were all first proposed based solely on fossil and archaeological evidence. However, it proved impossible for such evidence to resolve the fierce debates over the merits of the three hypotheses; instead, genetic evidence provided the answer. This is not surprising, because the three hypotheses make different predictions about the genetic contributions of non-African archaic humans to modern humans. MRE explicitly rejects any special role for Africa in the origin of modern humans, and instead posits that all archaic human groups, both in and outside Africa, contributed to the process of becoming modern; replacement posits that all modern human ancestry is from Africa and there was no contribution from archaic non-Africans; and assimilation posits that the majority of modern human ancestry is from Africa, but there was still some (small) contribution from archaic non-Africans. Thus, to distinguish among the three hypotheses, one should turn to the genetic evidence.

The first genetic evidence to weigh in on the debate over modern human origins came from mitochondrial DNA (mtDNA). The seminal studies (Cann, Stoneking, and Wilson 1987; Vigilant et al. 1991) of human mtDNA variation found that the greatest diversity among mtDNA sequences occurred in Africa, with the variation outside Africa a subset of the variation within Africa; that phylogenetic trees indicated a likely African ancestor; and that a molecular clock approach estimated that it would have taken ~200 ky to produce all of the mtDNA variation present in contemporary populations. Although these initial studies were inevitably questioned, there is now little doubt that there was a recent African origin of modern human mtDNA (Ingman et al. 2000), and moreover there is no evidence of any contribution of archaic human mtDNA to contemporary populations (Krings et al. 1997). Studies of mtDNA were soon followed by high-resolution studies of Y-chromosome variation (Underhill et al. 2000), which came to the same conclusion, namely a recent African origin of human Y-chromosome variation. The single-locus studies were soon followed by genome-wide studies (Jakobsson et al. 2008; Li et al. 2008; Rosenberg et al. 2002); to make a long story short, genetic studies of contemporary populations have convincingly found an overwhelming signal of a recent African origin all across the genome. Genomic evidence thus firmly rejects MRE in favor of RAO. However, it was not possible to convincingly distinguish between the replacement vs. assimilation versions of RAO by studying only contemporary populations. Instead, it took technical advances in the extraction, sequencing, and analysis of ancient DNA from Neandertal bones to provide the final answer: there is a small (~ 2% on average) but definite contribution of Neandertal ancestry in all non-Africans (Green et al. 2010). Moreover, genomic analysis of the DNA obtained from an undiagnostic finger bone from Denisova Cave in southern Russia identified a previously unknown hominin group (Denisovans) who contributed ancestry to groups throughout Asia and Oceania, with the highest amounts (~ 4%–6%) observed in Australians and New Guineans (Reich et al. 2010, 2011). Thus, the genomic evidence from archaic humans firmly rejects RAO with replacement in favor of RAO with assimilation, which is now the generally accepted model for the origin of our species.

As is so often the case in science, once this supposedly important question about modern human origins was answered, the answer was thought to be trivial, and other questions that were deemed even more important were raised. One of these questions is: How many times did our ancestors leave Africa? While there are a variety of lines of evidence that could be used to address this question, the discussion here is limited to what the genomic evidence tells us.

Single vs. Multiple Dispersals

Rather than focusing on the actual number of dispersals, the question is usually phrased as follows: Was there a single dispersal of modern humans from Africa, or more than one dispersal? At first glance, this seems to be a rather ridiculous question, because, after all, if our ancestors could leave Africa once, surely they did it more than once. And yet, the initial genetic evidence pointed to just a single dispersal. The first genetic evidence for a single dispersal came from mtDNA; a wide variety of mtDNA lineages are found in Africa, whereas all mtDNA types found outside Africa fall into just two major lineages, called M and N, which in turn are derived from a single sublineage of a major African lineage called L3 (Behar et al. 2008; Torroni et al. 2006). Given that there are many branches of L3 in Africa, not to mention many other major mtDNA lineages in Africa (L0, L1, L2, etc.), each with many sublineages, then if there were multiple dispersals of modern humans from Africa, we should expect to see additional mtDNA lineages outside Africa. Moreover, there was rapid diversification of sublineages of M and N as modern humans spread outside Africa, which also suggests a single dispersal (Macaulay et al. 2005; Thangaraj et al. 2005). The Y chromosome shows a similar pattern (i.e., one major founding lineage for non-African Y chromosomes with a rapid radiation of non-African lineages) and therefore is also considered to support a single dispersal (Hallast et al. 2015; Karmin et al. 2015; Lippold et al. 2014). However, it should be pointed out that one Y-chromosome lineage, called E-M215, is thought to represent a late Pleistocene migration from Africa to the Middle East and southern Europe (Karmin et al. 2015; Lippold et al. 2014; Semino et al. 2004); the fact that there is no real counterpart to E-M215 in the mtDNA variation suggests that this was a primarily male-mediated dispersal.

Of course, studies of single genetic loci, such as mtDNA or the Y chromosome, suffer from the drawback that the history of a single locus can differ from that of a population or species because of selection or because of chance events. Nevertheless, initial studies of genome-wide variation also suggested a single dispersal of modern humans from Africa (Jakobsson et al. 2008; Li et al. 2008). For example, there is a decline in genetic diversity of populations that is closely correlated with their distance from East Africa, which is consistent with a model of a single dispersal of modern humans from Africa, followed by serial bottlenecks as humans moved further and further from Africa (Prugnolle, Manica, and Balloux 2005).

Still, there is a difference between stating that genetic evidence is "consistent" with a single dispersal and actually testing the predictions of single vs. multiple dispersal models to see if the data do provide significant support for one model over the other. One way to carry out such a test is to simulate genomic data, assuming a particular demographic history, and then to compare the simulated data to the observed data to see how closely they match. This can be done for different models of demographic history (e.g., single vs. multiple dispersals), as well as for different combinations of demographic parameters (divergence times, population-size changes, migration rates, etc.). The history that provides the best match between the simulated and observed data is then taken as the best model of the actual history of the populations. While this is a powerful approach that goes beyond storytelling to allow testing the predictions of different models, there is an important caveat to keep in mind: all models are by definition simplifications of reality, and therefore all models are, by definition, wrong. But some models are more wrong than others, and if we can identify the least wrong model, then we call that progress (with apologies to George Box).

One of the first attempts to apply this modeling approach to the question of single vs. multiple dispersals analyzed genome-wide SNP data from several human populations (Wollstein et al. 2010). The focus of the study was on Oceanian populations, and one of the questions that was addressed was whether New Guinea was initially colonized by people descended from the same dispersal from Africa as all other non-Africans (e.g., a single dispersal out of Africa), or instead by a separate, earlier migration from Africa (e.g., multiple dispersals of modern humans from Africa), as some had suggested (Lahr and Foley 1994). Three models were analyzed: (1) one dispersal of modern humans from Africa, followed by one migration to Asia and New Guinea; (2) one dispersal of modern humans from Africa, followed by separate migrations from this non-African source population to Asia and to New Guinea; and (3) an earlier dispersal of modern humans who went to New Guinea, followed by a later dispersal of the ancestors of other modern human populations in Europe and Asia. The modeling approach employed by the authors also accounted for the ascertainment bias that plagues SNP data obtained from commercial SNP arrays, which arises from how the SNPs are chosen for inclusion on the array. This is typically decided from data obtained from populations of European ancestry (e.g., European-Americans) and so the data overestimate diversity in European-related populations and underestimate diversity in more distantly related populations.

The results of the modeling procedure, based on a few million simulations, were that model 2 (single dispersal from Africa followed by separate dispersals to Asia and to New Guinea) received the highest support (~ 74%), model 1 (single dispersal from Africa to Asia and New Guinea) received the next highest support (~ 24%), and model 3 (separate dispersals from Africa to New Guinea and to Asia) received practically no support (~ 2%). So, model 3 is rejected, and while model 2 is the best model, model 1 still receives some support. Furthermore, at around the same time as the study by Wollstein et al., the first genome sequence from Neandertals was obtained, and it indicated that all non-Africans share a common signal of Neandertal ancestry (Green et al. 2010). This in turn suggests a single dispersal of modern humans from Africa, followed by admixture with Neandertals, followed by dispersals from this non-African source population to Europe, Asia, New Guinea, etc.—findings that are the same as the results suggested by Wollstein et al. (2010).

The Plot Thickens

Thus, by around 2010, analyses of uniparental markers, genome-wide data, and Neandertal ancestry were converging on the same story, namely a single dispersal of modern humans from Africa that took place around 50–80 kya. However, later two studies appeared that challenged this view, and instead found evidence for multiple dispersals from Africa, based on analyses of cranial data and of genome-wide SNP data (Reyes-Centeno et al. 2014; Tassi et al. 2015). The genomic data analyzed, assembled from a variety of published sources, indicated that the ancestors of Australians and New Guineans diverged from East Africans at least 88 kya, while the ancestors of Europeans diverged from East Africans at most 79 kya. These results cannot be explained by ascertainment bias, by a single dispersal followed by genetic drift in the ancestors of Australians and New Guineans, or by back-migration from Europe to Africa; moreover, the authors controlled for the extra archaic ancestry in Australians and New Guineans by masking Denisovan ancestry (more on this later). The authors thus concluded that their results support multiple dispersals of modern humans from Africa, with an early dispersal to Australia/New Guinea.

This was shortly followed by three studies that analyzed whole genome sequence (WGS) data (so, no issues with SNP ascertainment bias) from a variety of human populations and addressed the topic of single vs. multiple dispersals from Africa. One study produced WGS data from 83 aboriginal Australians and 25 Papuans (Malaspinas et al. 2016) and found that, when only modern human populations are analyzed, the best-fitting model invoked an earlier dispersal (~ 81 kya) of the ancestors of Australo-Papuans from Africa, followed by a later dispersal (~ 67 kya) of the ancestors of Eurasians. However, when archaic ancestry from Neandertals and Denisovans was added to the model, then the best-fitting model was a single dispersal (~ 72 kya) of the ancestors of all modern non-Africans from Africa. Thus, the extra archaic ancestry that Australo-Papuans have from Denisovans makes them appear more genetically different from Africans than Eurasians are, which then leads to an erroneous inference of an earlier divergence time if the archaic ancestry is not considered properly.

Another study, which produced WGS data for 300 individuals from around the world (Mallick et al. 2016), also included Neandertal and Denisovan ancestry in the modeling approach and came to the same conclusion, namely that the best-fitting model involved a single dispersal of the ancestors of all modern non-Africans from Africa. The third study, which produced WGS data for 379 individuals from around the world (Pagani et al. 2016), masked segments of archaic ancestry from Denisovans and also found strongest support for a single dispersal from Africa— plus something extra in Papuans. That is, they found that the ancestors of Papuans diverged from Africans ~ 15 kya before the ancestors of Eurasians diverged from Africans, but upon further investigation they did not find support for a model of multiple dispersals. Instead, they suggested that Papuans derive ~ 2% of their ancestry from a putative earlier migration of modern humans from Africa, one that left Africa ~ 120 kya and subsequently went extinct, but not before contributing to the ancestry of Papuans.

To further complicate matters, another study appeared that utilized yet another approach to investigate single vs. multiple dispersals (Wall 2017). This study analyzed WGS data from Pagani et al. (2016) and restricted the analyses to sites in the genome where archaic humans and Africans were homozygous for the ancestral allele, while the derived allele was restricted to non-Africans. It is quite likely that such variants arose after modern humans left Africa, and hence should not be confounded by archaic ancestry. The author then modeled a single dispersal from Africa, following Malaspinas et al. (2016), as well as a single dispersal from Africa with additional ancestry from an earlier dispersal from Africa in Papuans, following Pagani et al. (2016). The simulated data under the Malapsinas model was a significantly better match to the observed data, leading the author to conclude that the apparent extra ancestry in Papuans identified by Pagani et al. (2016) was residual Denisovan ancestry that was not adequately masked.

Other Evidence

The genomic evidence for single vs. multiple dispersals, as summarized above, seems to favor a single dispersal from Africa that occurred around 50–80 kya (give or take a few thousand years). By contrast, there is abundant and growing evidence from paleontology and archaeology for the presence of modern humans outside Africa well before this time (Bae, Douka, and Petraglia 2017; Groucutt et al. 2015). While some of this evidence is debatable, such as evidence for modern humans in China before 80,000 years ago (Liu et al. 2015), or even in the New World 130,000 years ago (Holen et al. 2017), some of it is not, such as evidence for modern humans in Israel by at least 177,000 years ago (Hershkovitz et al. 2018). Yet, with the possible exception of the Pagani et al. (2016) study, there is no evidence that these earlier non-Africans contributed to the ancestry of contemporary humans. However, a signal from these earlier dispersals does potentially exist in the genomes of Neandertals. The first high-coverage Neandertal genome sequence (Prufer et al. 2014) included about 1% modern human DNA sequence, which was originally attributed to contamination from humans during excavation, handling, or processing of the sample. However, subsequent in-depth analysis of the human DNA sequence in this Neandertal showed that it was not closely related to any contemporary human population, as would be expected of modern DNA contamination. Instead, it appears to be derived from a population that diverged from Africans before the out-of-Africa dispersal of the ancestors of contemporary non-Africans (Kuhlwilm et al. 2016). This "ghost" population mixed with Neandertals and subsequently went extinct, without leaving any genetic trace in contemporary populations. Recently, a second high-coverage Neandertal genome sequence confirmed the signal from the ghost population (Prufer et al. 2017). This is, to date, the best genomic evidence for modern humans outside Africa prior to the main out-of-Africa dispersal of the ancestors of all non-Africans.

Some Additional Thoughts

When it comes to using genomic evidence to address the question of single vs. multiple dispersals of modern humans out of Africa, one must take into consideration the Neandertal and (especially) Denisovan ancestry in different populations. Even so, there is a marked discrepancy between studies that either explicitly model or account for archaic admixture (e.g., Malaspinas et al. 2016; Mallick et al. 2016; Wall 2017), vs. those that attempt to mask archaic ancestry by identifying archaic genomic segments and removing them prior to analysis (e.g., Pagani et al. 2016; Tassi et al. 2015). The former approaches invariably support a single dispersal out of Africa, while the latter support multiple dispersals (or a single dispersal plus something else). A likely explanation for this discrepancy is that masking may fail to adequately identify and remove all the segments of Denisovan ancestry in Australo-Papuans; what remains is enough to cause Australo-Papuans to be more divergent from Africans than Eurasians are, leading to an inference of an older divergence time for Australo-Papuans and hence multiple dispersals.

What makes this explanation even more likely is the fact that the Denisovan genome sequence that we have in hand is quite different from the Denisovan DNA sequences that are found in contemporary Australo-Papuans; the divergence time between the Denisovan genome sequence and the admixed Denisovan DNA sequence is on the order of 300,000 years (Prufer et al. 2014). A vivid demonstration of the impact of relatedness on masking ability comes from the second high-coverage Neandertal genome sequence that was obtained (Prufer et al. 2017), from Vindjia Cave in Croatia. This Neandertal was more closely related to the Neandertals that admixed with the ancestors of contemporary non-Africans than was the first Neandertal to be sequenced to high coverage (from Denisova Cave in the Altai region of southern Russia); when the Vindjia Neandertal genome sequence was used to identify segments of Neandertal DNA in contemporary humans, an additional 10% to 20% Neandertal DNA was found, compared to the amount of Neandertal DNA inferred when the Altai Neandertal genome sequence was used for this purpose (Prufer et al. 2017). Thus, the more distant the relationship between the introgressing archaic humans and the actual archaic genome sequence in hand, the more difficult it is to reliably identify and mask the archaic DNA sequences in contemporary humans. It therefore seems quite likely that insufficient masking of the archaic DNA sequence explains the evidence for multiple dispersals in studies that attempted to mask Denisovan ancestry.

Conclusions

To summarize, there are several aspects of genomic variation that support a single dispersal from Africa of the ancestors of contemporary non-Africans. The limited mtDNA and Y-chromosome variation outside Africa, compared to the enormous variation within Africa, is not expected with multiple dispersals, as one would expect multiple dispersals to carry additional African mtDNA and Y-chromosome lineages. All non-African populations experienced a bottleneck (reduction in population size) associated with dispersal from Africa, and this bottleneck seems to be shared, i.e., all non-African populations experienced the same bottleneck, not independent bottlenecks (Lippold et al. 2014; Malaspinas et al. 2016; Mallick et al. 2016). All non-Africans have a shared signal of Neandertal ancestry (Green et al. 2010; Prufer et al. 2014, 2017); while there were undoubtedly additional contributions by Neandertals to the ancestors of specific populations (Vernot et al. 2016), underlying these different contributions is a signal of some specific Neandertal ancestry that is shared by all non-Africans. This shared ancestry is best explained by a single dispersal of modern humans from Africa, to some location where they met and admixed with Neandertals, followed by subsequent dispersals from this non-African source population to the rest of Eurasia, Oceania, and (ultimately) the New World.

At the same time, while the genomic evidence seems best explained by a single dispersal from Africa around 50 to 80 kya, there is growing, unequivocal fossil and archaeological evidence for the presence of modern humans outside Africa beginning at least ~180 kya. And yet, with the possible exceptions of a genetic contribution to Neandertals—and maybe to contemporary Papuans, although that doesn't seem so likely—we see no genomic evidence of these earlier dispersals of modern humans. Which begs the question: Why not? After all, our ancestors left Africa, met up with archaic humans like Neandertals and Denisovans, and mixed with them, and so we find traces of archaic DNA in contemporary humans today. Surely, if our ancestors also met up with the descendants of these earlier migrations from Africa, they would have also mixed with them, so why don't we see traces of these earlier modern human dispersals in the genomes of contemporary non-Africans? Potential explanations are either they met up and didn't mix, or the descendants of the earlier migrations had gone extinct by the time our ancestors were dispersing out of Africa; neither of these explanations seems very likely. Another possible explanation is that traces of the earlier modern human dispersals do exist in our genomes, but they are too subtle to detect with our current methods and data. After all, definitive proof of archaic human ancestry in contemporary genomes required having archaic human genome sequences in hand to analyze; maybe, if we are ever able to obtain genome sequences from the skeletal remains of these earlier modern humans, we will find that they actually did contribute some ancestry to contemporary populations. Until such time, we are left with the less-than-satisfactory conclusion that genomic evidence currently best supports a single dispersal of our ancestors out of Africa, but we can't completely rule out multiple dispersals.

Acknowledgments

The author thanks the organizers of the Human Migration Conference for the opportunity to present this work, and the members of the Department of Evolutionary Genetics at the Max Planck Institute for Evolutionary Anthropology for valuable discussion.

References

Bae, C. J., K. Douka, and M. D. Petraglia. 2017. "On the origin of modern humans: Asian perspectives." *Science* 358 (6368):eaai9067. doi: 10.1126/science.aai9067.

Behar, D. M., R. Villems, H. Soodyall, et al. 2008. "The dawn of human matrilineal diversity." *American Journal of Human Genetics* 82: 1130–1140.

Cann, R. L., M. Stoneking, and A. C. Wilson. 1987. "Mitochondrial DNA and human evolution." *Nature* 325: 31–36.

Green, R. E., J. Krause, A. W. Briggs, et al. 2010. "A draft sequence of the Neandertal genome." *Science* 328: 710–722.

Groucutt, H. S., M. D. Petraglia, G. Bailey, et al. 2015. "Rethinking the dispersal of *Homo sapiens* out of Africa." *Evolutionary Anthropology* 24: 149–164.

Hallast, P., C. Batini, D. Zadik, et al. 2015. "The Y-chromosome tree bursts into leaf: 13,000 high-confidence SNPs covering the majority of known clades." *Molecular Biology and Evolution* 32: 661–673.

Hershkovitz, I., G. W. Weber, R. Quam, et al. 2018. "The earliest modern humans outside Africa." *Science* 359: 456–459.

Holen, S. R., T. A. Demere, D. C. Fisher, et al. 2017. "A 130,000-year-old archaeological site in southern California, USA." *Nature* 544: 479–483.

Ingman, M., H. Kaessmann, S. Paabo, and U. Gyllensten. 2000. "Mitochondrial genome variation and the origin of modern humans." *Nature* 408: 708–713.

Jakobsson, M., S. W. Scholz, P. Scheet, et al. 2008. "Genotype, haplotype and copy-number variation in worldwide human populations." *Nature* 451: 998–1003.

Karmin, M., L. Saag, M. Vicente, et al. 2015. "A recent bottleneck of Y chromosome diversity coincides with a global change in culture." *Genome Research* 25: 459–466.

Krings, M., A. Stone, R. W. Schmitz, H. Krainitzki, M. Stoneking, and S. Paabo. 1997. "Neandertal DNA sequences and the origin of modern humans." *Cell* 90: 19–30.

Kuhlwilm, M., I. Gronau, M. J. Hubisz, et al. 2016. "Ancient gene flow from early modern humans into Eastern Neanderthals." *Nature* 530: 429–433.

Lahr, M. M., and R. A. Foley. 1994. "Multiple dispersals and modern human origins." *Evolutionary Anthropology* 3: 48–60.

Li, J. Z., D. M. Absher, H. Tang, et al. 2008. "Worldwide human relationships inferred from genome-wide patterns of variation." *Science* 319: 1100–1104.

Lippold, S., H. Xu, A. Ko, M. Li, G. Renaud, A. Butthof, R. Schroder, and M. Stoneking. 2014. "Human paternal and maternal demographic histories: insights from high-resolution Y chromosome and mtDNA sequences." *Investigative Genetics* 5: 13.

Liu, W., M. Martinon-Torres, Y. J. Cai, et al. 2015. "The earliest unequivocally modern humans in southern China." *Nature* 526: 696–699.

Macaulay, V., C. Hill, A. Achilli, et al. 2005. "Single, rapid coastal settlement of Asia revealed by analysis of complete mitochondrial genomes." *Science* 308: 1034–1036.

Malaspinas, A. S., M. C. Westaway, C. Muller, et al. 2016. "A genomic history of Aboriginal Australia." *Nature* 538: 207–214.

Mallick, S., H. Li, M. Lipson, et al. 2016. "The Simons Genome Diversity Project: 300 genomes from 142 diverse populations." *Nature* 538: 201–206.

Pagani, L., D. J. Lawson, E. Jagoda, et al. 2016. "Genomic analyses inform on migration events during the peopling of Eurasia." *Nature* 538: 238–242.

Prufer, K., C. de Filippo, S. Grote, et al. 2017. "A high-coverage Neandertal genome from Vindija Cave in Croatia." *Science* 358: 655–658.

Prufer, K., F. Racimo, N. Patterson, et al. 2014. "The complete genome sequence of a Neanderthal from the Altai Mountains." *Nature* 505: 43–49.

Prugnolle, F., A. Manica, and F. Balloux. 2005. "Geography predicts neutral genetic diversity of human populations." *Current Biology* 15: R159–R160.

Reich, D., R. E. Green, M. Kircher, et al. 2010. "Genetic history of an archaic hominin group from Denisova Cave in Siberia." *Nature* 468: 1053–1060.

Reich, D., N. Patterson, M. Kircher, et al. 2011. "Denisova admixture and the first modern human dispersals into Southeast Asia and Oceania." *American Journal of Human Genetics* 89: 516–528.

Reyes-Centeno, H., S. Ghirotto, F. Detroit, D. Grimaud-Herve, G. Barbujani, and K. Harvati. 2014. "Genomic and cranial phenotype data support multiple modern human dispersals from Africa and a southern route into Asia." *PNAS* 111: 7248–7253.

Rosenberg, N. A., J. K. Pritchard, J. L. Weber, H. M. Cann, K. K. Kidd, L. A. Zhivotovsky, and M. W. Feldman. 2002. "Genetic structure of human populations." *Science* 298: 2381–2385.

Semino, O., C. Magri, G. Benuzzi, et al. 2004. "Origin, diffusion, and differentiation of Y-chromosome haplogroups E and J: Inferences on the neolithization of Europe and later migratory events in the Mediterranean area." *American Journal of Human Genetics* 74: 1023–1034.

Stoneking, M. 2008. "Human origins. The molecular perspective." *EMBO Reports* 9 Suppl 1: S46–S50.

Tassi, F., S. Ghirotto, M. Mezzavilla, S. T. Vilaca, L. De Santi, and G. Barbujani. 2015. "Early modern human dispersal from Africa: genomic evidence for multiple waves of migration." *Investigative Genetics* 6: 13.

Thangaraj, K., G. Chaubey, T. Kivisild, A. G. Reddy, V. K. Singh, A. A. Rasalkar, and L. Singh. 2005. "Reconstructing the origin of Andaman Islanders." *Science* 308: 996.

Torroni, A., A. Achilli, V. Macaulay, M. Richards, and H. J. Bandelt. 2006. "Harvesting the fruit of the human mtDNA tree." *Trends in Genetics* 22: 339–345.

Underhill, P. A., P. Shen, A. A. Lin, et al. 2000. "Y chromosome sequence variation and the history of human populations." *Nature Genetics* 26: 358–361.

Vernot, B., S. Tucci, J. Kelso, et al. 2016. "Excavating Neandertal and Denisovan DNA from the genomes of Melanesian individuals." *Science* 352: 235–239.

Vigilant, L., M. Stoneking, H. Harpending, K. Hawkes, and A. C. Wilson. 1991. "African populations and the evolution of human mitochondrial DNA." *Science* 253: 1503–1507.

Wall, J. D. 2017. "Inferring human demographic histories of non-African populations from patterns of allele sharing." *American Journal of Human Genetics* 100: 766–772.

Wollstein, A., O. Lao, C. Becker, S. Brauer, R. J. Trent, P. Nurnberg, M. Stoneking, and M. Kayser. 2010. "Demographic history of Oceania inferred from genome-wide data." *Current Biology* 20: 1983–1992.

2

Unangan (Aleut) Migrations

Causes and Consequences

Michael H. Crawford, Sarah Alden, Randy E. David, and Kristine Beaty

This research is dedicated to the memory of Alice Petrivelli, Aleut elder, who introduced the authors to the Aleut communities and accompanied the research teams.

Introduction

Human populations have experienced a series of migrations and geographic expansions, with a genesis in Africa more than 200,000 years BP, to the far corners of Eurasia, the Americas, and Oceania. The causes and consequences of such migrations depended on ecological, climatic, social, demographic, and biological factors. This chapter focuses on the causes and consequences of Aleut (endonym: Unangan) migrations and settlements from Siberia, across Beringia, and ultimately to an archipelago consisting of 200 islands distributed over 1,800 km² of the Pacific, between the North American and Asian continents (Figure 2.1). The chronology of this migration was reconstructed using a synthesis of archaeological, geological, and molecular genomic evidence.

The earliest evidence of human habitation on the Aleutian archipelago is based on stratigraphy and radiocarbon (^{14}C) dating to approximately 9,000 to 8,000 years BP from sites located, respectively, on the eastern Fox Islands at Anagula and various Hog Island sites in Unalaska Bay (Davis and Knecht 2010). The more central islands of the Aleutian archipelago (Andreanof Islands) contain archeological sites dating to 6,000 years BP. The westernmost Aleutian Islands (Rat Islands, Attu, and Shemya) were settled comparatively later, some 3,000 to 2,000 years BP. According to evidence, the earliest Aleuts traversed Beringia more than 9,000 years BP and colonized what became known as the Aleutian Islands, starting from the Alaska Peninsula, in a westward direction, toward Kamchatka Peninsula, reaching Attu only ~ 3,000 years ago (West et al. 2007). Aleut populations failed to settle the westernmost islands of the archipelago, independently known as the Commander Islands (Bering and Mednii), and the northeastern Pribilof islands (St. George and

Michael H. Crawford, Sarah Alden, Randy E. David, and Kristine Beaty, *Unangan (Aleut) Migrations* In: *Human Migration*. Edited by: Maria de Lourdes Muñoz-Moreno and Michael H. Crawford, Oxford University Press. © Oxford University Press 2021. DOI: 10.1093/oso/9780190945961.003.0003

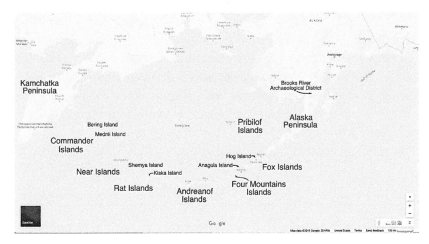

Figure 2.1 Map of the Aleutian Islands, eastern Siberia, and western Alaska (Crawford et al. 2010).

St. Paul). Subsequently, however, Russian nationals relocated Aleut hunters (ca. 1825–1830) from Umnak and Unalaska to the Pribilof and Commander Islands to harvest fur pelts for sale in Europe.

Causes of Migration

Archaeological evidence supports the westward expansion of the Unangan people from 9,000 years BP in the eastern islands, 6,000 years BP in the central islands, and 3,000 years BP in the westernmost islands of the Aleutian island chain. What drove the migration of Siberian peoples into the Americas? Why did the Unangans cross Beringia and then move in a westerly direction from island to island beginning at the Alaska Peninsula rather than migrating from the Kamchatka Peninsula in an easterly direction?

Population Dynamics

Given a relative scarcity of food and limited technological advancements, most bands of hunters and gatherers tend to exist in small subsistence units made up of extended families of 50 to 60 individuals (Crawford 1998). In populations migrating into new econiches with abundant resources, however, mortality rates decrease and fertility increases, resulting in group fission and further migration. There is evidence to suggest that historical Aleut migratory expansion fits the premise put forth by optimal foraging theory, which posits that organisms migrate according to the efficient identification, capture/harvest, and consumption of various local food resources (Winterhalder and Smith 1981). Aleuts maximized their net energy intake per unit time by colonizing new islands. Evidence of this tactic's being employed by migrating Aleuts is inferred from the earliest archaeological site on Shemya Island, where the largest fish were initially caught by the first migrants, followed by smaller fish and harder to catch birds. When Aleuts depleted their ideal resources, they relied on smaller, harder to obtain, and less calorically rich resources. Populations increased in size, fission occurred, and ultimately migration to other islands resulted, in an effort to acquire previously untapped hunting/foraging grounds.

Volcanic Eruptions

Evidence from the Anagula Blade archaeological site suggests that periodic volcanic eruptions destroyed food sources and forced the Aleut population to emigrate to adjacent islands. Blade tools excavated from Anagula were covered with 10 to 20 cm of volcanic ash, associated with an eruption that created the Okmok Caldera (McCartney et al. 1996). McCartney et al. have suggested that Anagula inhabitants fled to the Four Mountains Islands to escape the Okmok explosion. More recent volcanic activity from 2008, the Kasatochi Island eruption, demonstrated the destructive effects that major volcanic activity can have on the greater Bering Sea ecosystem, resulting in forced human relocations.

Climatic Changes

Climatic variation has played a significant role in Aleut migrations. Aleuts migrated during cold intervals from the eastern to the central region of the archipelago, 6,000 years BP (West and Crockford 2010). Furthermore, Aleuts migrated from the central islands to the western Near Islands of the archipelago ~ 4,000 years BP (West et al. 2007). Savinetsky et al. (2010) analyzed diatoms from natural peat bog deposits on Shemya Island and demonstrated that the western Aleutian Islands were settled during the coldest period of the Holocene, ca. 4,000 years BP. Peat bog analyses on Shemya Island revealed a climatic shift to a cooler and drier climate, again causing decreased cyclonic activity, less wind, and calmer seas. Cold climatic intervals coincided with increased biodiversity, bringing species that are not commonly seen in the region, including saffron cod (*Eleginus gracilis*), an extremely cold-tolerant fish. These colder but calmer conditions enabled long-distance interisland travel by umiaks to new islands in the far west (e.g., from Rat Island to the Near Islands, a distance of 221 km).

Culture, Contact, and Colonization

The earliest Russian contact with the Aleuts began in the 18th century with voyages of exploration by Vitus Bering and Aleksei Cherikov. The Commander Islands were accidentally "discovered" by Vitus Bering and crew when they were shipwrecked during a return voyage from the Americas (Jochelson 1933). After spending the winter on Bering Island, survivors returned to Siberia with a harvest of fur pelts, triggering a rush of Russian *promyshlenniki* (fur hunters) to the Commander Islands. Between 1824 and 1828, Aleuts from central and western islands were forcibly relocated to the Commander Islands to be employed by the Russian-American Company. As a result of this contact, the Aleut population was reduced via epidemics and warfare from an estimated 8,000 to 20,000 in the 17th century, to fewer than 2,000 (Reedy-Maschner 2010). The population of Aleuts in the Commander Islands fluctuated from 45 residents in 1825 to a maximum of 626 residents in 1892. In addition, Aleut males primarily from Unalaska were forcibly transplanted from 1825 to 1830 to the Pribilof Islands (St. George and St. Paul) to hunt fur seals.

In 1867, the Alaskan territory of Russia was purchased by the United States. All of the Aleutian Islands, except for the Commander Islands (Bering and Mednii) were included in the sale and came under U.S. political jurisdiction. The purchase consequently isolated western Aleuts from central and eastern groups, creating a political barrier to population movement and fluidity.

In 1943, during World War II, the Japanese Imperial Navy bombed Dutch Harbor in the eastern Aleutians and invaded the islands of Attu and Kiska to the west. Aleut residents and U.S. Naval weather observers were imprisoned in a prisoner-of-war internment camp on Hokkaido Island, Japan. Upon their release, the few Attuans who survived internment returned to the island of Atka, as well as mainland Alaska, leaving Attu unoccupied. Moreover, the American military relocated Aleuts living on six Aleutian islands to evacuation camps in southeastern Alaska. As a consequence of the aforementioned incidents, most of the islands of the Aleutian archipelago are currently uninhabited.

DNA Sampling and Analyses

In 1999, under the sponsorship of the National Science Foundation, we began an initial 8-year research program on the population genomics of the Aleutian Islands and Kamchatka, Siberia. Our research team sampled 267 Aleut volunteers from 11 island populations: Akutan, Atka, Bering, False Pass, King Cove, Nelson Lagoon, Nikolski, Pribilof Islands (St. George and St. Paul), Sand Point, and Umnak and from relocated Aleuts residing in Anchorage. Buccal swabs and/or whole blood samples were collected. DNA was then extracted, isolated, and analyzed for mitochondrial DNA (mtDNA) haplotypes and sequences, as well as nonrecombining Y-chromosome (NRY) markers (Crawford et al. 2010; Rubicz et al. 2003; Zlojutro et al. 2009).

Sampling of Aleut populations was further expanded in October of 2015 during a yearly meeting of Aleutian Trustees in Anchorage, Alaska. More than 200 representatives from 11 of the Aleutian Islands attended this meeting, and 115 volunteered to participate in the study. Through the support of a National Geographic Society grant and the availability of GENO 2.0 SNP chips, we were able to extract DNA and test for 750,000 single-nucleotide polymorphisms (SNPs) distributed throughout the genome. Gene by Gene Ltd. laboratories genotyped the DNA of the most recent study volunteers, plus 30 individuals from the earlier investigations on Bering Island.

Consequences of Migration

Maternal Migration

Of the five mtDNA haplogroups observed in Siberia and among indigenous Americans, only two, A and D, were detected in contemporary Aleut populations. The A7 subclade (A2a1A) observed in Aleut populations differs from the A3 subclade, commonly found among Na-Dene and Inuit populations, by a substitution specific to Aleuts, 16212 G (Zlojutro et al. 2006). The geographic distribution of mitochondrial haplogroups among Aleuts and others from the region is displayed in Figure 2.2 and is suggestive of a west to east gradient. In part, this gradient is exaggerated by the relocation of specific families bearing only D haplogroups to Bering and Mednii Islands in 1825, which was not part of the original settlement of the Aleutian archipelago. The relationship between geography and genetic markers was measured using Mantel tests (Mantel 1967) and was found to be highly significant ($r = 0.72$; $P < 0.000$), reflecting the maternal Aleut genetic structure derived from migration and settlement (Crawford 2007). Thus, the maternal genetic structure in Aleuts has indeed been preserved in mtDNA, as evidenced by the highly significant correlation between genetics and geographic distance.

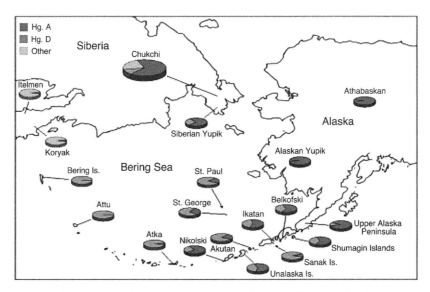

Figure 2.2 The Aleutian Islands showing the distribution of mtDNA haplogroup frequencies (Crawford 2007).

Reduction in Genetic Variability

Raff et al. (2010) found the B2 mitochondrial subclade to be present in 25% of skeletal remains from the Brooks River Archaeological District (Alaska Peninsula; see Figure 2.1). DNA transitions at np 16189C and np 16217C were diagnostic of the B2 lineage and suggestive of greater genetic diversity in prehistoric populations from the region. The presence of B2, as well as A7 and D2a1a in prehistoric samples, was followed by population fission, kin migration, founder effect, and an overall loss of the B2 subclade in contemporary populations throughout the archipelago.

Genetic Evidence of Kin Migration

Based on hypervariable region I and II mtDNA sequences, the spatial autocorrelation statistical method of Bertorelle and Barbujani (1995) was utilized to further elucidate the relationship between geography and genetic markers. In Figure 2.3, the *x* axis contains the lag geographic distances (km), while the *y* axis displays product moment coefficients analogous to Moran's I. As predicted by the isolation-by-distance model, the populations closest geographically have the greatest significant correlations. The populations beyond approximately 1,300-km lag geographic distance, however, have negative correlations, suggestive of kin migration following population fission and resettlement. Spatial autocorrelation plots are suggestive of kin/family migration from island to island and indicate a rapid genetic differentiation of island populations and greater interisland genetic differentiation.

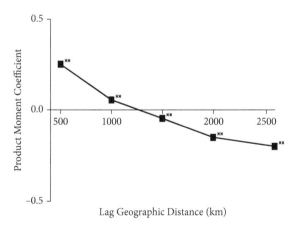

Figure 2.3 Spatial autocorrelation plot of Aleutian Island populations using mtDNA, illustrating a significant relationship between geography and genetics (Crawford et al. 2010).

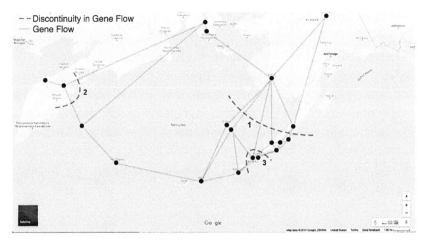

Figure 2.4 Triangulation plot using Monmonier's algorithm for identifying genetic barriers. The numbers in the plot identify the locations of genetic discontinuity, based on the proportion of the total genetic variance resulting from differences between groups (Crawford et al. 2010).

Climatic Fluctuations and Migration

Given the documented pattern of population migration, fission, and settlement from eastern to western islands, we expected the genetics to reflect an isolation-by-distance model. However, as shown in Figure 2.3, spatial autocorrelation does not fully support this model, as it displays negative Moran's I values between populations that are more than 1,000 km apart. We further probed, using Monmonier's algorithm, for regions of genetic discontinuity, or the existence of potential mating/genetic barriers (see Figure 2.4). Triangulation-based Monmonier boundaries indicate the existence of a genetic barrier (#3) in the central Aleutian Islands. This discontinuity may be explained by a temporal pause in the initial migration to the eastern islands (ca. 9,000 years BP), followed by more turbulent climatic conditions and seas that brought a virtual cessation to interisland crossings. Archaeological evidence suggests that the regional climate became drier and the seas less turbulent approximately 6,000 years BP, permitting genetically distinct families to migrate into the central islands. The other two barriers displayed in Figure 2.4 represent expected Aleut–Inuit genetic discontinuity (#1) and Siberian–Aleut discontinuity (#2).

Migration of Russians into the Aleutian Islands

Russian cultural contact and colonization of the Aleutian Islands resulted in a number of demographic and genetic repercussions: drastic population reduction due to disease (smallpox, tuberculosis, measles, etc.) and warfare; the forcible relocation of Aleuts from westerly islands to the Commander Islands, and from the central islands to the Pribilof Islands; asymmetric gene flow, i.e. Russian men reproducing coercively or otherwise with Aleut women, resulting in the relatively high frequency of indigenous American mtDNA (A and D haplogroups) throughout the western and central islands.

Aleut Ancestry

Principal component analysis (PCA) is a useful approach to evaluating ancestral trends, since changes among genetic markers often mirror the evolutionary and demographic histories of human populations. PCA employs a dimension-reduction methodology, and in this study PCA was performed on 170,636 SNPs from the 750,000 characterized by the GENO 2.0 chip (Patterson et al. 2006). SNPs were pruned using the software PLINK, which removed individuals whose genotypes were either in linkage disequilibrium, missing more than 10% of candidate SNP positions, or had a relative who shared an IBD (identical by descent) value of 0.2 (only one individual was retained). PCA captures variation in multidimensional data and reorganizes it along orthogonal axes, ordered by the variances that they capture (Price et al. 2006). The 170,636 SNPs in each of 2,499 individuals were aggregated, providing a glimpse into both the genetic structure and ancestry of Aleuts, by comparing them to Africans, East Asians, Europeans, Native Americans, and South Asians compiled from the 1000 Genomes Project (The 1000 Genomes Project Consortium, 2015). The first principal component (PC #1) primarily separated Africans from all other populations, while the second principal component (PC #2) primarily separated East Asians from all other populations. A few of the Aleuts clustered with Europeans, indicating considerable Russian or western European admixture. Overall, Aleuts were distributed between East Asians and European populations along the y axis (PC #2) and adjacent to Native American and South Asian populations along the x axis (PC #1; see Figure 2.5).

Admixture Analysis

There is considerable variation in the proportion of Aleut versus Russian/European ancestry depending on methods used and genetic markers employed. If uniparental markers are used for a diallelic estimate, 100% of sampled Aleuts display only indigenous American haplogroups—A and D according to mtDNA—while only 15% of Aleuts display NRY markers indicative of indigenous American ancestry. Admixture estimates based on autosomal STR (short tandem repeat) markers suggest that 60% of Aleut ancestry-informing sites are indigenous American and 40% are the result of gene flow from European sources (see Table 2.1). Admixture estimates for Bering Island (the least admixed of all islands in the archipelago)

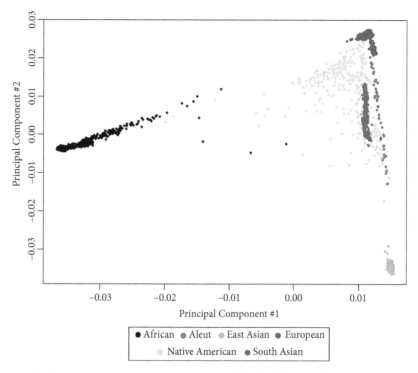

Figure 2.5 Principal component analysis using 170,636 SNPs and based on data from 1000 Genomes Project, with the following sample sizes: Africans, $N = 652$; East Asians, $N = 495$; Europeans, $N = 501$; Native Americans, $N = 297$; South Asians, $N = 482$; Aleuts $N = 72$.

Table 2.1 Admixture/ancestry estimates based on uniparental markers, autosomal short tandem repeats (STRs), and genomic single-nucleotide polymorphisms (SNPs)

Genetic Marker(s)	Percentage Aleut (Indigenous American)	Percentage Russian/ European
Mitochondrial DNA	100	0
Nonrecombining Y STRs	15	85
Autosomal STRs	60	40
Genomic SNPs	60	40

based on 750,000 SNPs provide similar ancestry proportions of 60% Aleut and 40% Russian gene flow.

The Admixture 1.3 program was utilized to estimate ancestry in a model-based analysis from a large autosomal SNP data set using unrelated individuals (Alexander et al. 2009). The K-means algorithm sorted individuals into an increasing number of clusters and provided a Bayesian information criterion (BIC) value for K. The K value with the lowest BIC provides evidence of the best fit for the data. Equal sample-size populations of genomes were used to represent six geographic regions

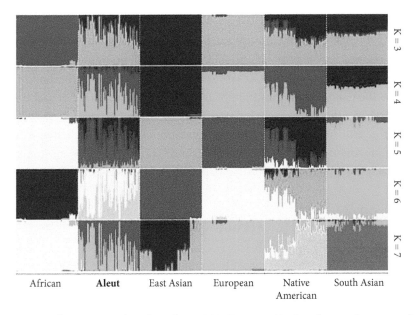

K = 3
K = 4
K = 5
K = 6
K = 7

African   **Aleut**   East Asian   European   Native   South Asian
American

Figure 2.6 Admixture analysis based on 1000 Genomes Project data and compared to Aleut autosomal SNPs. Individuals are represented by vertical lines partitioned into segments corresponding to their membership in genetic clusters, indicated by color.

(downloaded from the 1000 Genomes Project) and were compared with Aleuts. The best fit for the admixture analysis (with the lowest CV error) was K = 5. These results suggest that Aleut ancestry is a combination of Native American, European, and East Asian peoples (see Figure 2.6).

Conclusion

The evolutionary consequences of the settlement of the Aleutian Islands include loss of genetic variability resulting from population fission and founder effect. This conclusion is corroborated by the absence of mitochondrial haplogroup B in contemporary populations, despite its being found in 25% of prehistoric populations.

Evidence for kin migration is supported by spatial autocorrelation analyses. This form of familial migration resulted in increased genetic drift, and hence rapid genetic differentiation of distant island populations, and further reduced genetic variation in Aleut populations of the archipelago. Genetic discontinuity, as illustrated by the presence of mating/genetic barriers, reflects historical climatic variation. Long-distance migrations occurred principally during periods of extreme cold and aridity that resulted in calm seas. By contrast, periods of cyclonic activity and tempestuous seas coincided with relatively less population movement and greater genetic differentiation of island group populations. mtDNA sequences preserved interisland maternal population structure and migration patterns. Evidence from mtDNA supports the high correlation of genetic distance with geographic distance ($r = 0.72$; $P < 0.000$). This intimate relationship between genetics and geography was maintained in contemporary populations despite numerous sociopolitical and demographic disruptions, such as World War II and the forced relocation of Aleut males to the Commander and Pribilof Islands for the harvesting of seal furs. No statistically significant correlation between genomics and geography was observed for NRY genetic markers. Only 15% of contemporary Aleut populations exhibited Indigenous American NRY haplogroups, Q and Q3. This finding is due to the clear pattern of gene flow between Russian male migrants and Aleut females.

Acknowledgments

This research was supported in part by grants from NSF, OPP-9905090 and OPP-0327676 and National Geographic Society GENO 2.0. Graduate assistants Rohina Rubicz, Mark Zlojutro, and Liza Mack assisted in data collection. The authors thank the Aleutian and Pribiloff Island Corporation and the Unangan volunteers for their participation in the research.

Literature Cited

Alexander, D. H., J. Novembre, and K. Lange. 2009. "Fast model-based estimation of ancestry in unrelated individuals." *Genome Research* 19: 1655–1664.

Bertorelle, G., and G. Barbujani. 1995. "Analysis of DNA diversity by spatial autocorrelation." *Genetics* 140: 811–819.

Crawford, M. H. 2007. "Genetic structure of circumpolar populations: A synthesis." *American Journal of Human Biology* 19: 203–217.

Crawford, M. H. 2010. "Origins of Aleuts and the genetic structure of populations of the archipelago: Molecular and archaeological perspectives." *Human Biology* 82, no. 5-6: 695–717.

Davis, R. S., and R. A. Knecht. 2010. "Continuity and change in the Eastern Aleutian archaeological sequence." *Human Biology* 82, no. 5-6: 507–524.

Jochelson, V. I. 1933. *History, Ethnology, and Anthropology of the Aleut* (Carnegie Institution of Washington Publication 432). Washington DC: Carnegie Institution.

Map Data: Google/Zenrin. 2019. "Aleutian Islands." https://www.google.com/maps/@58.5305411,-172.4689024,4.93z

McCartney, A. P., and D. W. Veltre. 1996. "Anagula Core and Blade Site." In *American Beginnings: The Prehistory and Paleoecology of Beringia,* edited by F. H. West. Chicago: University of Chicago Press.

Patterson, N., A. L. Price, and D. Reich. 2006. "Population structure and eigenanalysis." *PLOS Genetics* 2, no. 12: e190.

Price, A. L., N. J. Patterson, R. M. Plenge, M. E. Weinblatt, N. A. Shadick, and D. Reich. 2006. "Principal component analysis corrects for stratification in genome-wide association studies." *Nature Genetics* 38, no. 8: 904–909.

Raff, J., J. Tackney, and D. H. O'Rourke. 2010. "South from Alaska: A pilot aDNA study of genetic history on the Alaska Peninsula and the eastern Aleutians." *Human Biology* 82, no. 5-6: 677–694.

Reedy-Maschner, K. 2010. "Where did all the Aleut men go? Aleut male attrition and related patterns in Aleutian historical demography and social organization." *Human Biology* 82, no. 5-6: 583–611.

Rubicz, R. C., T. G. Schurr, P. I. Babb, and M. H. Crawford. 2003. "Mitochondrial DNA variation and the origins of the Aleuts." *Human Biology* 75: 809–835.

Savinetsky, A. B., N. K. Kiseleva, and B. F. Khassanov. 2010. "Paleoenvironment—Holocene Deposits from Shemya Island." In *The People at the End of the World: The Western Aleutians Project and Archaeology of Shemya Island,* edited by Debra Corbett, Dixie West, and Christine Lefevre, 71–82. Anchorage: Aurora.

The 1000 Genomes Project Consortium. 2015. "A global reference for human genetic variation." *Nature* 526: 68–74.

West, D. L., and S. Crockford. 2010. "Conclusions." In *The People Before: The Geology, Paleoecology, and Archaeology of Adak Island, Alaska,* edited by D. West, V. Hatfield, E. Wilmerding, C. Lefevre, and L. Gualtieri. Santa Fe: Aurora Press.

West, D., A. Savinitsky, and M. H. Crawford. 2007. "Aleutian Islands: Archaeology, molecular genetics and ecology." *Transactions of the Royal Society of Edinburgh: Earth and the Environmental Sciences* 98: 47–57.

Winterhalder, B., and E. A. Smith. 1981. "Optimal Foraging Strategies and Hunter-Gatherer Research in Anthropology." In *Hunter-Gatherer Foraging Strategies: Ethnographic and Archeological Analyses*, 1–35. Chicago: University of Chicago Press.

Zlojutro, M. R., R. C. Rubicz, and M. H. Crawford. 2009. "Mitochondrial DNA and Y-chromosome variation in five eastern Aleut communities: Evidence for genetic substructure in the Aleutian population." *American Journal of Human Biology* 36, no. 5: 511–526.

Zlojutro, M. R., R. C. Rubicz, E. J. Devor, and M. H. Crawford. 2006. "Genetic structure of the Aleuts and circumpolar populations based on mitochondrial DNA sequences: A synthesis." *American Journal of Physical Anthropology* 129, no. 3: 446–464.

3

Early Peopling of the Americas

A Palaeogenetics Perspective

Constanza de la Fuente, J. Víctor Moreno-Mayar, and Maanasa Raghavan

The peopling of the Americas has been a subject of debate in the anthropological, archaeological, and genetic literature for several decades. Some of the primary research questions relating to this process focus on the origin of the founding population, timing of entry into the Americas, the number of migration events, migration routes, and the processes of early population diversification. While some of these questions remain contentious and investigations of others have opened further research avenues, it is imperative to acknowledge that several fields from both the social and natural sciences, including bioanthropology, archaeology, linguistics, genetics, and geology, have contributed immensely to our current understanding of the early peopling of the Americas. This review focuses on studies of ancient DNA (palaeogenetics/palaeogenomics) set in the Americas, with the understanding that the genetic data sets have been contextualized and interpreted together with evidence from the diverse aforementioned fields to obtain more nuanced models of the population history of this region.

Characterizing the Present-Day Genetic Diversity of Indigenous Populations in the Americas

Prior to discussing the contributions of ancient DNA (aDNA) research to the reconstruction of human population history of the Americas, it is imperative to briefly review key genetic research on present-day populations in order to better contextualize the former. Biomarkers like microsatellites, blood types, and, in particular, mitochondrial DNA (mtDNA) have been used in the past to characterize molecular diversity in the Americas. The definition of pan-American haplogroups A, B, C, and D (Forster et al. 1996; Schurr et al. 1990; Torroni et al. 1993) rapidly followed a more detailed catalog of several lineages widely distributed in the continents, including A2, B2, C1b, C1c, C1d, C1d1, D1, and D4h3 (Perego et al. 2009; Tamm et al. 2007). The analysis of these lineages, as well as lineages that are geographically more restricted (e.g., X2a, C4c, and D4e1), has fueled an intense discussion over the origins of the indigenous populations in the continents as well as the time and the number

Constanza de la Fuente, J. Víctor Moreno-Mayar, and Maanasa Raghavan, *Early Peopling of the Americas* In: *Human Migration*. Edited by: Maria de Lourdes Muñoz-Moreno and Michael H. Crawford, Oxford University Press. © Oxford University Press 2021. DOI: 10.1093/oso/9780190945961.003.0004 of migration events that could better explain the observed present-day diversity. Analyses of the defined mtDNA lineages in the Americas has supported an East Asian origin (Perego et al. 2009; Tamm et al. 2007). Nonetheless, the presence and distribution of rare mDNA haplogroups have contributed to alternate hypotheses for the origin of present-day indigenous populations in the Americas. For example, the presence of mtDNA haplogroup X in North America and Europe has been used to support a Pleistocene trans-Atlantic migration to the Americas, proposed by the Solutrean hypothesis (Bradley and Stanford 2004; Oppenheimer, Bradley, and Stanford 2014). Primarily based on archaeological data, this hypothesis has little support from genetics. As argued by Raff and Bolnick (Raff et al. 2015), the absence of mtDNA lineages that are ancestral to haplogroup X2a in Siberia does not provide unquestionable evidence to support this hypothesis. Furthermore, the identification of this lineage in the ~ 8.5 kya (thousand years ago) individual from Washington state (US) referred to as The Ancient One (Kennewick Man), who was found to be genetically most closely related to present-day Native Americans (Rasmussen et al. 2015), places the most basal X2a lineage found to date in the Pacific Northwest and not in northeastern Canada, as Oppenheimer et al. (Oppenheimer, Bradley, and Stanford 2014) highlighted as proof of a trans-Atlantic migration (Raff et al. 2015). Moreover, mtDNA variation across the Americas has consistently supported the Beringian Standstill or Incubation model, based on the high frequency of unique and widespread mtDNA lineages in America (Tamm et al. 2007). According to this model, most Native American variation is explained by the migration of people from Asia via Beringia (the land bridge that connected Northeast Asia and America during the last ice age). The founding population would have occupied Beringia and remained isolated in the region for around 15,000 years, accumulating specific variants that separate Asian lineages from the Native American ones, hence explaining the appearance of unique haplogroups widespread in the Americas. The hypothesis further postulates that the ancestors of Native Americans would have entered the Americas around 15 kya, moving southward rapidly, which explains, in turn, the wide and weakly structured distribution of founding mtDNA lineages across the continents (Brandini et al. 2018; Tamm et al. 2007).

More recently, important contributions to our understanding of the present-day genetic variation in the Americas have been made through the analysis of genome-wide data. One of the first studies to analyze autosomal data at a continental scale reported 678 microsatellites in 422 individuals from 29 Native American populations (Wang et al. 2007). By analyzing the intracontinental population structure and the geographic distribution of genetic variation, this work found support for a single major migration event, most likely of Asian origin, that moved through Beringia and subsequently followed a coastal route southward (Wang et al. 2007). More recently, the analysis of ~ 364k single-nucleotide polymorphisms (SNPs) from 52 Native American and 17 Siberian populations proposes a different scenario regarding the number of founding migrations that would fit the observed present-day diversity more accurately (Reich et al. 2012). Although most indigenous populations in the Americas derive the majority of their ancestry from a single source population in Asia (termed "First Americans"), this study suggests two additional gene-flow events from Asia are needed to explain the genetic diversity found among present-day Inuit and Aleutian Islanders (Eskimo-Aleut speakers) in the North American Arctic and the Chipewyan Na-Dene speakers, respectively. In contrast, another study analyzing present-day whole genomes from across the Americas and Siberia found support for a single migration wave for all Native Americans, including the Na-Dene speakers, from Siberia around 23 kya, reaching South America by at least 14.6 kya to account for archaeological evidence of humans in Monte Verde (Chile), and subsequently diversifying into two basal branches—"Northern Native Americans" (NNA) and "Southern Native Americans" (SNA)—about 13 kya (Raghavan et al. 2015). Some populations affiliated with the northern branch, particularly in northwestern North America, have been shown to have received additional gene flow from East Asia after the split (Raghavan et al. 2015).

There is no doubt that inferences based on present-day genetic data have contributed to a better understanding of the demographic processes that shaped Native American diversity. However, it is becoming increasingly clear that modern genetic data sets are unable to accurately capture past demographic events. Particularly in the Americas, it is well known that colonization led to a massive reduction in the indigenous genetic diversity and certain lineages might not, in fact, be represented in present-day populations. In this regard, ancient genomics has the power to provide unprecedented nuance to models of the peopling of the Americas by allowing researchers to sample genetic diversity at different time points. In what follows, we review the evidence for early migrations, structure, and admixture as informed by aDNA studies.

Genetic Origins of Early Founding Populations

Describing the genetic diversity of the Native American founding population(s) has been a major focus of aDNA research. Applying aDNA techniques to some of the oldest human remains in the Americas, such as the ~ 14.3 kya coprolites from the Paisley 5 Mile Point Cave in Oregon, revealed the presence of the Native American founding mtDNA lineages A2 and B2 in these and younger remains from the site (Gilbert, Jenkins et al. 2008; Jenkins et al. 2012). Hence, results from these studies support an East Asian origin for early populations in the Americas. More recently, by sequencing 92 whole-mitochondrial genomes from ancient remains across the Americas dating from 8.6 to 0.5 kya, Llamas et al. (2016) confirmed previous East Asian affinities, time estimates, and population size changes included in the Beringian Standstill model and highlighted a substantial mtDNA lineage loss following European colonization.

Further evidence for the Eurasian genetic origins of Native American populations have been contributed by the analysis of genome-wide data. In 2012, Patterson and colleagues proposed a "ghost" (i.e., unsampled) ancient northern Eurasian (ANE) population that contributed to the gene pool of both Indigenous American and northern European populations (Patterson et al. 2012). A few years later, in 2014, the genome of a 24,000-year-old individual representing this lineage was sequenced, providing further insights into the makings of the ancestral population of the Americas (Raghavan et al. 2014b). This individual (MA-1), from the Mal'ta site in south-central Siberia, was shown to be a member of a basal western Eurasian lineage, which contributed ~ 14% to 38% of present-day Native American ancestry. Based on the uniform distribution of ANE-related ancestry across the Americas, this gene-flow event was inferred to have occurred after the ancestral population split from its East Asian ancestors but to have predated the diversification of Native American populations.

Early Structure Within the Americas

Ancient genomics has provided the opportunity to document ancestral Native American populations that have not been characterized based on present-day genomic data (Figure 3.1). In 2011, Potter et al. (2011) described the Upward Sun River (USR) archaeological site in interior Alaska, where the remains of three individuals were identified and dated to the terminal Pleistocene (11.5 kya). A high-quality, high-depth genome was sequenced from one of the individuals buried in the site (Xach'itee'aanenh T'eede Gaay/Sunrise Child-Girl, or USR1; Moreno-Mayar et al. 2018b). Genomic data revealed that USR1 was a member of a population termed Ancient Beringians (AB), which fell within Native American genetic variation, yet was not a member of the previously characterized NNA or SNA branches. Instead, AB are an outgroup to all present-day and ancient Native Americans for which genetic data is available today.

Demographic modeling including the USR1 genome suggested that the ancestors of Native Americans (including AB) became isolated from Eurasian populations ~ 25 kya and confirmed that the NNA–SNA split took place 17.5 to 14.6 kya. These results support the notion that most Native American groups can be traced to a single Siberian or Northeast Asian Late Pleistocene founding population. In contrast to present-day groups that inhabit northern North America today, individuals buried at the USR site, as well as a second AB individual from the Seward Peninsula who lived ~9 kya (Moreno-Mayar et al. 2018a), did not bear Asian and Siberian ancestry found in present-day Na-Dene speakers, the Inuit, and Iñupat groups or the ancient Arctic culture predating the Inuit (Saqqaq/Palaeo-Eskimos). Thus, the ages of the AB individuals (~ 11.5–9 kya) provide an upper bound for the timing of Eurasian gene flow that contributed to the ancestry of northern present-day groups that crossed the Bering Strait.

Once AB were isolated from other Native Americans, the latter split into NNA and SNA 17.5 to 14.6 kya, around the estimated time of initial entry south of the

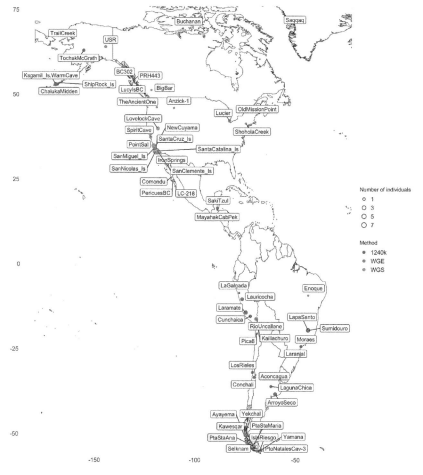

Figure 3.1 Location of ancient samples from the Americas with genome-wide data higher than 0.1X. References: de la Fuente et al. 2018; Flegontov et al. 2019; Lindo et al. 2017, 2018; Moreno-Mayar et al. 2018a, 2018b; Posth et al. 2018; Raghavan et al. 2014a, 2015; Rasmussen et al. 2010, 2014, 2015; Scheib et al. 2018.

glacial ice (Moreno-Mayar et al. 2018b). The ancient human remains from the Americas that have been dated closest to this time (12.8 kya; Becerra-Valdivia et al. 2018) are those of an infant buried at the Anzick site in Western Montana. Whole-genome sequencing and analysis of the individual, referred to as Anzick1, showed that he was one of the earliest members of the SNA clade (Rasmussen et al. 2014). Moreover, comparison to other present-day Native Americans showed that Anzick1 was "directly ancestral" to other members of the SNA branch, suggesting that Anzick1 lived close to the time in which the NNA–SNA split took place. As such, results based on the radiocarbon age of Anzick1 and the analysis of his genome confirm that the NNA–SNA split occurred early, soon after Native Americans moved from Eastern Beringia and south of the ice.

Ancient genomics has also contributed to building a more detailed narrative of the demographic events following the basal splits of the AB, NNA, and SNA branches. Most interpretations of the available data indicate that present-day Native American groups inhabiting regions that were once south of the continental ice carry SNA-related ancestry almost exclusively. However, Scheib et al. (2018) suggested that both the NNA and SNA branches admixed substantially early on, and that traces of both ancestries can be found in present-day individuals across the continents. However, a simpler model has been favored by more recent studies. These studies, based on whole-genome (Moreno-Mayar et al. 2018a) and SNP capture (Posth et al. 2018) sequencing data from Holocene human remains, showed that, after splitting from NNA, SNA underwent a rapid radiation ~ 14 kya (Moreno-Mayar et al. 2018a) as they spread south of the ice, ultimately reaching southern South America. This radiation gave rise to multiple early populations that were genetically similar and that extended over thousands of kilometers as they explored an essentially unrestricted territory. Moreover, both studies suggest that more recent population movements and admixture followed the initial exploration and settlement. Moreno-Mayar et al. (2018a) observed that present-day South Americans could be best modeled as a mixture of ancestry represented by Late Pleistocene and Early Holocene individuals and a more recently introduced Mesoamerican-related ancestry component. A similar result was presented by Posth et al. (2018), where ~ 4.2 kya Andean individuals carry ancestry that is found in ~ 4.9 kya individuals from California.

While both studies reached similar broad conclusions, they differed in a number of aspects. On the one hand, Moreno-Mayar et al. (2018a) inferred that Mesoamericans—as well as present-day South Americans, through admixture with the former—bear additional ancestry from an unsampled population that is most closely related to Native Americans but that is neither AB, NNA, nor SNA. On the other hand, Posth et al. (2018) inferred that the ancestry in Late Pleistocene and Early Holocene Native Americans was further structured with the oldest populations from Brazil and Chile (10.9–9.6 kya) being more closely related to the Anzick1 individual than are other South Americans postdating 9 kya.

Similarly, ancient genomics has provided evidence that the population history of North America was more complex than previously understood, with continuity, structure, and admixture occurring in different regions and time periods. In northern North America, ABs persisted in the region until ~ 9 kya (Moreno-Mayar et al. 2018a). In the Pacific Northwest, ancient genomes from individuals who lived after ~ 6 kya in coastal British Columbia (Lindo et al. 2017; Raghavan et al. 2015) and in the interior Fraser Plateau suggest that population structure remained in place for an extended period of time, as the former are more closely related to the NNA branch, while the latter could best be modeled as an outgroup to both NNA and SNA (Moreno-Mayar et al. 2018a). Based on shallow sequencing data (< 0.1 X) from a ~ 10.3 kya individual from Southeastern Alaska, referred to as Shuká

Káa ("Man Ahead of Us"), broad genetic continuity between Shuká Káa, the afore-mentioned 6–1.7 kya NNA-related individuals from coastal British Columbia, and present-day groups in the region has been suggested.

However, long-term structure was not geographically widespread and admixture between early lineages potentially occurred, as evidenced by the genome from the ~ 8.5 kya individual known as The Ancient One (Kennewick Man; Rasmussen et al. 2015). The Ancient One was most closely related to present-day Native Americans, and he had a greater genetic affinity to present-day populations from the same geographic region, in particular, the Colville, Ojibwa, and Algonquin (Rasmussen et al. 2015). Whereas "his affinity to the SNA and NNA lineages" had been considered "ambiguous" (Skoglund and Reich 2016), further analyses showed that the ancestry of this individual was best modeled as a mixture of both NNA and SNA (Moreno-Mayar et al. 2018a).

This diverse array of demographic histories is further displayed across the conti-nent. Ancient genomes from the North American Great Basin suggest genetic con-tinuity through the Holocene, with a population that remained largely isolated until the last millennium, when Mesoamerican-related admixture was introduced in the region (Moreno-Mayar et al. 2018a). Importantly, it remains to be resolved how or whether this Mesoamerican-related admixture is expressed culturally both in this region and in South America. Further south, recent studies from the Patagonian archipelago also present evidence for genetic continuity through time, from early maritime hunter-gatherer individuals dated to 6.2 kya to present-day populations from the region (de la Fuente et al. 2018; Moreno-Mayar et al. 2018a).

Eurasian Gene Flow into the Americas

Evidence of early splits and structure in the Americas has provided further support for a rapid process of peopling of the continents, followed by the diversification of ge-ographically restricted lineages. However, an intriguing body of evidence suggests a connection between Amazonian groups from Brazil, particularly Surui and Karitiana, and present-day Australasian populations represented by Papuans, Australians, and Andaman Islanders (Raghavan et al. 2015; Skoglund et al. 2015). Such a connec-tion is evidenced by excess allele and haplotype sharing between Amazonian and Australasian groups, when contrasted with Mesoamerican populations. Several hypotheses have been proposed to explain this connection, mainly involving addi-tional migrations into the Americas or the presence of a structured, nonhomogeneous population in Siberia during the founding event (Raghavan et al. 2015; Skoglund et al. 2015). While the craniofacial morphology-based model, or the Palaeoamerican Model, which purports two separate migration waves into the Americas, has been consistently rejected using genomic data (Moreno-Mayar et al. 2018a; Raghavan et al. 2015), the fact remains that this genetic signal is present in present-day and ancient individuals from South America (Moreno-Mayar et al. 2018a).

Recently, the analysis of ancient genomes from the Lapa do Sumidouro site in the Lagoa Santa region in Brazil, dated to ~ 10.4 kya, reproduced the observation based on present-day data (Moreno-Mayar et al. 2018a). However, this was not the case for genetically similar ancient individuals from the nearby site of Lapa do Santo, dated to ~ 9.6 kya (Posth et al. 2018). Therefore, if this signal is not a false-positive result, the geographic and temporal distribution of the Australasian signature suggests that early populations were structured and not evenly spread across the continent. Currently, population structure in the ancestral Native American population (Skoglund and Reich 2016) stands as the most parsimonious explanation for the Australasian ancestry in a reduced number of ancient (Moreno-Mayar et al. 2018a) and present-day Native Americans (Raghavan et al. 2015; Skoglund et al. 2015). However, the absence of this signature in ancient or present-day individuals in North and Central America as well as the rest of South America hints that migration south of the ice from the structured ancestral population might have also occurred in a structured fashion over both space and time.

Inferring Routes of Migrations into and Across the Americas

The routes followed by the first inhabitants of the Americas have been a topic of ongoing debate. For many years, an entry through an ice-free corridor (IFC) between the Cordilleran and Laurentide ice sheets was considered the earliest and only route into the Americas. This particular hypothesis was linked to the "Clovis First" paradigm, under which Clovis culture, characterized by a fluted projectile lithic technology, represented the first people entering the continents following the IFC route (Meltzer 2009; Potter et al. 2017). However, the presence of several pre-Clovis archaeological sites, some of them as far south as 41ºS latitude (Monte Verde, Chile), has favored a Pacific coastal route scenario (Dillehay et al. 2008; Lesnek et al. 2018). The use of one route over the other relies on several environmental factors: the presence of physical barriers to migration (glaciers or lakes), and the biological productivity and availability of resources to support human migrations (Lesnek et al. 2018; Potter et al. 2017). Recent research has suggested that the IFC was physically open ~ 14 to 15 kya, but it became biologically viable only around 12.6 kya (Pedersen et al. 2016). Considering these dates for the opening and biological viability of the IFC, older sites south of the ice sheets likely result from an alternate route that followed the Pacific coast (Lesnek et al. 2018; Pedersen et al. 2016).

In addition to environmental and archaeological evidence, genetic data have also contributed to this debate by reconstructing the patterns of genetic diversity on a continent-wide basis. The wide distribution of the founding mtDNA lineages, for example, supports a rapid movement southward followed by an early diversification of the lineages. According to the analysis of present-day and ancient mitogenomes, the incubation period in Beringia was followed by a rapid lineage diversification and an increase in the female effective population size starting ~ 16 kya (Llamas et al. 2016). Assuming this to be the time of the initial entry into the continent, a migration model involving the movement southward through the IFC is not compatible, while a coastal route was more likely to have been available around this time (Llamas et al. 2016; Lesnek et al. 2018). Additionally, the rapid diversification of the mtDNA lineages and its distribution across the continents would also imply a rapid movement southward. Similarly, the patterns of genome-wide genetic diversity have also supported the coastal route, suggesting a greater correlation between genetic diversity and geographical distance when the coast is taken as facilitator of migration (Reich et al. 2012; Wang et al. 2007).

Migrations into the North American Arctic

The peopling of the North American Arctic postdates the Pleistocene founding migrations by several thousands of years. Archaeological research has broadly characterized three distinct cultural waves in the North American Arctic: Pre-Dorset/Saqqaq (ca. 5–2.8 kya), Dorset (ca. 2.8–0.6 kya), and Thule/Inuit (~ 1 kya; Raghavan et al. 2014a). The extent of their genetic origins, relatedness, and continuity over time has, however, remained a topic of discussion in the literature. The complete mtDNA and nuclear genomes of a 4,000-year-old Saqqaq individual revealed that this first culture in the Arctic resulted from a migration independent of those that gave rise to Native Americans and the Thule (Gilbert, Kivisild et al. 2008; Rasmussen et al. 2010). Ancient mtDNA and genome-wide data have further shown that Saqqaq and Dorset share the same mtDNA haplogroup, hg D2a1, and autosomal data from individuals affiliated with these cultures show high genetic affinity, indicating genetic continuity over a period of ~ 4,000 years (Hayes, Coltrain, and O'Rourke 2005; Raghavan et al. 2014a). The Thule, on the other hand, have a very different mtDNA haplogroup profile, primarily carrying A2a, A2b, and D4b1a (formerly D3), and their autosomal signatures also support a new migration into the North American Arctic, ultimately giving rise to present-day Inuit (Gilbert et al. 2007; Hayes, Coltrain, and O'Rourke 2005; Raff et al. 2015; Raghavan et al. 2014a; Tackney et al. 2019).

Recent research suggests that while the Thule constituted a new migration into the North American Arctic, largely replacing the Dorset, present-day populations, such as Aleutian Islanders and the Inuit and related groups, may still harbor ancestry from a source related to the Saqqaq (Flegontov et al. 2019). Importantly, Na-Dene speakers in North America have been shown to bear Siberian ancestry introduced during the Holocene (Moreno-Mayar et al. 2018a; Reich et al. 2012). It has been suggested that the source of such ancestry also contributed to other present-day Siberian populations, such as the Koryaks (Moreno-Mayar et al. 2018b) and the Saqqaq (Flegontov et al. 2019), and is best represented by a ~ 9.8 kya individual from the Duvanny Yar site at the Kolyma River in northeastern Siberia (Sikora et al. 2019).

Final Perspectives

From a palaeogenetics perspective, the increasing numbers of analyzed ancient individuals and genetic markers have contributed to a deeper understanding of the complexity of the early human population structure and migratory processes in the Americas. While many key questions about the peopling of the Americas remain disputed, either due to disagreements between different disciplines or ambiguous genetic evidence, a broad model suggests a single source population stemming from East Asia around 25 kya. Evidence of early structure in Beringia and within the Americas, at least by ~ 14.6 kya, is strongly supported by ancient genomics (Moreno-Mayar et al. 2018a; Posth et al. 2018). Going forward, palaeogenomics studies of later diversification and migration events, with a focus on testing regional hypotheses, will be critical to ongoing research efforts in the field.

While aDNA research has undoubtedly contributed to the reconstruction of the population history of the Americas, largely thanks to the development of new sampling strategies, sequencing technologies, and computational tools, fewer efforts have been devoted to assessing and promoting sustainability and inclusivity in the research undertakings. Although multidisciplinary approaches have partially contributed to ameliorating this issue, further efforts must be made to reduce the current gap between archaeology, genetics, and indigenous communities. Drawing from personal experiences, actionable points, and opinions shared by other researchers (e.g., Claw et al. 2018; Fox and Hawks 2019; Tackney and Raff 2019), we should guide our research plans in ancient genomics by considering important questions, such as: Are local researchers, museums, and communities active participants in the project? Does the research satisfy ethical concerns of key stakeholders, including communities, archaeologists, curators, and geneticists? What is being done to ensure the morphological preservation of samples (e.g., are CT scanning, casts, etc., being used?) in light of destructive sampling for aDNA studies, especially with scarce samples? Are there efforts to build local capacity? Is there a dedicated strategy to disseminate results, particularly to indigenous communities, beyond just the scientific publication? Careful examination of these and other relevant questions, as well as measures to address them, should become the norm and should be considered as integral to the design and execution of a research project within the field as its scientific merit.

References

Becerra-Valdivia, Lorena, Michael R. Waters, Thomas W. Stafford, Sarah L. Anzick, Daniel Comeskey, Thibaut Devièse, and Thomas Higham. 2018. "Reassessing the chronology of the archaeological site of Anzick." *PNAS* 115, no. 27: 7000–7003. https://doi.org/10.1073/pnas.1803624115

Bradley, Bruce, and Dennis Stanford. 2004. "The North Atlantic ice-edge corridor: A possible palaeolithic route to the New World." *World Archaeology* 36, no. 4: 459–478. https://doi.org/10.1080/0043824042000303656

Brandini, Stefania, Paola Bergamaschi, Marco Fernando Cerna, Francesca Gandini, Francesca Bastaroli, Emilie Bertolini, Cristina Cereda, et al. 2018. "The Paleo-Indian entry into South America according to mitogenomes." *Molecular Biology and Evolution* 35, no. 2: 299–311. https://doi.org/10.1093/molbev/msx267

Claw, Katrina G., Matthew Z. Anderson, Rene L. Begay, Krystal S. Tsosie, Keolu Fox, and Nanibaa' A. Garrison. 2018. "A framework for enhancing ethical genomic research with indigenous communities." *Nature Communications* 9, no. 1: 2957. https://doi.org/10.1038/s41467-018-05188-3

de la Fuente, Constanza, María C. Ávila-Arcos, Jacqueline Galimany, Meredith L. Carpenter, Julian R. Homburger, Alejandro Blanco, Paloma Contreras, et al. 2018. "Genomic insights into the origin and diversification of late maritime hunter-gatherers from the Chilean Patagonia." *PNAS* 115, no. 17: E4006–E4012. https://doi.org/10.1073/pnas.1715688115

Dillehay, Tom D., C. Ramírez, M. Pino, M. B. Collins, J. Rossen, and J. D. Pino-Navarro. 2008. "Monte Verde: Seaweed, food, medicine, and the peopling of South America." *Science* 320, no. 5877: 784–786. https://doi.org/10.1126/science.1156533

Flegontov, Pavel, N. Ezgi Altınışık, Piya Changmai, Nadin Rohland, Swapan Mallick, Nicole Adamski, Deborah A. Bolnick, et al. 2019. "Palaeo-Eskimo genetic ancestry and the peopling of Chukotka and North America." *Nature* 570, no. 7760: 236–240. https://doi.org/10.1038/s41586-019-1251-y

Forster, P., R. Harding, A. Torroni, and H. J. Bandelt. 1996. "Origin and evolution of Native American mtDNA variation: A reappraisal." *American Journal of Human Genetics* 59, no. 4: 935–945.

Fox, Keolu, and John Hawks. 2019. "Use ancient remains more wisely." *Nature* 572, no. 7771: 581–583. https://doi.org/10.1038/d41586-019-02516-5

Gilbert, M. Thomas P., Durita Djurhuus, Linea Melchior, Niels Lynnerup, Michael Worobey, Andrew S. Wilson, Claus Andreasen, and Jørgen Dissing. 2007. "MtDNA from hair and nail clarifies the genetic relationship of the 15th century Qilakitsoq Inuit mummies." *American Journal of Physical Anthropology* 133, no. 2: 847–853. https://doi.org/10.1002/ajpa.20602

Gilbert, M. Thomas P., Dennis L. Jenkins, Anders Götherstrom, Nuria Naveran, Juan J. Sanchez, Michael Hofreiter, Philip Francis Thomsen, et al. 2008. "DNA from pre-Clovis human coprolites in Oregon, North America." *Science* 320, no. 5877: 786–789. https://doi.org/10.1126/science.1154116

Gilbert, M. Thomas P., Toomas Kivisild, Bjarne Grønnow, Pernille K. Andersen, Ene Metspalu, Maere Reidla, Erika Tamm, et al. 2008. "Paleo-Eskimo mtDNA genome reveals matrilineal discontinuity in Greenland." *Science* 320, no. 5884: 1787–1789. https://doi.org/10.1126/science.1159750

Hayes, M. G., J. B. Coltrain, and Dennis O'Rourke. 2005. "Molecular Archaeology of the Dorset, Thule, and Sadlermiut: Ancestor–Descendant Relationships in Eastern North American Arctic Prehistory." In *Contributions to the Study of the Dorset Palaeo-Eskimos*, edited by Patricia D. Sutherland, 11–32. Ottawa: University of Ottawa Press.

Jenkins, Dennis L., Loren G. Davis, Thomas W. Stafford, Paula F. Campos, Bryan Hockett, George T. Jones, Linda Scott Cummings, et al. 2012. "Clovis age Western stemmed projectile points and human coprolites at the Paisley Caves." *Science* 337, no. 6091: 223–228. https://doi.org/10.1126/science.1218443

Lesnek, Alia J., Jason P. Briner, Charlotte Lindqvist, James F. Baichtal, and Timothy H. Heaton. 2018. "Deglaciation of the Pacific Coastal corridor directly preceded the human colonization of the Americas." *Science Advances* 4, no. 5: eaar5040. https://doi.org/10.1126/sciadv.aar5040

Lindo, John, Alessandro Achilli, Ugo A. Perego, David Archer, Cristina Valdiosera, Barbara Petzelt, Joycelynn Mitchell, et al. 2017. "Ancient individuals from the North American Northwest Coast reveal 10,000 years of regional genetic continuity." *PNAS* 114, no. 16: 4093–4098. https://doi.org/10.1073/pnas.1620410114

Lindo, John, Randall Haas, Courtney Hofman, Mario Apata, Mauricio Moraga, Ricardo A. Verdugo, James T. Watson, et al. 2018. "The genetic prehistory of the Andean highlands 7000 years BP through European contact." *Science Advances* 4, no. 11: eaau4921. https://doi.org/10.1126/sciadv.aau4921

Llamas, Bastien, Lars Fehren-Schmitz, Guido Valverde, Julien Soubrier, Swapan Mallick, Nadin Rohland, Susanne Nordenfelt, et al. 2016. "Ancient mitochondrial DNA provides high-resolution time scale of the peopling of the Americas." *Science Advances* 2, no. 4: e1501385. https://doi.org/10.1126/sciadv.1501385

Meltzer, David J. 2009. *First Peoples in a New World: Colonizing Ice Age America*. Berkeley: University of California Press.

Moreno-Mayar, J. Víctor, Lasse Vinner, Peter de Barros Damgaard, Constanza de la Fuente, Jeffrey Chan, Jeffrey P. Spence, Morten E. Allentoft, et al. 2018a. "Early human dispersals within the Americas." *Science* 362, no. 6419: eaav2621. https://doi.org/10.1126/science.aav2621

Moreno-Mayar, J. Víctor, Ben A. Potter, Lasse Vinner, Matthias Steinrücken, Simon Rasmussen, Jonathan Terhorst, John A. Kamm, et al. 2018b. "Terminal Pleistocene Alaskan genome reveals first founding population of Native Americans." *Nature* 553, no. 7687: 203–207. https://doi.org/10.1038/nature25173

Oppenheimer, Stephen, Bruce Bradley, and Dennis Stanford. 2014. "Solutrean hypothesis: Genetics, the mammoth in the room." *World Archaeology* 46, no. 5: 752–774. https://doi.org/10.1080/00438243.2014.966273

Patterson, Nick, Priya Moorjani, Yontao Luo, Swapan Mallick, Nadin Rohland, Yiping Zhan, Teri Genschoreck, Teresa Webster, and David Reich. 2012. "Ancient admixture in human history." *Genetics* 192, no. 3: 1065–1093. https://doi.org/10.1534/genetics.112.145037

Pedersen, Mikkel W., Anthony Ruter, Charles Schweger, Harvey Friebe, Richard A. Staff, Kristian K. Kjeldsen, Marie L. Z. Mendoza, et al. 2016. "Postglacial viability and colonization in North America's ice-free corridor." *Nature* 537, no. 7618: 45–49. https://doi.org/10.1038/nature19085

Perego, Ugo A., Alessandro Achilli, Norman Angerhofer, Matteo Accetturo, Maria Pala, Anna Olivieri, Baharak Hooshiar Kashani, et al. 2009. "Distinctive Paleo-Indian migration routes from Beringia marked by two rare mtDNA haplogroups." *Current Biology* 19, no. 1: 1–8. https://doi.org/10.1016/j.cub.2008.11.058

Posth, Cosimo, Nathan Nakatsuka, Iosif Lazaridis, Pontus Skoglund, Swapan Mallick, Thiseas C. Lamnidis, Nadin Rohland, et al. 2018. "Reconstructing the deep population history of Central and South America." *Cell* 175, no. 5: 1185–1197.e22. https://doi.org/10.1016/j.cell.2018.10.027

Potter, Ben A., Joel D. Irish, Joshua D. Reuther, Carol Gelvin-Reymiller, and Vance T. Holliday. 2011. "A terminal Pleistocene child cremation and residential structure from Eastern Beringia." *Science* 331, no. 6020: 1058–1062. https://doi.org/10.1126/science.1201581

Potter, Ben A., Joshua D. Reuther, Vance T. Holliday, Charles E. Holmes, D. Shane Miller, and Nicholas Schmuck. 2017. "Early colonization of Beringia and northern North America: Chronology, routes, and adaptive strategies." *Quaternary International* 444, no. Part B: 36–55. https://doi.org/10.1016/j.quaint.2017.02.034

Raff, Jennifer A., Margarita Rzhetskaya, Justin Tackney, and M. Geoffrey Hayes. 2015. "Mitochondrial diversity of Iñupiat people from the Alaskan North Slope provides evidence for the origins of the paleo- and neo-Eskimo peoples." *American Journal of Physical Anthropology* 157, no. 4: 603–614. https://doi.org/10.1002/ajpa.22750

Raghavan, Maanasa, Michael DeGiorgio, Anders Albrechtsen, Ida Moltke, Pontus Skoglund, Thorfinn S. Korneliussen, Bjarne Grønnow, et al. 2014a. "The genetic prehistory of the New World Arctic." *Science* 345, no. 6200: 1255832. https://doi.org/10.1126/science.1255832

Raghavan, Maanasa, Pontus Skoglund, Kelly E. Graf, Mait Metspalu, Anders Albrechtsen, Ida Moltke, Simon Rasmussen, et al. 2014b. "Upper Palaeolithic Siberian genome reveals dual ancestry of Native Americans." *Nature* 505, no. 7481: 87–91. https://doi.org/10.1038/nature12736

Raghavan, Maanasa, Matthias Steinrücken, Kelley Harris, Stephan Schiffels, Simon Rasmussen, Michael DeGiorgio, Anders Albrechtsen, et al. 2015. "Genomic evidence for the Pleistocene and recent population history of Native Americans." *Science* 349, no. 6250: aab3884. https://doi.org/10.1126/science.aab3884

Rasmussen, Morten, Sarah L. Anzick, Michael R. Waters, Pontus Skoglund, Michael DeGiorgio, Thomas W. Stafford, Simon Rasmussen, et al. 2014. "The genome of a Late Pleistocene human from a Clovis burial site in Western Montana." *Nature* 506, no. 7487: 225–229. https://doi.org/10.1038/nature13025

Rasmussen, Morten, Yingrui Li, Stinus Lindgreen, Jakob Skou Pedersen, Anders Albrechtsen, Ida Moltke, Mait Metspalu, et al. 2010. "Ancient human genome sequence of an extinct Palaeo-Eskimo." *Nature* 463, no. 7282: 757–762. https://doi.org/10.1038/nature08835

Rasmussen, Morten, Martin Sikora, Anders Albrechtsen, Thorfinn Sand Korneliussen, J. Víctor Moreno-Mayar, G. David Poznik, Christoph P. E. Zollikofer, et al. 2015. "The ancestry and affiliations of Kennewick Man." *Nature* 523, no. 7561: 455–458. https://doi.org/10.1038/nature14625

Reich, David, Nick Patterson, Desmond Campbell, Arti Tandon, Stéphane Mazieres, Nicolas Ray, Maria V. Parra, et al. 2012. "Reconstructing Native American population history." *Nature* 488, no. 7411: 370–374. https://doi.org/10.1038/nature11258

Scheib, C. L., Hongjie Li, Tariq Desai, Vivian Link, Christopher Kendall, Genevieve Dewar, Peter William Griffith, et al. 2018. "Ancient human parallel lineages within North America contributed to a coastal expansion." *Science* 360, no. 6392: 1024–1027. https://doi.org/10.1126/science.aar6851

Schurr, T. G., S. W. Ballinger, Y. Y. Gan, J. A. Hodge, D. A. Merriwether, D. N. Lawrence, W. C. Knowler, K. M. Weiss, and D. C. Wallace. 1990. "Amerindian mitochondrial DNAs have rare Asian mutations at high frequencies, suggesting they derived from four primary maternal lineages." *American Journal of Human Genetics* 46, no. 3: 613–623.

Sikora, Martin, Vladimir V. Pitulko, Vitor C. Sousa, Morten E. Allentoft, Lasse Vinner, Simon Rasmussen, Ashot Margaryan, et al. 2019. "The population history of northeastern Siberia since the Pleistocene." *Nature* 570, no. 7760: 182–188. https://doi.org/10.1038/s41586-019-1279-z

Skoglund, Pontus, Swapan Mallick, Maria Cátira Bortolini, Niru Chennagiri, Tábita Hünemeier, Maria Luiza Petzl-Erler, Francisco Mauro Salzano, Nick Patterson, and David Reich. 2015. "Genetic evidence for two founding populations of the Americas." *Nature* 525, no. 7567: 104–108. https://doi.org/10.1038/nature14895

Skoglund, Pontus, and David Reich. 2016. "A genomic view of the peopling of the Americas." *Current Opinion in Genetics & Development* 41 (December): 27–35. https://doi.org/10.1016/j.gde.2016.06.016

Tackney, Justin, Anne M. Jensen, Caroline Kisielinski, and Dennis H. O'Rourke. 2019. "Molecular analysis of an ancient Thule population at Nuvuk, Point Barrow, Alaska." *American Journal of Physical Anthropology* 168, no. 2: 303–317. https://doi.org/10.1002/ajpa.23746

Tackney, Justin, and Jennifer A. Raff. 2019. "A different way: Perspectives on human genetic research from the Arctic." *The SAA Archaeological Record* 19, no. 2: 20–24.

Tamm, Erika, Toomas Kivisild, Maere Reidla, Mait Metspalu, David Glenn Smith, Connie J. Mulligan, Claudio M. Bravi, et al. 2007. "Beringian Standstill and spread of Native American founders." *PLOS One* 2, no. 9: e829. https://doi.org/10.1371/journal.pone.0000829

Torroni, A., T. G. Schurr, M. F. Cabell, M. D. Brown, J. V. Neel, M. Larsen, D. G. Smith, C. M. Vullo, and D. C. Wallace. 1993. "Asian affinities and continental radiation of the four founding Native American mtDNAs." *American Journal of Human Genetics* 53, no. 3: 563–590.

Wang, Sijia, Cecil M. Lewis, Jr., Mattias Jakobsson, Sohini Ramachandran, Nicolas Ray, Gabriel Bedoya, Winston Rojas, et al. 2007. "Genetic variation and population structure in Native Americans." *PLOS Genetics* 3, no. 11: e185. https://doi.org/10.1371/journal.pgen.0030185

PART II
ANCIENT DNA AND MIGRATION

4

An Arctic Lens for American Migration

Integrating Genomics, Archaeology, and Paleoecology

Dennis H. O'Rourke, Justin Tackney, and Lauren Norman

Introduction

The peopling of the Western Hemisphere has intrigued scholars of migration for centuries. The Jesuit friar Jose de Acosta was the first to suggest that the indigenous people of the Western Hemisphere derived from early Asian populations based on patterns of morphological similarity (de Acosta 1592). de Acosta's early insight was essentially confirmed by the middle of the last century through more detailed morphological studies (see Scott et al. 2018), documented patterns of genetic variation across the Americas, the similarity to such patterns in Asian populations (McComb et al. 1995; Torroni et al. 1993; for a recent review, see Bolnick et al. 2016), and the documentation of the late Pleistocene land connection between Asia and North America, Beringia (Hoffecker et al. 2014, 2016; Hopkins 1967; Hultén 1937). The route of entry via Beringia was established, as de Acosta had speculated, but the timing of initial entry into the hemisphere and the routes of dispersal into the interior of North America and, ultimately, South America have remained elusive and controversial (Goebel et al. 2008; Llamas et al. 2016, 2017; Moreno-Mayar et al. 2018; Mulligan and Szathmary 2018; Nei and Roychoudhury 1993; O'Rourke 2014; and Posth et al. 2018; and references therein).

   The long-held view of the initial entry into the American continents consisted of "waves" of small migrating bands from Asia (Siberia), each of which moved eastward across the Beringian land bridge in sequence at the end of the Last Glacial Maximum (LGM). The migrant waves filtered southward through an ice-free corridor (IFC) between the retreating Laurentide and Cordilleran glacial masses and ultimately dispersed throughout the continents (Figure 4.1A). According to the traditional view, the migration was rapid, occurred after 14 thousand years ago (kya), and has typically been associated with a subsistence strategy focused on big-game hunting, resulting in the extinction of megafaunal species in North America (the Overkill Hypothesis).

Dennis H. O'Rourke, Justin Tackney, and Lauren Norman, *An Arctic Lens for American Migration* In: *Human Migration*. Edited by: Maria de Lourdes Muñoz-Moreno and Michael H. Crawford, Oxford University Press. © Oxford University Press 2021. DOI: 10.1093/oso/9780190945961.003.0005

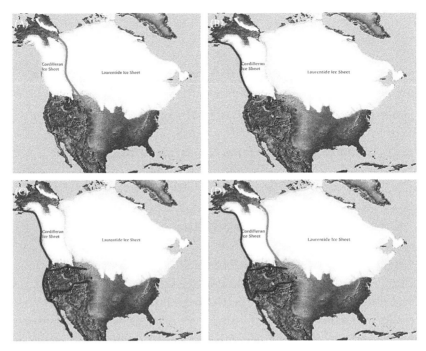

Figure 4.1   A: migration path from Beringia to the midcontinent via the Ice Free Corridor (IFC), traces the traditionally proposed route into the Western hemisphere. Panels (B), (C), and (D) show an alternative migration model. B: alternative route supported by some molecular genetic and paleoenvironmental data suggests a slightly earlier southward migration along the Pacific Coast. C: coastal migrants could have begun to exploit riverine resources that would draw them toward the interior of the continent, where, D: the model suggests, they met other migrants traveling southward via the IFC.

Source: Base map modified from Cosmographic Research 2009 (http://www.cosmographicresearch.org/prelim_glacial_maximum.htm).

An Alternative Migration Model

Fladmark (1979) proposed that an earlier movement of people along the Pacific Coast may have predated migration through the IFC (see also Erlandson et al. 2007; Scheib et al. 2018). While this contrasting migration scenario remains subject to debate, recent research in molecular genetics and isotopic analysis has given considerable support to the early coastal entry model. Lesnek et al. (2018) and Misarti et al. (2012) provide independent evidence that a Pacific coastal route was open and available by at least 17 kya. Ancient DNA analysis of lake cores in the former IFC (Pedersen et al. 2016) and ancient mitochondrial DNA (mtDNA) from late Pleistocene bison fossils (Heintzman et al. 2016) independently demonstrate that the IFC was not available for human transit until at least 13 kya, millennia after the opening of a Pacific coastal route south from Beringia (but, for a contrasting view, see Potter et al. 2018). The late opening of the IFC places it after well-documented archaeological sites in both North and South America (Goebel et al. 2008; Hoffecker et al. 2016; Llamas et al. 2017).

Molecular genetic data, coupled with paleoecological data that indicate when dispersal routes into the Americas became available, suggest a more complex model of initial peopling. Populations moving from (see below) or through Beringia first moved southward along the Pacific Coast (Figure 4.1B) and dispersed eastward into the interior of North America once south of the retreating Cordilleran ice sheet (Figure 4.1C). Later, after the IFC was available for human transit, populations that remained in eastern Beringia moved south and ultimately met populations that had dispersed some millennia earlier along the coastal route (Figure 4.1D). The archaeological record is also clear that by 12 kya, interior North American populations were moving north through the corridor (Clark 1984; Hoffecker 2011; Smith et al. 2013; Smith and Goebel 2018), presumably following large game. Several archaeological sites in the north are considered evidence of mid-latitude populations that had migrated north by this time (see Smith and Goebel 2018). Under such a scenario, the northward-migrating big-game hunters of the interior might well have encountered and interacted with Beringian populations inhabiting the IFC, or similar populations making their way southward through the IFC.

This more complex view of American dispersals indicates at least two points of entry into North America, a Pacific coastal route and an interior path through the IFC. Both routes may have been used repeatedly by formerly isolated groups over long periods of time, rather than acting as conduits for discrete waves of migration as traditionally proposed. Such a perspective on American peopling also implies early admixture in interior North America between groups that may have been separated by several centuries, perhaps millennia. Such early admixture has not been formally analyzed, nor have the outcomes been modeled, using available genetic/genomic data.

Paleoecology and the Beringian Standstill

Geneticists have long speculated that the original migrants to the hemisphere derived from an isolated population on the Beringian land mass (Mulligan and Szathmary 2017; Salzano and Bonatto 1997; Szathmary 1993; Tamm et al. 2007). Paleoenvironmental reconstructions of Beringia, especially south-central and southeast Beringia, lend support to this hypothesis, generally referred to as the Beringian Standstill model or the Beringian Pause model. Beringia was long characterized as a cold, barren, steppe tundra that supported megafaunal grazers but little else. Its climate was seen as too harsh to be a place where human populations tarried. In more recent Beringian paleoecological reconstructions (Elias et al. 1996; Elias and Crocker 2008; Hoffecker et al. 2016), southern Beringia is characterized by a mesic tundra biome, with substantial resources, including wood from small trees in river valleys, birds, and small, as well as some large, game. This would have been an appropriate refugium environment for one or more subdivided human populations to have existed in, in isolation from Asian/Siberian populations, throughout the LGM. If a population in Beringia was, in fact, isolated for sufficient time, the emergence of an identifiable genome, a Native American genome, distinct from ancestral Asian populations, would have been possible.

Such an isolated Beringian population would also have occupied a geographic area immediately adjacent to the opening corridor along the Pacific Coast by 17 kya. Thus, a plausible source population for an early coastal entry, residing much closer to the opening of the Pacific coastal route than the interior IFC, is possible and perhaps likely. Certainly, there is evidence from other medium to large vertebrate species of survival in multiple refugia in northern North America, especially along the Pacific Coast, and redispersal after the LGM (Shafer et al. 2010). It is not clear why humans would have been an exception.

A primary difficulty is dating the initial dispersal, an area that has traditionally been the domain of archaeology. However, the precise area most likely to provide archaeological evidence of a Beringian refugium population is inaccessible to archaeological investigation—the low-lying mesic tundra of south-central Beringia that now forms the northern Bering Sea shelf. Alternatively, molecular estimates of lineage coalescences can be used as proxies. In this case, as more genomic data become available, the estimates of the timing of the origin of the source population genome are increasingly older than the ~ 14 kya (e.g., Llamas et al. 2016; Moreno-Mayar et al. 2018; Raghavan et al. 2015) traditionally obtained from mainland archaeological research. An obvious question is: How reliable are the molecular dates?

Potential Bias in Molecular Dating Estimates?

In a recent analysis, Llamas et al. (2017) sequenced 92 ancient mitochondrial genomes from across South America. Over 80 of those mitochondrial lineages had never been observed in contemporary populations. Additionally, one of the most common mtDNA lineages in living southern South American populations, D4h3a, was not observed at all in the large ancient sample. This clearly indicates that the pre-Columbian indigenous population was more variable than previously assumed and that, in some cases, contemporary patterns of genetic variation are poor indicators of past patterns of variation. Why might this be so?

One obvious reason is the postcolonial demographic history of indigenous communities in the Americas. The demographic collapse of many Indigenous populations and cultures at European contact is well known and documented (Crosby 1986; Kelton 2015; Koch et al. 2019). This collapse meant that surviving populations suffered an extreme, and often protracted, bottleneck. The genetic result is a dramatic loss of variation; thus, frequently (but not universally) modern populations do not reflect precontact levels of genetic diversity. For example, the lowland South American populations of the Karitiana and Surui were among the earliest to have whole genomes made available. They have routinely been used as baseline populations for studying patterned diversity and population history in the Americas. However, it is now known that they are characterized by very high levels of homozygosity, reflecting decreased variation after contact. For the Surui, the mean identity by descent (IBD) value, a measure of reduced variability in two or more people due to inheritance of alleles from a common ancestor, is one to two orders of magnitude higher than for other populations around the world, including many populations that also underwent historical bottlenecks (Henn et al. 2012). Using microsatellite data on over 60 Indigenous American populations, Hunley et al. (2016) also found that three isolated lowland populations in southern Amazonia, the Surui, the Karitiana, and the Ache, had very low levels of genetic variation, coupled with virtually no European admixture. These results made them extreme genetic outliers relative to other Indigenous American populations' extent of genetic variation. Moreover, Szpiech et al. (2013) found that the Karitiana and Surui were characterized by much longer runs of homozygosity (RoH) detected by exome sequencing. Most populations have long RoH comprising less than 10% of the genome, while the two South American populations have genomes that are characterized by ≥ 40% RoH, reflecting a dramatic decrease in variation relative to other populations. The effects of basing population genetic and phylogenetic parameter estimates, such as coalescent times or source population sizes, on such samples has not been effectively studied. Dramatic underestimates of the extent of population variation, coupled with nonstationary population histories, could result in underestimations of coalescent times and population sizes (Cox 2009). It is equally important to note that the effects of genetic drift in small populations, both prehistorically and in postcolonial times, could erroneously highlight or minimize unique genetic signatures in these populations present earlier in their history, leading to "ghost" signals of ancestry.

Another issue to be confronted is undersampling. There are sparse molecular data actually available for the whole of North America south of the Arctic (Bolnick et al. 2016; Scheib et al. 2018). There are only a few contemporary populations molecularly characterized at a genomic level in Canada, e.g., the Chipewyan, Cree, and Ojibwa. In an analysis of microsatellite variation in 63 Indigenous American populations, Hunley et al. (2016) showed that the three Canadian groups were characterized by higher levels of European ancestry and therefore exhibited lower levels of pan-Indian ancestry than other populations in the analysis. Thus, these North American populations appear divergent from other Indigenous American populations in some genetic analyses, but for different reasons than seen in the lowland South American populations (Hunley et al. 2016). South of the United States–Mexico border, a number of indigenous populations have been studied molecularly, and the number of groups with genomic data grows throughout South America. This curious distribution means most estimates of the timing of the origin of the Native American genome and possible initial dispersal routes based on molecular data derive from the genetic study of South American populations. This undersampling of North American indigenous populations may also result in bias in estimates of timing of migration events. Marin and Hedges (2018) have recently shown that undersampling of genomes has a negative effect on the accuracy of estimation of time and rates of change in phylogenetic studies, such as the diversification and dispersal of populations in the Western Hemisphere considered here (see also Navascués and Emerson 2009).

Implications

The implications of the potential bias in evolutionary parameter estimates may have substantial ramifications for our views of hemispheric colonization and the way that we balance the archaeological, genetic, and paleoecological records. It is now clear that the Arctic was not the impediment for human occupation it was once thought. The extensive archaeological record at the Yana Rhinoceros Horn site in the Yana-Indigurka plain in northeastern Siberia (northwestern Beringia) demonstrates that, at least by 32 kya, populations were occupying very-high-latitude environments, and did so throughout the LGM (Pitulko et al. 2004, 2012). Bourgeon and colleagues (2017) reexamined the faunal material recovered from the Bluefish Caves in northern Yukon and found cut marks had been made on fresh bone that dated between 18 kya and 28 kya, substantially before most archaeologists have assumed such high-latitude environments were occupied in eastern Beringia. Although there is no evidence the Bluefish Caves were occupation centers (rather, they appear to be temporary hunting/butchery sites), the fact that people were operating this far north during the height of the LGM is surprising. Not surprisingly, many archaeologists have been reluctant to accept the early dates for Bluefish Caves, much like the case a generation ago for the sites of Monte Verde and Meadowcroft Rock Shelter, sites that are now widely accepted as predating 14 kya. More recently, Vachula et al. (2019) present evidence from an LGM lake core in north Alaska containing charcoal of presumed anthropogenic origin along with coprastanols that indicate the presence of humans on the landscape. The deposits containing these biogeochemical markers date to 32 kya, coeval with the Yana site further west.

If the new reports of human presence in Arctic North America during and before the LGM are ultimately confirmed and accepted, it will require a major reassessment of traditional views of both the timing of entry into the Western Hemisphere and subsequent dispersal throughout the Americas. As more genomic-scale data become available, the continued reliance on genomes that are highly homozygous and depauperate of variation will decrease, minimizing any bias in time estimates from molecular data in contemporary populations with reduced genetic variability. Moreover, if, as suggested by recent genomic research (Llamas et al. 2016, 2017; Moreno-Mayar et al. 2018; Raghavan et al. 2015), the origin of indigenous genomes in the Western Hemisphere is placed during or before the LGM (e.g., 20 kya to 25 kya), the molecular and newly emerging archaeological records seem likely to become more concordant with each other, and with the paleoecological record as well.

It seems increasingly likely that a case can be made for human habitation in Beringia, isolated from Eurasian populations to the southwest during the LGM, and for multiple pathways from Beringia over several thousands of years into the interior of the American continents. Discrete waves of migration directly from Asia now seem inadequate to accommodate an isolated and extended human occupation in one or more Beringian refugia during the LGM. The more complex histories of refugium populations and their dispersal are consistent with the most recent molecular genomic studies emphasizing complex patterns of genetic diversity and the demographic scenarios that likely gave rise to them (Moreno Mayar et al. 2018a, 2018b; Pinotti et al. 2019; Posth et al. 2018).

References

Bolnick, D. A., J. A. Raff, L. C. Springs, A. W. Reynolds, and A. T. Miró-Herrans. 2016. "Native American genomics and population histories." *Annual Review of Anthropology* 45: 20.1–20.22.

Bourgeon, L., A. Burke, and T. Higham. 2017. "Earliest human presence in North America dated to the Last Glacial Maximum: New radiocarbon dates from Bluefish Caves, Canada." *PLOS One* 12, no. 1: e0169486. doi:10.1371/journal.pone.0169486

Clark, D. W. 1984. "Northern fluted points: Paleo-Eskimo, Paleo-Arctic, or Paleo-Indian." *Canadian Journal of Anthropology* 4, no. 1: 65–81.

Cox, M. 2009. "Accuracy of molecular dating with the rho statistic: Deviations from coalescent expectations under a range of demographic models." *Human Biology* 81, no. 5/6: 911–933.

Crosby, A. W. 1986. *Ecological Imperialism: The Biological Expansion of Europe, 900-1900.* Cambridge: Cambridge University Press.

Dillehay, T. D., S. Goodgred, M. Pino, V. F. Vasquez Sanchez, T. R. Tham, J. Adovasio, M. B. Collins, P. J. Netherly, C. A. Hastorf, K. L. Chiou, D. Piperno, I. Rey, and N. Velchoff. 2017. "Simple technologies and diverse food strategies of the Late Pleistocene and Early Holocene at Huaca Prieta, Coastal Peru." *Science Advances* 3: e1602778.

Dillehay, T. D., C. Ocampo, J. Saavedra, A. O. Sawakuchi, R. M. Vega, M. Pino, M. B. Collins, L. S. Cummings, I. Arregui, X. S. Villagran, G. A. Hartmann, M. Mella, A. Gonzalez, and George Dix. 2015. "New archaeological evidence for an early human presence at Monte Verde, Chile." PLOS One e0141923. doi:10.1371/journal.pone.0141923

Elias, S. A., and B. Crocker. 2008. "The Bering land bridge: A moisture barrier to the dispersal of steppe-tundra biota?" *Quaternary Science Reviews* 27: 2473–2483.

Elias, S. A., S. K. Short, C. H. Nelson, et al. 1996. "Life and times of the Bering land bridge." *Nature* 382: 60–63.

Erlandson, J. M., M. H. Graham, B. J. Bourque, D. Corbett, J. A. Estes, and R. S. Steneck. 2007. "The kelp highway hypothesis: Marine ecology, the coastal migration theory, and the peopling of the Americas." *Journal of Island and Coastal Archaeology* 2, no. 2: 161–174.

Fladmark, K. R. 1979. "Routes: Alternate migration corridors for early man in North America." *American Antiquity* 44, no. 1: 55–69.

Goebel, T., M. Waters, and D. H. O'Rourke. 2008. "The Late Pleistocene dispersal of modern humans in the Americas." *Science* 319: 1497–1502.

Heintzman, P. D., D. Froese, J. W. Ives, A. E. R. Soares, G. D. Zazula, B. Letts, T. D. Andrews, J. C. Driver, E. Hall, P. G. Hare, C. N. Jass, G. MacKay, J. R. Southon, M. Stiller, R. Woywitka, M. A. Suchard, and B. Shapiro. 2016. "Bison phylogeography constrains dispersal and viability of the ice free corridor in western Canada." *PNAS* 113, no. 29: 8057–8063.

Henn, B. M., L. Hon, J. M. Macpherson, N. Eriksson, S. Saxonov, I. Pe'er, and J. L. Mountain. 2012. "Cryptic distant relatives are common in both isolated and cosmopolitan genetic samples." *PLOS One* 7, no. 4: 1–13.

Hlusko, L. J., J. P. Carlson, G. Chaplin, S. A. Elias, J. F. Hoffecker, M. Huffman, N. G. Jablonski, T. A. Monson, D. H. O'Rourke, M. A. Pilloud, and G. R. Scott. 2018. "Environmental selection during the last ice age on the mother-to-infant transmission of vitamin D and fatty acids through breast milk." *PNAS* 115, no. 19: E4426–E4432.

Hoffecker, J. F. 2011. "Assemblage Variability in Beringia: The Mesa Factor." In *From the Yenisei to the Yukon: Interpreting Lithic Assemblage Variability in Late Pleistocene/Early Holocene Beringia*, edited by T. Goebel and I. Buvit. College Station: Texas A&M University Press.

Hoffecker, J. F., S. A. Elias, and D. H. O'Rourke. 2014. "Out of Beringia?" *Science* 343: 979–980.

Hoffecker, J. F., S. A. Elias, D. H. O'Rourke, G. R. Scott, and N. H. Bigelow. 2016. "Beringia and the global dispersal of modern humans." *Evolutionary Anthropology* 25: 64–78.

Hopkins, D. M., ed. 1967. *The Bering Land Bridge*. Palo Alto: Stanford University Press.

Hunley, K., K. Gwin, and B. Liberman. 2016. "A reassessment of the impact of European contact on the structure of Native American genetic diversity." *PLOS One* 11, no. 8: e0161018. doi:10.1371/journal.pone.0161018

Kelton, P. 2015. *Cherokee Medicine, Colonial Germs: An Indigenous Nation's Fight Against Smallpox, 1518–1824*. Norman: Oklahoma University Press.

Kock, A., C. Brierley, M. M. Maslin, and S. L. Lewis. 2019. "Earth system impacts of the European arrival and Great Dying in the Americas after 1492." *Quaternary Science Reviews* 207: 13–36.

Llamas, B., L. Fehren-Schmitz, G. Valverde, J. Soubrier, S. Mallick, et al. 2016. "Ancient mitochondrial DNA provides high-resolution time scale of the peopling of the Americas." *Science Advances* 2: e1501385.

Llamas, B., K. M. Harkins, and L. Fehren-Schmitz. 2017. "Genetic studies of the peopling of the Americas: What insights do diachronic mitochondrial genome datasets provide?" *Quaternary International* 444: 26–35.

Lesnek, A. J., J. P. Briner, C. Lindqvist, J. F. Baichtal, and T. H. Heaton. 2018. "Deglaciation of the Pacific coastal corridor directly preceded the human colonization of the Americas." *Science Advances* 4: eaar5040.

Marin, J., and S. B. Hedges. 2018. "Undersampling genomes has biased time and rate estimates throughout the tree of life." *Molecular Biology and Evolution* 35, no. 8: 2077–2084.

McComb, J., N. Blagitko, A. G. Comuzzie, M. S. Schanfield, R. I. Sukernik, W. R. Leonard, and M. H. Crawford. 1995. "VNTR DNA variation in Siberian indigenous populations." *Human Biology* 67: 217–229.

Misarti, N., B. P. Finney, J. W. Jordan, H. D. G. Maschner, J. A. Addison, M. D. Shapley, A. Krumhardt, and J. E. Beget. 2012. "Early retreat of the Alaska Peninsula glacier complex and the implications for coastal migrations of First Americans." *Quaternary Science* 48: 1–6.

Moreno-Mayar, J. V., B. Potter, L. Vinner, M. Steinbrücken, S. Rasmussen, et al. 2018a. "Terminal Pleistocene Alaskan genome reveals first founding population of Native Americans." *Nature* 553: 203–208.

Moreno-Mayar, J. V., L. Vinner, P. de B. Damgaard, C. de la Fuente, J. Chan, et al. 2018b. "Early human dispersals within the Americas." *Science* 362: eaav2621. doi:10.1126/science.aav2621

Mulligan, C. J., and E. J. E. Szathmáry. 2017. "The peopling of the Americas and the origin of the Beringian occupation model." *American Journal of Physical Anthropology* 162: 403–408. doi:10.1002/ajpa.23152.

Navascués, M., and B. C. Emerson. 2009. "Elevated substitution rate estimates from ancient DNA: Model violation and bias of Bayesian methods." *Molecular Ecology* 18: 4390–4397.

Nei, M., and A. K. Roychoudhury. 1993. "Evolutionary relationships of human populations on a global scale." *Molecular Biology and Evolution* 10, no. 5: 927–943.

O'Rourke, D. H. 2011. "Contradictions and concordances in American colonization models." *Evolution: Education & Outreach* 4, no. 2: 244–253.

Pedersen, M. W., A. Ruter, C. Schweger, H. Friebe, R. A. Staff, K. K. Kjeldsen, M. L. Z. Mendoza, A. B. Beaudoin, C. Zutter, N. K. Larsen, B. A. Potter, R. Nielsen, R. A. Rainville, L. Orlando, D.

J. Meltzer, K. H. Kjaer, and E. Willerslev. 2016. "Postglacial viability and colonization in North America's ice-free corridor." *Nature* doi:10.1038/nature19085

Pinotti, T., A. Bergström, M. Geppert, M. Bawn, D. Ohasi, et al. 2019. "Y chromosome sequences reveal a short Beringian Standstill, rapid expansion, and early population structure of Native American founders." *Current Biology* 29: 1–9.

Pitulko, V. V., P. A. Nikolsky, E. Y. Girya, A. E. Basilyan, V. E. Tumskoy, S. A. Koulakov, S. N. Astoakhov, E. Y. Pavlova, and M. A. Anisimov. 2004. "The Yana RHS site: Humans in the Arctic before the last glacial maximum." *Science* 30-3, no. 5654: 52–56.

Pitulko, V. V., E. Y. Pavlova, P. A. Nikolskiy, and V. V. Ivanova. 2012. "The oldest art of the Eurasian Arctic: Personal ornaments and symbolic objects from Yana RHS, arctic Siberia." *Antiquity* 86, no. 333: 642–659.

Posth, C., N. Nakatsuka, I. Lazaridis, P. Skoglund, S. Mallick, et al. 2018. "Reconstructing the deep population history of Central and South America." *Cell* 175: 1–13.

Roos, C. I., M. N. Zedeno, K. L. Hollenback, and M. M. H. Erlick. 2018. "Indigenous impacts on North American Great Plains fire regimes of the past millennium." *PNAS* 115, no. 32: 8143–8148.

Scheib, C. L., H. Li, T. Desai, V. Link, C. Kendall, et al. 2018. "Ancient human parallel lineages within North America contributed to a coastal expansion." *Science* 360, no. 6392: 1024–1027.

Scott, G. R., K. Schmitz, K. N. Heim, K. S. Paul, R. Schomberg, and M. A. Pilloud. 2018. "Sinodonty, sundadonty, and the Beringian Standstill model: Issues of timing and migrations into the New World." *Quaternary International* 466: 233–246.

Shafer, A. B. A., C. I. Cullingham, S. D. Côté, and D. W. Coltman. 2010. "Of glaciers and refugia: A decade of study sheds new light on the phylogeography of northwestern North America." *Molecular Ecology* 19, no. 21: 4589–4621.

Smith, H. L., and T. Goebel. 2018. "Origins and spread of fluted-point technology in the Canadian ice-free corridor and eastern Beringia." *PNAS* 115: 4116–4121.

Smith, H. L., J. T. Rasic, and T. Goebel. 2013. "Biface Traditions of Northern Alaska and Their Role in the Peopling of the Americas." In *PaleoAmerican Odyssey*, edited by K. E. Graf, C. V. Ketron, and M. R. Waters. College Station: Center for the Study of the First Americans.

Szathmary, E. J. E. 1993. "Genetics of aboriginal North Americans." *Evolutionary Anthropology* 1: 202–220.

Tamm, E., T. Kivisild, M. Reidla, M. Metspalu, D. G. Smith, C. J. Mulligan, C. M. Bravi, O. Rickards, C. Martinez-Labarga, E. K. Khusnutdinova, S. A. Federova, M. V. Gouibenko, V. A. Stepanov, M. A. Gubina, S. I. Zhadanov, L. P. Ossipova, L. Damba, M. I. Voevoda, J. E. Dipierri, R. Villems, and Ripan Malhi. 2007. "Beringian Standstill and spread of Native American founders." *PLOS One* 2, no. 9: e829. doi:10.1371/journal.pone.0000829

Torroni, A., R. I. Sukernik, T. G. Schurr, Y. B. Starikovskaya, M. F. Cabell, M. H. Crawford, A. G. Comuzzie, and D. C. Wallace. 1993. "Mitochondrial DNA variation of aboriginal Siberians reveals distinct genetic affinities with Native Americans." *American Journal of Human Genetics* 53: 591–608.

Vachula, R. S., Y. Huang, W. M. Longo, S. G. Dee, W. C. Daniels, and J. M. Russell. 2019. "Evidence of Ice Age humans in eastern Beringia suggests early migration to North America." *Quaternary Science Reviews* 205: 35–44.

5

Mitochondrial DNA Analysis and Pre-Hispanic Maya Migrations

Languages and Climate Influence

*Maria de Lourdes Muñoz-Moreno, Mirna Isabel Ochoa-Lugo,
Gerardo Pérez-Ramírez, Kristine G. Beaty, Adrián Martínez Meza,
and Michael H. Crawford*

Introduction

The Maya civilization is one of the best-known classical civilizations of Mesoamerica, and it developed into a society equipped with astronomy, mathematics, a calendar system, hieroglyphic writing, astrology, and agricultural innovations (Houston et al. 2000). It was during the Classic period that the Maya constructed most of the extensive cities and pyramids that have made them famous. The present-day indigenous populations of Mexico and Central America are descendants of the people who built these great cities, and some of them are still residing on the same lands. Based on historical documentation and recent studies, it is accepted that the Maya are the descendants of the earliest settlers of the Americas who eventually came to the Yucatan Peninsula and south to Lake Petén, Guatemala, where they established a kingdom with their capital and sacred city of Flores Island (Ochoa-Lugo et al. 2016). The Yucatan Peninsula became the principal region of a new culture, called Toltec/Maya, which formed when Toltec migrated from the north and integrated with the Maya people. The inscriptions that have been translated provide detailed accounts of their myths and political events, including conflicts between city-states (Law 2013).

Some archeologists have suggested that the Maya were Olmec descendants and/or trading partners, who settled along the Gulf of Mexico, including La Venta, between 1400 and 1150 BCE (Rust and Sharer 1988). However, based on architectural features and carbon-14 dating, Inomata et al. (2013) suggested that the development of lowland Maya civilization resulted from interactions with groups in the southwestern Maya lowlands, Chiapas, the Pacific Coast, and the southern Gulf Coast.

María de Lourdes Muñoz-Moreno, Mirna Isabel Ochoa-Lugo, Gerardo Pérez-Ramírez, Kristine G. Beaty, Adrián Martínez Meza, and Michael H. Crawford, *Mitochondrial DNA Analysis and Pre-Hispanic Maya Migrations* In: *Human Migration*. Edited by: Maria de Lourdes Muñoz-Moreno and Michael H. Crawford, Oxford University Press. © Oxford University Press 2021. DOI: 10.1093/oso/9780190945961.003.0006

Factors Influencing Migration and Population Movements

Maize Cultivation and Human Migration

In the transition from hunting and gathering to agriculture in Mexico, the population subsisted on corn cultivation. Basic pottery and stone tools, excavated in archaeological sites, date back to the Holocene epoch in 7200 BCE (Ranere et al. 2009). A phytolith analysis of sediments from San Andres, Tabasco, confirmed the diffusion of maize cultivation from the tropical Mexican Gulf Coast > 7,000 years ago (~ 5800 BCE), followed by the fast spread of the crop to South and North America by the Maya ancestors (da Fonseca et al. 2015; Piperno et al. 2009).

Maya Language

Migration had a role in diversifying languages within Mesoamerica, in contrast to language's being the driving force in dispersal through the Americas. Simultaneously, language and culture could also have played a crucial role in Mesoamerica during the colonization of the American continent. Proto-Mayan is the common ancestor of all modern Mayan languages today. Classic Maya languages are documented in the hieroglyphic inscriptions (Law 2013). According to the classification scheme developed by Campbell and Kaufman (1985), Proto-Mayan split into six main branches: K'iche'an, Mamean, Q'anjob'alan, Ch'olan-Tseltalan, Yukatekan, and Huastecan. The first division occurred around 2200 BCE, when Huastecan moved northwest along the Gulf Coast of Mexico. Subsequently, each subgroup was spawned, and at the present time the Maya language family includes approximately 31 languages spoken by more than 5 million people in Guatemala, Mexico, Belize, and Honduras, as well as in relocated communities in the United States and Canada (Law 2013).

Climate Change's Influence on Maya Population Migrations, Culture, and Languages Changes

Human societies around the world have had to either adapt to environmental changes and natural and anthropogenic disasters or be extinguished. Fluctuations in the environment have resulted in the disappearance of entire populations and cultures, which is of particular concern because it also results in the extinction of languages and traditional ecological knowledge (Maffi 2005). When environmental changes are gradual, they can lead people to migrate, which can cause changes in cultural practices, language integrity, and genetic diversity.

Overall, water management seemed to be a key to maintaining the political power in many historic cultures and states and to distributing settlements across landscapes (Lucero 2008). Several studies have shown that between the 9th and 15th centuries, regions like the Maya lowlands were highly susceptible to extreme variations in weather and to climatic shifts (Lucero et al. 2015). Among Maya cultures, water was managed and distributed via intricately linked systems, as in Tikal and the lowlands (Scarborough et al. 2012). Consequently, when environmental conditions changed, due in part to climate changes, the critical infrastructure could collapse, thus threatening survival of the human groups, who were forced to adapt to stringent conditions and/or eventually to migrate (Figueroa 2011).

Traditional explanations of the collapse of the Maya civilization have suggested that military conflicts with Aztec forces caused their collapse during the 9th to 7th centuries BCE. This was after years of Maya population increase and intensive land use. However, new evidence indicates that the collapse may have been the result of drought conditions that could not sustain the large Maya population (Evans et al. 2018; Kennet and Beach 2013; Turner 2012).

In summary, several weather anomalies and climatological factors (temperature, precipitation, humidity, and soil moisture) may have negatively affected agrarian and non-agrarian lifestyles and economies of ancient Maya (Hsiang et al. 2013). Consequently, the intersection of demographic, political, and climatic factors may have caused habitat destruction, resulting in cultural diversity and linguistic modification.

One response of local indigenous communities to the impacts of climate change is migration (López-Carr and Marter-Kenyon 2015). A good example of this is the study by Evans et al. (2018) that constrained the changing hydrological conditions at Lake Chichancanab using triple oxygen and hydrogen isotope data to independently deconvolve climate variables of precipitation, RHn (normalized relative humidity), and the amount effect. The study clarified how drought affected agriculture like maize production and produced the collapse of Maya civilization in the northern Maya lowlands during the Terminal Classic period. Furthermore, studies of the context of political and climatic change activity at Chichén Itzá around AD 1000, a century after the collapse of Puuc Maya cities and other interior centers in northern Yucatán, also showed a regional drying trend at the end of the Classic period and evidenced a population shift in the 11th century toward some coastal locations after monumental construction and art development. Chichén Itzá Maya may also have expanded their networks of political and economic interaction to increase access to declining natural resources during the drought at the end of the Classic period (Hoggarth et al. 2016).

Local migration results in minimal cultural and language change, while long-distance migration has the potential to diminish biological, cultural, and linguistic diversity (Pimm et al. 2014). Biological diversity can be impacted because evolution requires some level of genetic diversity on which to operate. If that genetic diversity is diminished owing to anthropogenic or other factors, the possibilities for future change and diversification are reduced. Therefore, it is important to maintain as much diversity as possible to ensure future diversity, in whatever direction evolution takes.

Maya Genetics and Migration

The magnitude of genetic admixture in the Maya population has been influenced by gene flow between different locations, including genetic exchange with people who adopted maize, other domesticates, and ceramic use but without a prolonged sedentary lifestyle (Inomata et al. 2015), as well as by gene flow introduced by other Maya groups. It also has been suggested the effect of admixture proportions of the migratory people not fully sedentary that contributed to the monumental constructions and public ceremonies in the lowland Maya site of Ceibal (Inomata et al. 2015). Ragsdale and Edgar (2015) also suggested that trade and political relationships affected the population structure among Postclassic Mexican populations, decreasing diversity between groups.

Biological Markers

Study of population structure using cranial (von Cramon-Taubadel et al. 2017) and dental morphology to reconstruct patterns of affinity among ancient and modern human groups has been used to understand migration and population history (Cucina et al. 2015; Ragsdale and Edgar 2015; Scherer 2007). Analysis of the odontometric variation of 827 skeletons of Maya from the Classic period (250 BCE to 900 CE) from 12 archaeological sites in Mexico, Guatemala, Belize, and Honduras indicated that the isolation-by-distance model is not applicable to the population structure of the Classic period of Maya (Scherer 2007). Additional studies examining dental morphology and $^{87}Sr:^{86}Sr$ ratios included many locations along the Gulf Coast of Campeche, many places inland in the Maya lowlands to the north, and inland sites in the Chenes region during the Maya Classic period and concluded that the different findings for the dental and isotopic indicators were consistent with proposed trade routes in the peninsula (Cucina et al. 2015).

Genomic Data from Contemporary Maya

Analyses of genomic data from modern and ancient people facilitate the understanding of the ancestral relationships between human populations and the identification of migration routes, diversification events, and genetic admixture among different groups (Ochoa-Lugo et al. 2016). Mitochondrial DNA (mtDNA) reveals only maternal inheritance because it does not recombine (Ingman et al. 2000). In addition, there are numerous copies of mtDNA in each cell, and databases of entire human mtDNA sequences are available for comparison, allowing mtDNA to be widely used both to elucidate the maternal evolutionary history of anatomically modern humans through the reconstruction of prehistoric human dispersals

(Goebel et al. 2008; Melton et al. 2013) and to study human genetic predispositions to some diseases (Falk et al. 2015; Saldaña-Maritnez et al. 2019).

Studies of contemporary Maya populations have provided some insight into the origins and movement of Maya. Phylogenetic studies using mtDNA data suggest moderate local endogamy and isolation of the contemporary Maya, combined with episodes of gene exchange with other ethnic groups, and that a recent adoption of an ethnic identity in the Guatemalan Maya stems from a cultural, rather than a biological, basis (Söchtig et al. 2015). However, the endogamy grade calculated for 619,205 markers in Mexican Maya populations from Yucatán was null (Domínguez-Cruz et al. 2018).

Maya History Through Ancient DNA

Ancient DNA (aDNA) methods have also been used to understand Maya population history. The origin and diversity of Maya maternal lineages in the Mexican region were assessed through the analysis of 38 pre-Hispanic Maya sequences from the mtDNA hypervariable region I (HVI) using samples from the archeological sites of Xcambo, Bonampak (Group Frey and Group Quemado), Palenque (Temple XIII, Temple XV, Group B), El Rey Quintana Roo, Comalcalco (Temple V, Temple III, Dren), Tenosique, Peje Lagarto Huimanguillo, Sueños de Oro, and Calicanto (Figure 5.1). In addition, haplogroup and genetic variation in these pre-Hispanic Maya and their relationship with contemporary Maya populations and other comparative groups were determined to better understand their relationship with other geographic regions and the effects of gene flow on contemporary Mayas (Ochoa-Lugo et al. 2016). mtDNA haplotypes identified were in lineages A (2.6%), A2 (44.73%), A2v (13.16%), C (2.6%), C1 (15.8%), C1b14 (15.8%), and D (5.3%). Lineage B was not identified in the pre-Hispanic Maya samples (Figure 5.1).

Maya History and Population Dynamics

Figure 5.1 displays the geographic distribution of the haplogroup frequencies of the pre-Hispanic Maya samples in the archaeological sites. The mismatch distribution of pairwise differences between sequences of mtDNA for pre-Hispanic and contemporary populations indicated population expansion (not shown), because the distribution was unimodal with main peaks for haplogroup A and bimodal with two peaks for haplogroups C and D.

Median joining networks of haplogroup A mtDNA HVS-I sequence analysis (Figure 5.2) showed the pattern of substitutions in the noncoding control HVS-I region from nucleotides 16106–16399 of 15 pre-Hispanic samples and 656 comparative samples from the GenBank database. The data indicate a genetically close relationship among the pre-Hispanic individuals from Tenosique and Comalcalco,

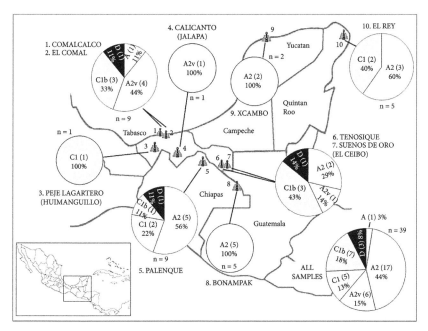

Figure 5.1 Geographic locations of the archaeological Maya sites where samples were collected. The pre-Hispanic samples were collected in the archaeological sites located in the states of Yucatan, Chiapas, and Tabasco. The number and percentage of individuals possessing each haplogroup are shown in the pie chart. The percentages of each haplogroup in the 38 samples are displayed in the pie chart at the bottom on the right side.

exhibiting haplogroup A2v and a shared common ancestor with native Mexican populations and Maya of Guatemala. The individuals from Palenque and Bonampak who share haplogroup A2 are genetically closely and geographically related to each other and with the populations from Peru and Mexican Maya.

Analysis of the pre-Hispanic Mexican Maya sequences containing haplogroup A showed haplotype sharing with the Mississipian archaeological site sample; contemporary Maya (Chiapas); Colombian, Brazilian, Venezuelan, and Beringia ethnic groups; and South Siberia populations (Figure 5.2). These results support the presence of a shared maternal genetic structure among Maya populations, Central American populations, and North and South American populations.

Sequences belonging to haplogroup C from 11 pre-Hispanic Maya populations and 346 populations from the American continent and Asia were also analyzed using reduced median networks (Figure 5.3). The results indicated that Mexican pre-Hispanic Maya groups spread to Central and South America. Interestingly, haplogroup C1b14 identified in five of the pre-Hispanic Maya from Comalcalco, Tenosique, and Sueños de Oro, and the Dren population from Tabasco (Figure 5.3) had been reported in only four contemporary individuals: two Mexican Americans, one Zapotec (Gómez-Carballa et al. 2015; Kumar et al. 2011; Rieux

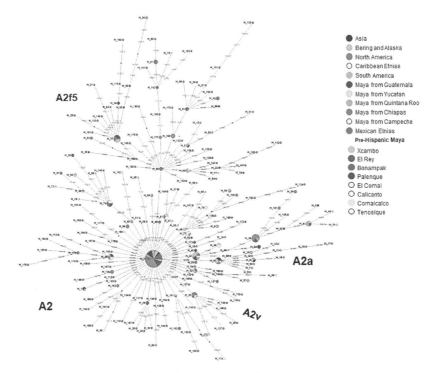

Figure 5.2 Haplotype network analysis of mtDNA haplogroup A of pre-Hispanic and contemporary groups from Maya, Asia, Beringia, and the American continent. A phylogenetic network was constructed with the mtDNA sequences from nucleotides 16106 to 16399. The size of each circle is proportional to the number of individuals in each haplotype present. The distances between the circles correspond to the number of mutations between haplotypes. Inferred missing haplotypes (single nucleotide changes) are marked with the black dots on the branches.

et al. 2014), and one Mexican patient with the mitochondrial disease Kearns-Sayre syndrome (KSS; Saldaña-Martínez et al. 2019). They are closely related to each other and with the main haplotype H10 containing sequences from North, Central, and South America. Decrease of this haplogroup in the contemporary population may be due to population reduction during the Spanish conquest or due to mitochondrial disease susceptibility. The sharing of haplogroup C1b14 among these individuals may suggest common ancestry of the pre-Hispanic Maya and contemporary Mexicans and Native Americans. The sample from Palenque having haplogroup C1d1c1 (H116) grouped with ancient samples from Patagonia, the Caribbean, and contemporary Peruvian, Argentinian, and Chilean populations.

Haplogroup D sequences (nucleotides 16153–16399) analyzed by a median network (Figure 5.4) included one sample from the two pre-Hispanic Maya recovered in Palenque and Comalcalco P3AE2, as well as 262 contemporary and ancient individuals from the Americas and Asia. The results showed a grouping of Maya

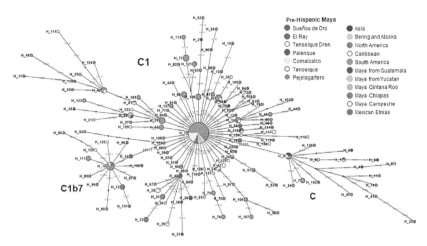

Figure 5.3 Haplotype network analysis of mtDNA haplogroup C of pre-Hispanic and contemporary groups from Maya, Asia, Beringia, and the American continent. A phylogenetic network was constructed with the mtDNA sequences from nucleotides as in Figure 6.2.

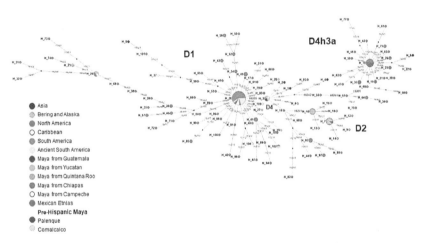

Figure 5.4 Haplotype network analysis of mtDNA haplogroup D of pre-Hispanic and contemporary groups from Maya, Asia, Beringia, and the American continent. A phylogenetic network was constructed with the mtDNA sequences from nucleotides as in Figure 6.2.

pre-Hispanic samples with one sequence from Ciboney and with contemporary individuals from northern Brazil, China, and northern Asia. This suggested a shared common ancestor of the pre-Hispanic individuals with individuals from China and northern Asia. This haplotype is also found in Cuba and South America. In addition, the pre-Hispanic Guane group from Colombia (Casas-Vargas et al. 2011) and two Mexican pre-Hispanic Maya individuals with haplogroup D have different haplotypes, suggesting different ancestral origins.

Haplotype sharing seen in this study supports the ethnogenesis of the Mexican Maya populations in pre-Hispanic times on a cultural and biological basis, in contrast to the etnogenesis in the contemporary Guatemalan Maya populations that is supported on a cultural rather than a biological basis (Söchtig et al. 2015). In addition, the present-day variation in the haplogroup C and D clades suggests that they expanded prior to the Last Glacial Maximum, with the oldest lineages present in eastern Asia, as suggested by these studies (Figure 5.5) and previously by Derenko et al. (2010), since haplotype H4 was shared among two of the pre-Hispanic Maya and sequences from eastern Asia. In addition, this haplotype was shared with pre-Hispanic samples from Ciboney and contemporary human samples from Brazil, supporting further dispersions of the pre-Hispanic Maya from Yucatan to Cuba or backward, as suggested by Lalueza-Fox et al. (2003), during the pre-Hispanic era mainly in successive migration movements from South America.

In summary, all these results demonstrated consistency with the hypothesis that direct ancestors of Native Americans were a hybrid of different Siberian groups that had migrated from eastern Beringia at different times, following different routes through the Americas (Figure 5.5) (Crawford et al. 2010; Kunz and Reanier 1994; Ochoa-Lugo et al. 2019; Perego et al. 2010; Starikovskaya et al. 2005). In addition, the median network analysis and haplotype sharing analysis (Figures 5.2–5.5) indicated: migration of the Maya ancestors from Asia across the Bering Strait land bridge, through the United States, and from northern Mexico to the Maya

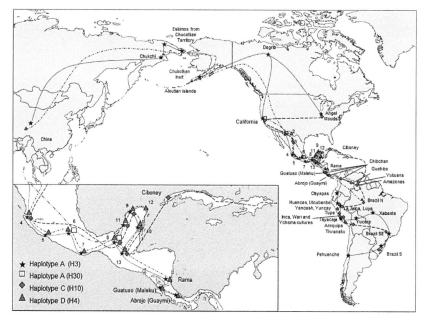

Figure 5.5 Geographic distribution of shared mtDNA haplotypes between the Maya pre-Hispanic populations in this study and those reported previously for contemporary and pre-Hispanic groups from Asia, Beringia, and America, showing the hypothetical migration path of the Maya ancestors.

region; continuity of pre-Hispanic Maya populations with contemporary Maya populations; some contemporary Maya populations' being descendants of both the original Maya population and other populations that migrated to these geographic areas; gene flow from North America to South America and common ancestors among some populations (Figure 5.5); and regional gene flow among the pre-Hispanic populations and regional interaction.

Conclusions

Weather and climate may have affected subsistence patterns of ancient Maya cultures. Consequently, the intersection of demographic, political, and climate factors may have resulted in population adaptation or migration and change in the mtDNA haplotype composition. This may also have been influenced by invasions of other populations, such as the Nahua of Uto-Aztecan affiliation. In addition, mtDNA analysis strongly supports that contemporary and pre-Hispanic Maya are descendants of a common ancestor. The analysis in this chapter and previous reports demonstrate a predominant unidirectional gene flow toward South America in the Mesoamerican area that probably occurred during the Preclassic (1800 BCE–200 CE) and Classic (200–1000 CE) eras of the Mesoamerican chronology. These findings are in agreement with the development of Maya civilization (Söchtig et al. 2015) and with the expansion of maize cultivation via the migration of archaic people from the Mexican Gulf Coast to the south (Malhi et al. 2003), since maize cultivation was dispersed from the lowland tropics of Tabasco to North and South America before 5,050 BCE (Merrill et al. 2009; Piperno et al. 2009) and into the U.S. Southwest before 2050 BCE (Kemp et al. 2010; Kohler and Reese 2014; Merrill et al. 2009), according to the Maya population expansion to North and South America (Figure 5.5). These studies also indicate migration of Native American ancestors from North to South and interregional migration among the Maya populations.

References

Campbell, L., and T. Kaufman. 1985. "Mayan linguistics: Where are we now?" *Annual Review of Anthropology* 14:187–198.

Casas-Vargas, A., I. Gómez, A. Briceño, et al. 2011. "High genetic diversity on a sample of pre-Columbian bone remains from Guane territories in northwestern Colombia." *American Journal of Physical Anthropology* 146: 637–649.

Crawford, M. H., R. C. Rubicz, and M. Zlojutro. 2010. "Origins of Aleuts and the genetic structure of populations of the archipelago: Molecular and archaeological perspectives." *Human Biology* 82: 695–717.

Cucina, A., T. D. Price, and E. Magaña-Peralta. 2015. "Crossing the peninsula: The role of Noh Bec, Yucatán, in ancient Maya Classic period population dynamics from an analysis of dental morphology and Sr isotopes." *American Journal of Human Biology* 27: 767–778.

da Fonseca, R. R., B. D. Smith, N. Wales, et al. 2015. "The origin and evolution of maize in the southwestern United States." *Nature Plants* 1: 15007.

Derenko, M., B. Malyarchuk, T. Grzybowski, et al. 2010. "Origin and post-glacial dispersal of mitochondrial DNA haplogroups C and D in northern Asia." *PLOS One* 5: e15214.

Domínguez-Cruz, M., M. Muñoz, A. Totomoch-Serra, et al. 2018. "Pilot genome-wide association study identifying novel risk loci for type 2 diabetes in a Maya population." *Gene* 677: 324–333.

Kennett, D. J., and T. P. Beach. 2013. "Archeological and environmental lessons for the Anthropocene from the classic Maya collapse." *Anthropocene*, 4, 88–100.

Evans, N. P., T. K. Bauska, F. Gázquez-Sánchez, et al. 2018. "Quantification of drought during the collapse of the classic Maya civilization." *Science* 361, no. 6401: 498–501.

Figueroa, R. M. 2011. "Indigenous peoples and cultural losses." In *Oxford Handbook of Climate Change and Society*, edited by J. S. Dryzek, R. B. Norgaard, and D. Schlosberg, 232–246. Oxford: Oxford University Press.

Falk, M. J., L. Shen, M. Gonzalez, et al. 2015. "Mitochondrial Disease Sequence Data Resource (MSeqDR): A global grass-roots consortium to facilitate deposition, curation, annotation, and integrated analysis of genomic data for the mitochondrial disease clinical and research communities." *Molecular Genetics and Metabolism* 114: 388–396.

Goebel, T., M. R. Waters, and D. H. O'Rourke. 2008. "The late Pleistocene dispersal of modern humans in the Americas." *Science* 319: 1497–1502.

Gómez-Carballa, A., L. Catelli, J. Pardo-Seco, et al. 2015. "The complete mitogenome of a 500-year-old Inca child mummy." *Scientific Reports* 5: 16462.

Hoggarth, J. A., S. F. M. Breitenbach, et al. 2016. "The political collapse of Chichén Itzá in climatic and cultural context." *Global and Planetary Change*, 138, 25–42.

Houston, S., J. Robertson, and D. Stuart. 2000. "The language of classic Maya inscriptions." *Current Anthropology* 41, no. 3: 321–356.

Hsiang, S. M., M. Burke, and E. Miguel. 2013. "Quantifying the influence of climate on human conflict." *Science* 341, no. 6151: 1235367.

Ingman, M., H. Kaessmann, S. Pääbo, et al. 2000. "Mitochondrial genome variation and the origin of modern humans." *Nature* 408: 708–713.

Inomata, T., J. MacLellan, D. Triadan, et al. 2015. "Development of sedentary communities in the Maya lowlands: Coexisting mobile groups and public ceremonies at Ceibal, Guatemala." *PNAS* 112: 4268–4273.

Inomata, T., D. Triadan, K. Aoyama, et al. 2013. "Early ceremonial constructions at Ceibal, Guatemala, and the origins of lowland Maya civilization." *Science* 340: 467–471.

Kemp, B. M., A. González-Oliver, R. S. Malhi, et al. 2010. "Evaluating the farming/language dispersal hypothesis with genetic variation exhibited by populations in the Southwest and Mesoamerica." *PNAS* 107: 6759–6764.

Kohler, T. A., and K. M. Reese. 2014. "Long and spatially variable Neolithic demographic transition in the North American Southwest." *PNAS* 111(28): 10101–10106.

Kuma, S., C. Bellis, M. Zlojutro, et al. 2011. "Large scale mitochondrial sequencing in Mexican Americans suggests a reappraisal of Native American origins." *BMC Evolutionary Biology* 11: 293.

Kunz, M. L., and R. E. Reanier. 1994. "Paleoindians in Beringia: Evidence from Arctic Alaska." *Science* 263: 660–662.

Lalueza-Fox, C., M. T. Gilbert, J. Martínez-Fuentes, et al. 2003. "Mitochondrial DNA from pre-Columbian Ciboneys from Cuba and the prehistoric colonization of the Caribbean." *American Journal of Physical Anthropology* 121: 97–108.

Law, D. 2013. "Mayan historical linguistics in a new age." *Language and Linguistics Compass* 7: 141–156.

López-Carr, D., and J. Marter-Kenyon. 2015. "Manage climate-induced resettlement." *Nature* 517: 265–267.

Lucero, L. J. 2008. "The collapse of the classic Maya: A case for the role of water control." *American Anthropologist* 104, no. 3: 814–826.

Lucero, L. J., R. Fletcher, and R. Coningham. 2015. "From 'collapse' to urban diaspora: the transformation of low-density, dispersed agrarian urbanism." *Antiquity* 89: 1139–1154.

Maffi, L. 2005. "Linguistic, cultural, and biological diversity." *Annual Review of Anthropology* 29: 599–617.

Melton, P. E., N. F. Baldi, R. Barrantes, et al. 2013. "Microevolution, migration, and the population structure of five Amerindian populations from Nicaragua and Costa Rica." *American Journal of Human Biology* 25: 480–490.

Merrill, W. L., R. J. Hard, J. B. Mabry, et al. 2009. "The diffusion of maize to the southwestern United States and its impact." *PNAS* 106: 21019–21026.

Ochoa-Lugo, M. I., M. L. Muñoz, G. Pérez-Ramírez, et al. 2016. "Genetic affiliation of pre-Hispanic and contemporary Mayas through maternal lineage." *Human Biology* 88, no. 2: 136–167.

Perego, U. A., N. Angerhofer, M. Pala, et al. 2010. "The initial peopling of the Americas: A growing number of founding mitochondrial genomes from Beringia." *Genome Research* 20: 1174–1179.

Pimm, S. L., C. N. Jenkins, R. Abell, et al. 2014. "The biodiversity of species and their rates of extinction, distribution, and protection." *Science* 344: 1246752.

Piperno, D. R., A. J. Ranere, I. Holst, et al. 2009. "Starch grain and phytolith evidence for early ninth millennium B.P. maize from the Central Balsas River Valley, Mexico." PNAS 106, no. 13: 5019–5024.

Ragsdale, C. S., and H. J. Edgar. 2015. "Cultural interaction and biological distance in Postclassic period Mexico." *American Journal of Physical Anthropology* 157: 121–133.

Ranere, A. J., D. R. Piperno, I. Holst, et al. 2009. "The cultural and chronological context of early Holocene maize and squash domestication in the Central Balsas River Valley, Mexico." *PNAS* 106: 5014–5018.

Rieux, A., A. Eriksson, M. Li, et al. 2014. "Improved calibration of the human mitochondrial clock using ancient genomes." *Molecular Biology and Evolution* 31: 2780–2792.

Rust, W. F., and R. J. Sharer. 1988. "Olmec settlement data from La Venta, Tabasco, Mexico." *Science* 242: 102–104.

Saldaña-Martínez, A., M. Muñoz, G. Pérez-Ramírez, et al. 2019. "Whole sequence of the mitochondrial DNA genome of Kearns Sayre syndrome patients: Identification of deletions and variants." *Gene*, 688: 171–181.

Scarborough, V. L., N. P. Dunning, K. B. Tankersley, et al. 2012. "Water and sustainable land use at the ancient tropical city of Tikal, Guatemala." *PNAS* 109, no. 31: 12408–12413.

Scherer, A. K. 2007. "Population structure of the Classic period Maya." *American Journal of Physical Anthropology* 132: 367–380.

Söchtig, J., V. Álvarez-Iglesias, A. Mosquera-Miguel, et al. 2015. "Genomic insights on the ethnohistory of the Maya and the 'Ladinos' from Guatemala." *BMC Genomics* 16: 131.

Starikovskaya, E. B., R. I. Sukernik, O. A. Derbeneva, et al. 2005. "Mitochondrial DNA diversity in indigenous populations of the southern extent of Siberia, and the origins of Native American haplogroups." *Annals of Human Genetics* 69: 67–89.

Turner, B. L., II, and J. A. Sabloff. 2012. "Classic period collapse of the Central Maya Lowlands: Insights about human–environment relationships for sustainability." *PNAS* 109, no. 35: 13908–13914.

von Cramon-Taubadel, N., A. Strauss, M. Hubbe. 2017. "Evolutionary population history of early Paleoamerican cranial morphology." *Science Advances* 3, no. 2: e1602289.

6

Mitochondrial DNA Haplotypes in Pre-Hispanic Human Remains from Puyil Cave, Tabasco, Mexico

María Teresa Navarro-Romero, María de Lourdes Muñoz-Moreno, and Enrique Alcalá-Castañeda

Introduction

The study of pre-Hispanic civilizations is important for understanding our past, which in turn allows us to understand our present. Such a study of pre-Hispanic civilizations would contribute to our understanding of human evolutionary history, population expansion worldwide, and enhance our understanding of cultural evolution that constitute the basis of modern societies. . Multiple pre-Hispanic civilizations, such as Olmec, Zoque, and Maya, among others, lived in Mesoamerica, a region located in south Mexico and countries of Central America, such as Guatemala, El Salvador, and west of Honduras, Nicaragua, and Costa Rica (Paul 1967; Figure 6.1). The pre-Hispanic era was divided mainly into four periods: the Archaic (8000–2000 BC), the Preclassic (2000 BC–200 AD), the Classic (200–800 AD), and the Postclassic (800–1521 AD; González-Martín et al. 2015; Serrano-Ursúa et al. 2005). Tabasco is in the Mesoamerican region, and it is one of 31 Mexican states located in the southeast region of the country, bordered by the states of Campeche, Chiapas, and Veracruz, with the Gulf of Mexico to the north and the country of Guatemala to the southeast. This state has a rich diversity of ecosystems and natural resources that favored the settlement of the Olmec (2000–400 BC), Zoque (1300 BC–400 AD), and Maya (250–1521 AD) pre-Hispanic civilizations (Blaha-Pfeiler et al. 2014; Wichmann, Beliaev, and Davletshin 2008). The topographic characteristics of the area, its climate, and its hydrology have favored the formation of large number of caves. The caves in Tabasco were used by the pre-Hispanic populations as tombs or for religious ceremonies (Blaha-Pfeiler et al. 2014), since caves had a sacred meaning and represented an important cosmogonic concept of the world as the origin of life and/or the entrance to the underworld (De-la-Paz-Monte and Linares-Villanueva 2017). The most significant event of the Preclassic period occurred with the emergence of the Olmecs, who are considered ancestral to Mesoamerican civilization. Olmecs influenced urbanism,

María Teresa Navarro-Romero, Maria de Lourdes Muñoz-Moreno, and Enrique Alcalá-Castañeda, *Mitochondrial DNA Haplotypes in Pre-Hispanic Human Remains from Puyil Cave, Tabasco, Mexico* In: *Human Migration*. Edited by: Maria de Lourdes Muñoz-Moreno and Michael H. Crawford, Oxford University Press. © Oxford University Press 2021. DOI: 10.1093/oso/9780190945961.003.0007

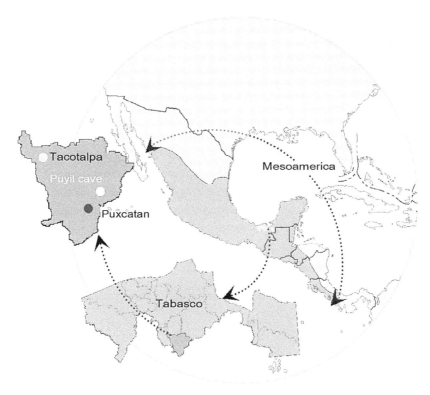

Figure 6.1 Geographic location of the Puyil cave in the Mesoamerican region, on San Felipe Mountain near to Puxcatan town in Tacotalpa municipality of Tabasco State, Mexico.

sculpture, mathematics, and astronomy of the Maya. Olmecs inhabited mainly the southeastern part of Veracruz and the west of Tabasco in Mexico. The Zoques are archaeologically and linguistically related to the Olmecs, who spoke the Proto-Mixe-Zoque language from 1800 to 1600 BC (Campa and Winter 2009). A variant of Proto-Mixe-Zoque language was Mixe-Zoque, which originated when the populations began to move to the southern regions of Tabasco in 1350 to 1300 BC (Campbell and Kaufman 1976; Navarrete 2010; Pye and Clark 2006; Terreros-Espinosa 2006), and Mixe-Zoque was spoken by the Zoques.

It has been suggested that ancient Maya populations have descended from the Olmecs. These populations were located in a large area in the southeast of Mexico in the states of Campeche, Chiapas, Quintana Roo, Tabasco, and Yucatán, as well as Belize, Guatemala, Honduras, and El Salvador. Among these populations were the Mayas Choles and Chontales, who inhabited in Tabasco in the Classic period. It is also accepted that the Maya civilization was one of the most important pre-Hispanic Mesoamerican cultures, due to their extensive knowledge of astronomy, architecture, mathematics, and sculpture, their development of a calendar, and their writing (Law 2013; Muñoz et al. 2012). The peak development of Maya civilization has been proposed to have been in ~ 250 AD, and it continued until the Spanish conquest in 1521.

In 2007, a group of archaeologists discovered archaeological material in a cave located in the middle of the jungle, ona mountain called San Felipe, near the town of Puxcatan, in Tacotalpa municipality of Tabasco State, Mexico (Figure 6.1). The Puyil cave was divided into six chambers according to the natural conformation of the vault, and each chamber was divided into sections according to divisions created by walls or calcareous structures. The greater amount of archaeological material and ~ 40 skeletal remains were distributed inside chambers 4, 5 and 6 (for more information about the Puyil cave, see Chapter 16).

The primary aim of the study of ancient bone samples found in the Puyil cave was to assess the maternal genetic origin through analysis of the mitochondrial DNA (mtDNA) sequences. This analysis included the identification and distribution of haplogroups, and the genetic relationship between contemporary and ancient native populations from the American continent and Asia. This genetic analysis contributes to the knowledge of pre-Hispanic populations and propose possible migratory routes.

Background

mtDNA has been used to study ancient samples because it is inherited from the mother (Nicholls and Minczuk 2014); it does not recombine, so any variation is the result of nucleotide changes accumulating over time; the noncoding region of the mitochondrial genome known as the control region or D-loop (1122 bp) has a high mutation rate; the highest variability in the D-loop is enclosed in hypervariable region I (HVR-I) from nucleotides 16024–16383 and HVR-II from the nucleotides 57–372 (Nicholls and Minczuk 2014), which are used for haplogroup classification (González-Martín et al. 2015). Molecular genetic studies of Native Americans have identified five haplogroups, A, B, C, D, and X (Gorostiza et al. 2012), which are distributed in different frequencies in populations of the American continent. Previous studies have shown that pre-Hispanic populations from Mesoamerica had haplogroups A and C at higher frequency, while B and D were present at low frequency (González-Oliver et al. 2001; Qi et al. 2013). Moreover, in pre-Hispanic Maya populations from different archaeological sites, haplogroups A, A2, A2v, C, C1, and D were identified, but not haplogroup B (Ochoa-Lugo et al. 2016). Nevertheless, haplogroup D1 was identified at a higher frequency in ancient populations from Monte Verde in South America (Fagundes et al. 2008; González-Martín et al. 2015; Gorostiza et al. 2012; Qi et al. 2013). Furthermore, this haplogroup was also identified in three pre-Hispanic Maya individuals of Palenque from Chiapas, Comalcalco, and Sueños de Oro from Tabasco (Ochoa-Lugo et al. 2016).

Materials

The study of aDNA was carried out with ten samples from different pre-Hispanic individuals collected from chambers 4, 5, and 6 in Puyil cave (geographical coordinates 17º27'38.04"N and 92º39'28.5"W). The ancient samples were recovered in 2007 by an archaeology group led by Luis Alberto Martos-López. The ages of the samples were determined using ^{14}C. The age at death was base on morphological features and reference tables; the haplogroups were identified using specific variants for each haplogroup (Navarro-Romero et al. 2020a); and gender was determined using a new methodology described by Navarro-Romero et al. (2020b). All these characteristics are shown in Table 6.1.

Methods

The molecular biology analyses were carried out in a clean room using the guidelines for working with ancient remains (Adler et al. 2011; Campos et al. 2012). The extraction of aDNA was extracted using two methods. The first method was developed in Mexico using magnetic bead technology (PerkinElmer, Waltham, Massachusetts, USA) in combination with the DNA Tissue10 Prepito Kit (PerkinElmer) used by the Prepito-D equipment (PerkinElmer), following the manufacturer's instructions. The second method was carried out in Germany (Lee et al. 2012) using the High Pure Viral Nucleotic Acid Kit (Roche). The libraries were prepared without treatment and with the enzyme uracil-DNA-glycosylase (UDG) used to repair the DNA following the protocol described by Meyer and Kircher (2010). Massive sequencing was carried out using the Illumina platform (Kiel University & Max Plank Institute for the Science of Human History, Germany). The bioinformatic analysis was carried out using the Eager pipeline (Peltzer et al. 2016). Phylogenetic networks were constructed using the Network 5.0.0.1 software (Bandelt, Forster, and Röhl 1999; Polzin and Daneshmand 2003), from nucleotides 16000 to 16569 (570 bp), 14767 to 16569 (1803 bp), and 16224 to 16569 (346 bp) for haplogroup A; 16224 to 16569 (346 bp) for haplogroup C; and 16064 to 16569 for haplogroup D (508 bp) of the mtDNA. The analysis included 992 sequences of contemporary native populations from all of the American continent and Asia, as well as ancient sequences from Peru, Chile, and Cuba (Navarro-Romero et al. 2020a).

Results

To identify the haplogroup of each sample, a fragment of 211 bp of HVR-I was amplified and sequenced by the Sanger method. All sequences obtained by both methods, Sanger and NGS, were aligned with the Cambridge Reference Sequence (CRs_human_hg38). The haplogroups identified were A, A2, C1, C1c, and D4,

Table 6.1 Features of the pre-Hispanic individuals found at the Puyil cave, as well as dating by ^{14}C, showing the conventional age and the calendar calibration

ID	Chamber*	Collection ID	Bone Sample	Age at Death (years)	Haplogroup	Gender**
PUXTABMEX001	6	ENT-6/Chamber 6	Femur	20–25	C1	Male
PUXTABMEX002	6	ENT-9/Chamber 6	Left rib	20–25	C1c	Male
PUXTABMEX003	5	ENT-11/ Chamber 5	5th or 6th cervical vertebrae	25–30	A2	Male
PUXTABMEX004	4	Skull-4/Point 50	Temporal cranium	5–10	C1	Male
PUXTABMEX005	4	ENT-2/Point 51	2nd or 3rd lumbar vertebrae	15–20	A	Male
PUXTABMEX006	4	ENT-11/Point 46	Third molar tooth	25–30	A	Male
PUXTABMEX007	4	Skull-1/Point 47	Mastoid bone	25–30	A	Male
PUXTABMEX008	4	Skull-1/Point 50	Mastoid bone	25–30	ND	Male
PUXTABMEX009	4	Skull-2/Point 50	Mastoid bone	25–30 yrs	D4	Male
PUXTABMEX010	4	Skull-3/Point 50	Mastoid bone	20–25	A	Male

* Chamber = Number of chamber inside Puyil cave. ** Gender = New molecular methodology using *TTTY2*, *TTTY7*, and *TSPY3* genes located on chromosome Y (Navarro-Romero et al. 2020a, 2020b). ND = Not determined.

while haplogroup B was not detected in any of the pre-Hispanic study individuals (Navarro-Romero et al. 2020a).

Haplotype network analysis was carried out to determine the relationship of the pre-Hispanic individuals with ancient and contemporary native populations of America and Asia, as described previously, using representative mtDNA sequences (Navarro-Romero et al. 2020a). The analysis for haplogroup A included five sequences from this study (PUXTABMEX003, PUXTABMEX005–PUXTABMEX007, and PUXTABMEX010) and 515 sequences previously reported. The individual PUXTABMEX003 containing haplogroup A2 was linked with contemporary native Maya of Kiche from Guatemala and Tzotzil from Mexico, as well as an Amazon tribe from Brazil (Figure 6.2A). The individuals PUXTABMEX005 and PUXTABMEX010 linked with individual PUXTABMEX007 were haplogroup A; the individual PUXTABMEX007 was linked with the contemporary population of Xikrin from Brazil and Mayas of Yucatán and Tzotzil from Mexico (Figure 6.2A). The individual PUXTABMEX006 containing haplogroup A was linked with the contemporary population of Mongolia and Mayas of Ladino, Verapaza, and La Tinta from Guatemala and Tzotzil from Mexico (Figure 6.2A).

The analysis for haplogroup C included three sequences from this study (PUXTABMEX001, PUXTABMEX002, and PUXTABMEX004) and 242 sequences previously reported. These three individuals were genetically closely related since they are in the same cluster. The individuals PUXTABMEX001 and PUXTABMEX004 contained haplogroup C1 and the individual PUXTABMEX002 contained haplogroup C1c. Samples PUXTABMEX002 and PUXTABMEX004 linked with individual PUXTABMEX001 and individual PUXTABMEX001 linked with native contemporary populations of Movima from Brazil, Mayas of Yucatán, Campeche, and Quintana Roo from Mexico, and populations from the Dominican Republic (Figure 6.2B).

The analysis for haplogroup D included one sequence from this study (PUXTABMEX009) and 235 sequences previously reported. This individual displaying haplogroup D4 was linked with populations from North of Brazil, Han from China, and Asian populations (Figure 6.2C).

Haplogroup could not be determined in individual PUXTABMEX008 because of a lower percentage of DNA.

Discussion

The archaeological, linguistic, and genetic evidence has indicated that the first human inhabitants of the Western Hemisphere probably arrived from Asia through the Beringia land mass located between the northeast of Siberia and Alaska during the Last Glacial Maximum, ~ 23,000 to 19,000 years ago (Kumar et al. 2011).

Of the ten individuals analyzed, the individuals PUXTABMEX004, PUXTABMEX006, and PUXTABMEX009 belonged to the Archaic period, and they contained the founding haplogroups C1, A, and D4, respectively (Table 6.2).

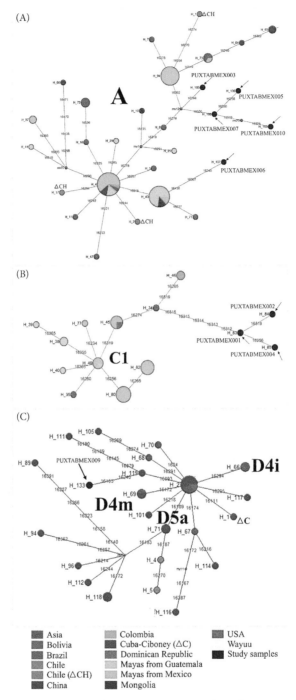

Figure 6.2 Haplotype networks of haplogroup A, C, and D of mtDNA HVR-I region. (A) Haplogroup A network analysis from nucleotide 16224 to 16569 (346 bp) of samples PUXTABMEX003 (H_105), PUXTABMEX005 (H_106), PUXTABMEX006 (H_107), PUXTABMEX007 (H_108), PUXTABMEX010 (H_109), and 515 previously reported. (B) Haplogroup C network analysis from nucleotide 16224 to 16569 (346 bp) of samples PUXTABMEX001 (H_83), PUXTABMEX002 (H_84), PUXTABMEX004 (H_85), and 242 previously reported sequences. (C) Haplogroup D network analysis from nucleotide 16064 to 16569 (508 bp) of sample PUXTABMEX009 (H_133) and 235 previously reported sequences. The triangles before C and CH indicate the sequence of ancient samples from Cuba and Chile, respectively.

The individuals PUXTABMEX004 and PUXTABMEX001 with haplogroup C1 (Table 6.2) and PUXTABMEX002 with haplogroup C1c are genetically closely related to each other and with the contemporary native populations from Mexico, Bolivia, and the Dominican Republic. These results support the theory that the American continental migration route went from North America to South America, probably following the route along the Pacific Ocean (Kumar et al. 2011; Tamm et al. 2007). Moreover, haplogroup C1 has been reported with an estimated age of 21.4 ± 2.7 kya/16.4 ± 1.5 kya (Kumar et al. 2011). This evidence suggests that the haplogroup dispersal followed a route from North America to Central America and, finally, to South America, and that, after several years, the populations followed the route from South America through the Gulf Coast to the Caribbean Islands.

The most common haplogroup previously identified in other pre-Hispanic populations from the Classic period was A2 (Kumar et al. 2011; Pierre et al. 2012; Table 6.2), and it was distributed in high frequency from North to South America. The estimated coalescence age for this haplogroup was 19.5 ± 1.3 to 16.1 ± 1.5 kya according to Kumar et al. (2011), and it has been identified with high frequency in Aleuts and Native American populations. This haplogroup was displayed by the individual PUXTABMEX003, who was genetically closely related to native populations from Brazil and Maya from Mexico and Guatemala, which suggested that this individual shared an ancestor with these contemporary native populations. In contrast, haplogroup A (Table 6.2) has been found in low frequency on the American continent. Previously, it was identified in a pre-Hispanic individual from Comalcalco in Tabasco (Ochoa-Lugo et al. 2016), and it was also identified in the individual PUXTABMEX006, who, it has been suggested, was genetically related to contemporary Maya populations from Mexico and Guatemala, as well as populations from Mongolia.

Table 6.2 Relationship between haplogroup and the conventional age and the calendar calibration

Haplogroup	ID	Age	Period (years)
A	PUXTABMEX005	ND	ND
A	PUXTABMEX006	Archaic	10,000–2,000 BP
A	PUXTABMEX007	Pre-Classic	2,000 BP–200C CE
A	PUXTABMEX010	Pre-Classic	2,000 BP–200C CE
A2	PUXTABMEX003	ND	ND
C1	PUXTABMEX001	Pre-Classic	2,000 BP–200C CE
C1	PUXTABMEX004	Archaic	10,000–2,000 BP
C1c	PUXTABMEX002	Pre-Classic	2,000 BP–200C CE
D4	PUXTABMEX009	Late-Archaic	10,000–2,000 BP
ND	PUXTABMEX008	ND	ND

ND = Not determined.

The individual PUXTABMEX009 contained haplogroup D4 (Table 6.2) and was genetically closely related to contemporary native populations from Brazil and populations from China. The genetic relationship with populations from South America and Asia, the high frequency of haplogroup D in South America (Fagundes et al. 2008; González-Martín et al. 2015; Qi et al. 2013) and the identification of the haplogroup in Monte Verde for the first time with an estimated age of 14.5 kya (Kumar et al. 2011) suggest a common ancestor who migrated from North to South America.

The individuals PUXTABMEX001, PUXTABMEX002, PUXTABMEX007, and PUXTABMEX010 belonged to the Classic period (200–800 AD; Table 6.2). The individuals PUXTABMEX001 and PUXTABMEX002 displaying haplogroup C1 and C1c, respectively, were genetically related to each other and with contemporary native populations from Bolivia linked by one mutational step with Mexican Maya, and the Dominican Republic population, suggesting a common ancestor of the pre-Hispanic individuals shared with Maya and Dominican Republic populations. In addition, because these two samples pertained to the Classic period, it is likely that they pertain to the same group. Haplogroup C1c has a coalescence estimated age of 15.8 ± 4.7 kya to 10.8 ± 2.0 kya (Achilli et al. 2008), suggesting its origin on the American continent. This is likely if the first Native Americans crossed the Beringia land bridge ~ 27 to 19 kya (Watson 2017).

The individuals PUXTABMEX005, PUXTABMEX007, and PUXTABMEX010 containing haplogroup A (Table 6.2) are genetically closely related and share a common ancestor with native contemporary Mexican Maya populations and contemporary populations from Brazil. All three individuals were genetically closely related and pertained to the Classic period (200–800 AD), suggesting that they belonged to the same pre-Hispanic group, but the temporality of individuals PUXTABMEX003, PUXTABMEX005, and PUXTABMEX008 were not determined.

Conclusion

In this study, mtDNA from pre-Hispanic individual predecessors of the Olmec, Zoque, or Maya Mesoamerican civilizations from the Archaic and Classic periods was sequenced and analyzed using haplotype networks. The results showed the close genetic relationship between the pre-Hispanic human individuals and the Maya from Mexico and Guatemala, and their relationship with the contemporary native populations from Bolivia, Brazil, the Dominican Republic, and Asia, with whom they shared common ancestors. In addition, the results support the theory of migratory routes and settlement of ancient populations from North to South America through the Caribbean, as well as the origin of Native American founding haplogroups A, C, and D from Asia. Because the archeological context of these ancient human bones and their different ^{14}C ages pertain to different eras, we also suggest that the caves were used as ceremonial centers at different periods of time.

References

Achilli, Alessandro, Ugo A. Perego, Claudio M. Bravi, Michael D. Coble, Qing-Peng Kong, Scott R. Woodward, Antonio Salas, Antonio Torroni, and Hans-Jürgen Bandelt. 2008. "The phylogeny of the four Pan-American mtDNA haplogroups: Implications for evolutionary and disease studies." *PLOS One* 3, no. 3: e1764. https://doi.org/10.1371/journal.pone.0001764

Adler, C. J., W. Haak, D. Donlon, and A. Cooper. 2011. "Survival and recovery of DNA from ancient teeth and bones." *Journal of Archaeological Science* 38, no. 5: 956–964. https://doi.org/10.1016/j.jas.2010.11.010

Bandelt, H. J., P. Forster, and A. Röhl. 1999. "Median-joining networks for inferring intraspecific phylogenies." *Molecular Biology and Evolution* 16, no. 1: 37–48. https://doi.org/10.1093/oxfordjournals.molbev.a026036

Blaha-Pfeiler, Barbara, B. Flor Canché-Teh, Cecilia García-Gómez, David Romero-Bernical, Mario Humberto Ruz-Sosa, Flora Leticia I. Salazar Ledesma, and Eladio Terreros Espinosa. 2014. *Tabasco serrano: miradas plurales: geografía, arqueología, historia, lingüística y turismo* [Possible Linguistic and Archaeological Correlations Linked to the Olmecs]. Mexico City: Universidad Nacional Autónoma de México.

Vazquez-Campa, Violeta, and Marcus Winter. 2009. "Mixes, Zoques y la arqueología del istmo sur de Tehuantepec [Mixes, Zoques and the archeology of the southern isthmus of Tehuantepec]."In *Medio ambiente, antropología, historia y poder regional en occidente de Chiapas y el Istmo de Tehuantepec*, edited by T. A. Lee Whiting, D. Domenici, V. M. Esponda-Jimeno y C. U. del Carpio Penagos, Tuztla Gutiérrez, Chiapas: Universidad de Ciencias y Artes de Chiapas, pp. 219–234.

Campbell, Lyle, and Terrence Kaufman. 1976. "A linguistic look at the Olmecs." *American Antiquity* 41, no. 1: 80–89. https://doi.org/10.2307/279044

Campos, Paula F., Oliver E. Craig, Gordon Turner-Walker, Elizabeth Peacock, Eske Willerslev, and M. Thomas P. Gilbert. 2012. "DNA in ancient bone—Where is it located and how should we extract it?" *Annals of Anatomy - Anatomischer Anzeiger*, Special Issue: Ancient DNA, 194, no. 1: 7–16. https://doi.org/10.1016/j.aanat.2011.07.003

Carracedo, A., W. Bär, P. Lincoln, W. Mayr, N. Morling, B. Olaisen, P. Schneider, et al. 2000. "DNA Commission of the International Society for Forensic Genetics: Guidelines for mitochondrial DNA typing." *Forensic Science International* 110, no. 2: 79–85.

De-la-Paz-Monte, Javier, and Eliseo Linares Villanueva. 2017. "Un Caso Prehispánico de Occipitalización Del Atlas. Estudio Antropofísico de Un Cráneo Humano de La Cueva El Tepesco Del Diablo, Chiapas [A Pre-Hispanic Case of Occipitalization of the Atlas. Anthropophysical Study of a Human Skull from El Tepesco Del Diablo Cave, Chiapas]." *Diario de Campo* no. 10–11: 46–54.

Fagundes, Nelson J. R., Ricardo Kanitz, Roberta Eckert, Ana C. S. Valls, Mauricio R. Bogo, Francisco M. Salzano, David Glenn Smith, et al. 2008. "Mitochondrial population genomics supports a single pre-Clovis origin with a coastal route for the peopling of the Americas." *American Journal of Human Genetics* 82, no. 3: 583–592. https://doi.org/10.1016/j.ajhg.2007.11.013

González-Martín, Antonio, Amaya Gorostiza, Lucía Regalado-Liu, Sergio Arroyo-Peña, Sergio Tirado, Ismael Nuño-Arana, Rodrigo Rubi-Castellanos, Karla Sandoval, Michael D. Coble, and Héctor Rangel-Villalobos. 2015. "Demographic history of indigenous populations in Mesoamerica based on mtDNA sequence data." *PLOS One* 10, no. 8: e0131791. https://doi.org/10.1371/journal.pone.0131791

González-Oliver, A., L. Márquez-Morfín, J. C. Jiménez, and A. Torre-Blanco. 2001. "Founding Amerindian mitochondrial DNA lineages in ancient Maya from Xcaret, Quintana Roo." *American Journal of Physical Anthropology* 116, no. 3: 230–235. https://doi.org/10.1002/ajpa.1118

Gorostiza, Amaya, Víctor Acunha-Alonzo, Lucía Regalado-Liu, Sergio Tirado, Julio Granados, David Sámano, Héctor Rangel-Villalobos, and Antonio González-Martín. 2012.

"Reconstructing the history of Mesoamerican populations through the study of the mitochondrial DNA control region." *PLOS One* 7, no. 9: e44666. https://doi.org/10.1371/journal.pone.0044666

Kirchhoff, Paul. 1967. "Mesoamérica, sus límites geográficos, composición étnica y caracteres culturales [Mesoamerica, its geographical limits, ethnic composition and cultural characteristics]." Escuela Nacional de Antropología e Historia, Sociedad de Alumnos.

Kumar, Satish, Claire Bellis, Mark Zlojutro, Phillip E. Melton, John Blangero, and Joanne E. Curran. 2011. "Large scale mitochondrial sequencing in Mexican Americans suggests a reappraisal of Native American origins." *BMC Evolutionary Biology* 11 (October): 293. https://doi.org/10.1186/1471-2148-11-293

Law, Danny. 2013. "Mayan historical linguistics in a new age." *Language and Linguistics Compass* 7, no. 3: 141–156. https://doi.org/10.1111/lnc3.12012

Lee, Esther J., Cheryl Makarewicz, Rebecca Renneberg, Melanie Harder, Ben Krause-Kyora, Stephanie Müller, Sven Ostritz, et al. 2012. "Emerging genetic patterns of the European Neolithic: Perspectives from a Late Neolithic bell beaker burial site in Germany." *American Journal of Physical Anthropology* 148, no. 4: 571–579. https://doi.org/10.1002/ajpa.22074

Meyer, Matthias, and Martin Kircher. 2010. "Illumina sequencing library preparation for highly multiplexed target capture and sequencing." *Cold Spring Harbor Protocols* no. 6: pdb.prot5448. https://doi.org/10.1101/pdb.prot5448

Muñoz, María de Lourdes, Eduardo Ramos, Alvaro Díaz-Badillo, María Concepción Morales-Gómez, Rocío Gómez, and Gerardo Pérez-Ramirez. 2012. "Migration of Pre-Hispanic and Contemporary Human Mexican Populations." In *Causes and Consequences of Human Migration: An Evolutionary Perspective*, edited by Benjamin C. Campbell and Michael H. Crawford, 417–435. Cambridge: Cambridge University Press. https://doi.org/10.1017/CBO9781139003308.023

Navarrete, Carlos. 2010. "Fuentes para la historia cultural de los Zoques [Sources for the cultural history of the Zoques]." *Anales de Antropología* 7: 207–246. https://repositorio.unam.mx/contenidos/37847

Navarro-Romero, María Teresa, María de Lourdes Muñoz, Enrique Alcala-Castañeda, Eladio Terreros-Espinosa, Eduardo Domínguez-de-la-Cruz, Normand García-Hernández, and Miguel Ángel Moreno-Galeana. 2020b. "A novel method of male sex identification of human ancient skeletal remains." *Chromosome Research*, 28 (July): 277–291. https://doi.org/10.1007/s10577-020-09634-1

Nicholls, Thomas J., and Michal Minczuk. 2014. "In D-loop: 40 years of mitochondrial 7S DNA." *Experimental Gerontology, Mitochondria, Metabolic Regulation and the Biology of Aging*, 56 (August): 175–181. https://doi.org/10.1016/j.exger.2014.03.027

Ochoa-Lugo, Mirna Isabel, María de Lourdes Muñoz, Gerardo Pérez-Ramírez, Kristine G. Beaty, Mauro López-Armenta, Javiera Cervini-Silva, Miguel Moreno-Galeana, et al. 2016. "Genetic affiliation of pre-Hispanic and contemporary Mayas through maternal lineage." *Human Biology* 88, no. 2: 136–167. https://doi.org/10.13110/humanbiology.88.2.0136

Peltzer, Alexander, Günter Jäger, Alexander Herbig, Alexander Seitz, Christian Kniep, Johannes Krause, and Kay Nieselt. 2016. "EAGER: Efficient Ancient Genome Reconstruction." *Genome Biology* 17 (March): 60. https://doi.org/10.1186/s13059-016-0918-z

Polzin, Tobias, and Siavash Vahdati Daneshmand. 2003. "On Steiner trees and minimum spanning trees in hypergraphs." *Operations Research Letters* 31, no. 1: 12–20. https://doi.org/10.1016/S0167-6377(02)00185-2

Pye, Mary, and John Clark. 2006. *Los Olmecas son Mixe-Zoques: Contribuciones de Gareth W. Lowe a la Arqueología del Formativo* [The Olmecs are Mixe-Zoques: Gareth W. Lowe's Contributions to the Archeology of the Formative], pp. 73–74. Guatemala: Ministerio de Cultura y Deportes.

Qi, Xuebin, Chaoying Cui, Yi Peng, Xiaoming Zhang, Zhaohui Yang, Hua Zhong, Hui Zhang, et al. 2013. "Genetic evidence of Paleolithic colonization and Neolithic expansion of modern humans on the Tibetan Plateau." *Molecular Biology and Evolution* 30, no. 8: 1761–1778. https://doi.org/10.1093/molbev/mst093

Saint Pierre, Michelle de, Francesca Gandini, Ugo A Perego, Martin Bodner, Alberto Gómez-Carballa, Daniel Corach, Norman Angerhofer, et al. 2012. "Arrival of Paleo-Indians to the Southern Cone of South America: New clues from mitogenomes." *PLOS One* 7, no. 12: e51311–e51311. https://doi.org/10.1371/journal.pone.0051311

Serrano-Ursúa, Félix Javier, Daniel Enrique Morales-Urbina, Claudia María Hernández-de-Diguero, Carlos Enrique Chian-Rodríguez, Luis Fernando Dubón-González, Thelma Zuly González-de-Dubón, and Mario Leonel Estrada-Furlán. 2005. "Mesoamérica." Editorial Galería Guatemala.

Tamm, Erika, Toomas Kivisild, Maere Reidla, Mait Metspalu, David Glenn Smith, Connie J. Mulligan, Claudio M. Bravi, et al. 2007. "Beringian standstill and spread of Native American founders." *PLOS One* 2, no. 9. https://doi.org/10.1371/journal.pone.0000829

Terreros Espinosa, Eladio. 2006. "Arqueología Zoque de La Región Serrana Tabasqueña. [Zoque Archeology of the Serrana Tabasqueña Region]" *Estudios Mesoamericanos* no. 7: 29–43.

Watson, Traci. 2017. "News feature: Is theory about peopling of the Americas a bridge too far?" *PNAS* 114, no. 22: 5554–5557. https://doi.org/10.1073/pnas.1705966114

Wichmann, Søren, Dmitri Beliaev, and Albert Davletshin. 2008. "Possibles Correlaciones Lingüísticas y Arqueológicas Vinculadas Con Los Olmecas [Possible Linguistic and Archaeological Correlations Linked to the Olmecs]." *Olmeca: Balance y Perspectivas*, January, 667–683.

PART III
REGIONAL MIGRATION

7

A Genetic Perspective on the Origin and Migration of the Samoyedic-Speaking Populations from Siberia

Tatiana M. Karafet, Ludmila P. Osipova, and Michael F. Hammer

Introduction

Siberia occupies the extensive geographic region across North Asia from the Ural Mountains in the west to the Pacific Ocean in the east, and from the Arctic Ocean in the north to Mongolia, Kazakhstan, and China in the south. Siberia is among the most sparsely populated regions on Earth, with only ~36 million people inhabiting 5.2 million square miles of territory (13.5 million km^2). Siberia, which has historically been a part of Russia since the 16th and 17th centuries, can be divided into Western Siberia, extending from the Ural to the Yenisey River, Central Siberia, extending from the Yenisey River to the Lena River, Eastern Siberia, extending from the Lena River to the mountain ranges of the Pacific Ocean watershed, and the Siberian Far East, which includes the Amur region, Sakhalin Island, Kamchatka, and Chukotka Peninsulas. Currently, 40 indigenous ethnic groups live in the territories of Siberia, and 24 of these groups are small nationalities with populations of less than 5,000 people. Their languages belong to four linguistic families—Uralic, Altaic, Eskimo-Aleut, and Chukchi-Kamchatkan—and two linguistic isolates—Ket and Nivkh. The archeological record suggests that the early settlement of Siberia was a lengthy and complex process, but the precise antiquity of anatomically modern humans in Siberia is still enigmatic. It is important to remember that as a human habitat Siberia was more favorable than Europe during glacial times (Chard 1974). The great boreal forest or taiga of Siberia was established early in the Pleistocene and has never been totally displaced.

While it is generally accepted that the earliest ^{14}C-dated Asian Upper Paleolithic industries occurred in the Altai Mountains of Southwest Siberia at 43,300 ± 1,600 years BP and spread to the east and north much later (Goebel, Derevianko, and Petrin 1993; Kuzmin et al. 1998), there are still open questions about the chronology of the initial peopling of northern Asia, the nature of its ancient human populations, and the origins and migrations of local prehistoric cultures. The archeological, anthropological, ethnographic, linguistic, and genetic data do not

Tatiana M. Karafet, Ludmila P. Osipova, and Michael F. Hammer, *A Genetic Perspective on the Origin and Migration of the Samoyedic-Speaking Populations from Siberia* In: *Human Migration*. Edited by: Maria de Lourdes Muñoz-Moreno and Michael H. Crawford, Oxford University Press. © Oxford University Press 2021. DOI: 10.1093/oso/9780190945961.003.0008 provide a consistent story, although there are some common themes in the main stages of the ethnic history of Siberia. Demographic events like migrations, admixture, and bottlenecks have a strong influence on genetic variation of populations. DNA-based investigations are particularly useful in tracing the genomic footprints of past historical events.

In this chapter, we describe genetic affinities of Samoyedic-speaking peoples with other Siberian populations to uncover migration events in western Siberia and Taymyr Peninsula and to test existing hypotheses on the origin of Samoyedic-speaking Siberian groups.

Samoyedic Populations

The Uralic linguistic family includes the Finno-Ugric and Samoyedic (or Samoyed) languages, which are spoken in a vast territory of northern Eurasia. Two linguistic branches are distinguished within the Samoyeds: the Northern Samoyedic branch includes the Nentsi, Entsi, and Nganasan languages. The Nentsi, the largest Samoyed population, are divided into two groups with different lifestyles. The nomadic reindeer-breeders of the tundra are called Tundra Nentsi, while the semi-nomadic hunters, fishers, and reindeer-breeders Forest Nentsi live in the taiga region north of the Middle Ob. They speak different dialects of a single language with low mutual intelligibility. The Nganasans, the most northerly people in Russia, are a small group of wild Arctic reindeer hunters and reindeer-breeders occupying the central tundra zone of the Taymyr Peninsula. The Entsi population, with less than 200 people (census 2010), is nearly extinct.

Several southern Samoyedic languages have been historically distributed over a wide territory in the Sayan-Baikal uplands of central Siberia. The only people who currently speak a southern Samoyedic language are Selkups. The ancestors of the Selkups are believed to have migrated to the taiga belt in the middle of the Ob' River during the early part of the first millennium AD (Prokof'yeva 1964). After the arrival of Russians in Siberia, a group of Selkups fled further to the North and occupied the Taz River and basins to avoid taxes and baptism. This group is known as the northern Selkups. The greatly assimilated southern Selkups live in Tomsk Oblast. All other southern-Samoyedic-speaking populations from the Sayan region were mixed with Turkic-speaking tribes and abandoned their own original languages in favor of Turkic (Helimsky 1990).

Population Samples

We examined autosomal genome-wide single nucleotide polymorphisms (SNPs) and Y-chromosome data from four Samoyed-speaking populations: the Nganasans, Tundra Nentsi, Forest Nentsi, and Northern Selkups. Samples were collected with

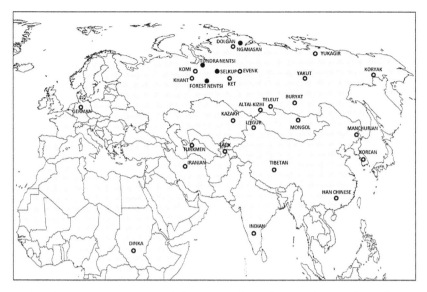

Figure 7.1 Approximate geographic locations of populations. Samoyed-speaking populations are shown as black circles. See Table 7.1 for sample sizes and linguistic affiliations.

informed consent of volunteers in traditional villages or tundra camps. Eighty-four samples from Samoyeds were tested for autosomal SNPs, while 292 samples from unrelated males were subjects of Y-chromosome analysis. An additional 11 Siberian and 12 reference populations were used for comparative analyses (Figure 7.1, Table 7.1). Samples were genotyped for 567,096 autosomal SNPs on the Affymetrix platform and 147 Y-chromosome polymorphic sites (Karafet et al. 2018). All samples analyzed here were described in our previous studies (Hammer et al. 2001; Karafet et al. 2002, 2018; Karafet, Osipova, and Hammer 2008).

The Pattern of Genetic Variation in Siberian Populations

Fixation Index (F_{ST}) and Multidimensional Scaling (MDS) Plots

Analyses of autosomal and Y-chromosome data reveal significant genetic variation among Siberian groups, which might be explained by founder effect, endogamy, and genetic drift due to small population sizes and limited gene flow. Differentiation among Siberian populations measured by F_{ST} is very high for autosomal data and for the Y chromosome. The mean F_{ST} (0.036) based on autosomal data for Siberia was almost identical to the F_{ST} (0.037) for 27 worldwide populations (Wilcoxon rank sum test P value = 0.810). Remarkably, a substantial level of variation among Samoyedic-speaking populations alone (F_{ST} = 0.029) does not differ significantly from the whole of Siberia (P value = 0.762). Y-chromosome F_{ST} statistics indicate an

Table 7.1 The list of population samples and the number of individuals analyzed (*N*)

Population	Language group	Code	Autosomal N	Y-chromosome N
Siberia				
Altai-Kizhi	Turkic	ALT	18	98
Buryat	Mongolic	BUR	16	80
Dolgan	Turkic	DOL	10	57
Evenk	Tungus-Manchu	EVK	13	91
Forest Nentsi	Samoyedic	FNE	26	82
Ket	Language isolate	KET	17	44
Khant	Ugric	KHA	12	165
Komi	Permic	KOM	13	78
Koryak	Chukotko-Kamchatkan	KOR	8	11
Mongol	Mongolic	MON	10	75
Nganasan	Samoyedic	NGA	18	34
Selkup	Samoyedic	SEL	29	129
Teleut	Turkic	TEL	10	40
Tundra Nentsi	Samoyedic	TNE	11	47
Yakut	Turkic	YAK	13	62
Yukagir	Uralic	YUK	7	10
Total			231	1103
Central Asia				
Kazakh	Turkic	KAZ	10	29
Tajik	Indo-Iranian	TAJ	11	15
Turkmen	Turkic	TUR	21	44
Uygur	Turkic	UYG	12	66
Total			54	154
East Asia				
Han Chinese	Sino-Tibetan	HAN	13	40
Korean	Korean	KRE	18	63
Manchurian	Tungus-Manchu	MAN	14	50
Tibetan	Sino-Tibetan	TIB	18	71
Total			63	224
Near East				
Iranian	Indo-Iranian	IRA	27	136
Europe				
German	Germanic	GER	22	37
South Asia				
Indian	Indo-Iranian	IND	27	51
Africa				
Dinka	Niger–Congo	YRI	17	
Totals			441	1705

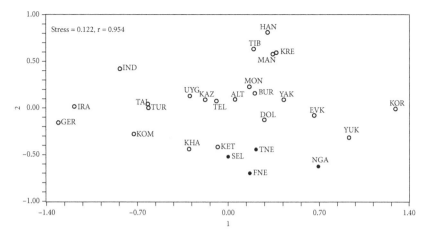

Figure 7.2 MDS plot of 27 populations, performed on autosomal SNP F_{ST} genetic distances. Samoyed-speaking populations are shown as black circles. See Table 7.1 for three-letter population codes.

even higher degree of population differentiation within Siberia (0.367). When all 27 populations were included, the F_{ST} value rose to 0.409. F_{ST} significantly increased to 0.544 when only Samoyedic populations were used for analysis.

To explore regional relationships, MDS based on autosomal and Y-chromosome SNPs was performed on 27 populations. The resulting MDS plot (Figure 7.2) constructed based on autosomal data to some extent recaps previously reported patterns of genetic structure across Eurasia from principal component analysis (PCA; Karafet et al. 2018; Pugach et al. 2016; Wong et al. 2016). Like F_{ST} statistics, MDS and PCA demonstrate high differentiation of the Siberian groups. The x-axis and PC1 correspond to west–east differentiation, while the y-axis and PC2 generally differentiate northern and southern populations. Consistent with their origin, Mongolic-speaking Buryats display genetic similarity with Mongols, and Turkic-speaking Altai-Kizhi and Teleuts are drawn toward Central Asian groups. The Tungusic-speaking Evenks are close to Yukagirs. Uralic-speaking Komi are drawn to Europe, while Khants show a closer affinity with Selkups and Tundra and Forest Nentsi. Yenisey-speaking Kets reveal close affinity to Selkups. Interestingly, Samoyedic-speaking Nganasans from the Taymyr Peninsula form a separate tight cluster closer to Evenks, Yukagirs, and Koryaks in PCA (Karafet et al. 2018) or occupy intermediate position between Evenks, Yukagirs, Koryaks, and Forest and Tundra Nentsi in MDS plots.

Similar to that for autosomal SNPs, a MDS plot based on Y-chromosome SNPs (Figure 7.3) demonstrates close relationships between Forest and Tundra Nentsi and between Sekups and Kets. Contrary to analysis of autosomal SNPs, Nganasans represent an outlier, not showing genetic similarity to Evenks, Yukagirs, and Koryaks. The result is not surprising because Nganasans differ significantly from Evenks and Yukagirs in their Y-chromosome haplogroup composition (Karafet et al. 2008,

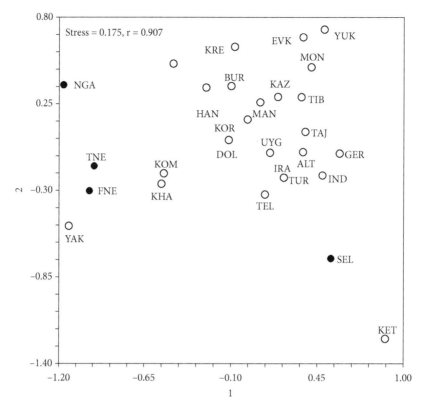

Figure 7.3 MDS plot of 27 populations, performed on Y-chromosome SNP F_{ST} genetic distances. Samoyed-speaking populations are shown as black circles. See Table 7.1 for three-letter population codes.

2018). Haplogroup C is dominant in Evenks and Yukagirs, reaching frequencies of 69.2% and 80%, respectively, whereas N-P63 is the major haplogroup in Nganasans (88.8%). No significant correlations were found between Y-chromosome and autosomal SNP structure with the Mantel test ($r = 0.028$, $P = 0.557$).

Inferring the Place of Admixture and Population Migrations: TreeMix, f3

To reconstruct patterns of population splits and mixtures in the history of the ancestors of present-day populations from Asia, we built a tree using TreeMix (Pickrell and Pritchard 2012) with the Dinka as an outgroup (Figure 7.4). The population tree splits most Siberian populations into two branches: one branch consists of Samoyed-speaking Tundra Nentsi, Forest Nentsi, and Selkups with Kets and Khants; and a second branch groups Nganasans, Evenks, Yukagirs, Dolgans, Koryaks, and Yakuts with Mongolians and East Asian populations. This branch leads to the Altai-Kizhi and Teleuts from the Altai-Sayan region. Allowing

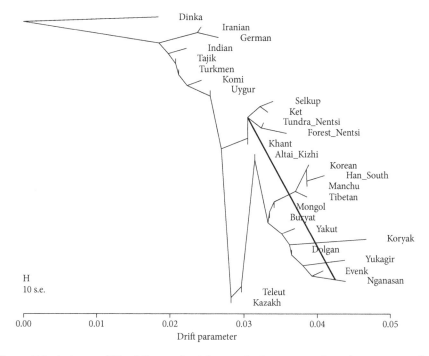

Figure 7.4 Autosomal TreeMix graph with one admixture event based on autosomal SNPs was built for 27 groups from Siberia, Asia, Europe, and Near East populations. Dinka population was used as an outgroup.

one admixture event finds evidence of admixture between the Nganasan and the Samoyed-speaking Tundra and Forest Nentsi, Selkups, and Ket branch.

We also calculated the population divergence time (*TF)* based on effective population size (*Ne*) and F_{ST} information among Siberian populations as described in McEvoy et al. (2011). In agreement with the phylogenetic tree, the smallest *TF* estimates within Siberian populations are found between Kets and Selkups (~ 1.7 kya) and between Forest and Tundra Nensti (~ 2 kya), while Nganasans represent the most diverged Samoyed population, with a time of separation that varied from the smallest of ~ 3.0 kya with Evenks to ~ 14.3 kya with the Komi (Table 7.2).

To infer evidence of admixture, we calculated the f3 statistic for all possible triple combinations of 27 modern populations with the test population (Karafet et al. 2018). A significantly negative value of the f3 statistic implies that the population is admixed. We found that several populations (i.e., Nganasans, Evenks, Yukagirs, and Koryaks) do not exhibit a history of admixture events, producing only positive f3 values (Karafet et al. 2018). For the remaining populations in our Siberian data set, the composition of mixing sources is strikingly similar: one source always includes Nganasans or Evenks, while the other is dominated by Germans or Iranians, which were used as representatives of the Europe and Near East. Among Siberian populations, the most negative statistics were found for the Altai-Kizhi and Teleuts

Table 7.2 *TF* divergence time estimates (in thousands of years) between populations from Siberia and Mongolia

	ALT	BUR	DOL	EVK	FNE	KET	KHA	KOM	KOR	MON	NGA	SEL	TEL	TNE	YAK	YUK
ALT	0															
BUR	2.7	0														
DOL	3.9	3.4	0													
EVK	6.2	4.6	2.4	0												
FNE	6.8	7.8	4.6	5.9	0											
KET	5.8	7.3	5.0	7.0	4.6	0										
KHA	6.7	8.4	5.7	8.1	4.5	4.1	0									
KOM	10.0	13.3	10.1	14.4	9.1	7.3	4.6	0								
KOR	10.7	10.8	8.0	7.2	8.2	9.6	10.0	15.9	0							
MON	2.6	0.9	3.8	5.3	8.0	7.6	8.4	12.9	10.7	0						
NGA	8.0	7.9	3.0	3.0	4.8	6.9	7.8	14.4	7.7	8.2	0					
SEL	6.0	6.9	4.5	5.9	4.3	1.7	4.2	8.3	8.6	7.4	5.8	0				
TEL	2.9	3.7	4.5	7.0	6.5	5.2	5.5	7.4	10.4	3.6	8.6	5.7	0			
TNE	5.1	5.7	2.9	4.5	2.0	3.4	2.8	7.5	7.8	6.3	3.7	3.2	4.9	0		
YAK	4.4	3.0	2.7	3.2	6.2	6.6	7.7	12.6	8.5	3.7	5.4	6.2	5.2	5.0	0	
YUK	7.6	7.4	4.7	4.0	6.0	7.6	8.1	13.6	6.3	7.5	4.7	6.6	8.2	5.5	5.2	0

See codes in Table 7.1.

from the Altai-Sayan region due to a multilayered history of admixture, and for the Dolgans because of their complex origin.

Identity by Descent (IBD) Analysis in Siberian Populations

The analysis of shared IBD segments complements analyses of population structure and gene flow. The mean IBD sharing between Siberian populations and reference groups is reported elsewhere (Karafet et al. 2018). Overall, the total genome-wide sharing for an average pair of samples from the same population varies greatly, from 7.54 cM in Buryats to 163.4 cM in Yukagirs. The majority of shared IBD segments were found within the same population, with some exceptions. Tundra Nentsi share more IBD with Forest Nentsi than with each other (83.96 versus 50.3, $P = 0.000055$) possibly due to the common origin and long-term gene flow. The Dolgan individuals share more segments with Nganasans than with other Dolgans (54.13 versus 41.72, Mann-Whitney test, $P < 10^{-10}$). The result is not surprising because the demographic data showed that the Nganasans were subjected to intense assimilation by the Dolgans in the second half of the 20th century (Goltsova et al. 2005). Selkup and Ket populations allocate significantly more IBD blocks between populations than within individuals from their own population (121.2 cM versus 85.9 cM for Kets, $P = 0.000008$; and 121.2 cM versus 114.9 cM for Selkups, $P = 0.043$).

The extensive IBD sharing within populations may be the result of isolation, limited gene flow, and rather small Siberian population sizes. We calculated the Pearson correlation coefficient between IBD sharing within populations and their census sizes. A negative, but not statistically significant, correlation ($r = -0.4248$, $P = 0.1$ for two-tailed probability, $P = 0.051$ for one-tailed probability) is observed between IBD sharing and census sizes. The distribution of length of shared IBD segments also might be a consequence of an isolation-by-distance process. A Mantel test was performed to compare pairwise geographic distances and genetic distances based on the mean IBD segments among populations normalized by the sample sizes. A statistically significant negative correlation was observed between geographic distances and the amount of shared IBD ($r = -0.349$, $P = 0.0006$). To infer the presence of a north–south or east–west gradient of shared IBD segments, we performed correlation analyses for each group between latitude or longitude and the average IBD sharing within all Siberian populations and Mongolia. Interestingly, we found a statistically significant north–south gradient ($r = 0.681$, $P = 0.002$), and no evidence of an east–west gradient ($r = -0.142$, $P = 0.3$), implying predominant north–south migrations.

Within Samoyedic-speaking populations, an unusual pattern of sharing IBD was observed. Only two Samoyedic-speaking populations, the Tundra and Forest Nentsi, shared long segments of IBD (84 cM). The Selkups reveal extremely high IBD sharing with the non-Samoyedic-speaking Kets (121.2 cM). Remarkably, the small Nganasan population from the remote Taymyr Peninsula demonstrated the greatest level of IBD sharing, not only with Samoyedic-speaking populations and neighboring Dolgans, but with nearly all other Siberian populations (average IBD = 17.84 cM). The highest sharing is especially evident for smaller IBD segments (< 6cM; Karafet et al. 2018).

Discussion

It is generally accepted by Russian ethnographers and anthropologists that all Samoyedic-speaking peoples in the vast territory of western Siberia and Taymyr Peninsula are descendants of autochthonous Paleo-Asiatic tribes that were assimilated by ancient Samoyeds who came from southern Siberia in the middle of the first millennium AD (Napolskikh 1995; Prokof'ev 1940; Vasil'ev 1983). The interactions between aboriginal peoples and the Samoyedic newcomers undoubtedly differed in various areas, resulting in cultural assimilation of the Samoyedic language and varied degrees of genetic mixing. Prokof'ef (1940) suggested that Nganasans, Nentsi, and Entsi were descendants of one aboriginal tribe, while Selkups shared distinct Paleo-Asiatic components with Ugric-speaking Khants and Mansi. However, the questions Who were those autochthonous tribes? and When did they inhabit these geographic areas? are unanswered.

A long independent evolution of the West Siberian population in Paleometal age was proposed by Molodin and Pilipenko (2016) based on genetic studies of ancient mitochondrial DNA (mtDNA). But the earliest evidence of human presence in the territory occupied by contemporary Samoyedic-speaking populations is sparse. A scarcity of early archaeological sites and human remains makes it difficult to clarify the history of Samoyedic-speaking populations and the relationships between ancient and modern populations. It was believed that in the Upper Paleolithic (since ca. 30,000 BP) people occupied just the periphery of the West Siberian Plain and Trans-Urals (Zenin 2002). Only in the early Holocene did people inhabit the whole of western Siberia. The initial human settlement of Taymyr was previously dated to not before 6,000 years BP (Khlobystin 2005). However, recent archaeological findings suggest that people lived across northern Siberia much earlier than previously thought. The oldest anatomically modern human genome, found recently at 57°N in western Siberia, produced a direct radiocarbon date of 45,000 years BP (Fu et al. 2014; Kuzmin et al. 2009). Pleistocene human remains (~ 31.6 kya) were found at the Yana RHS (~ 70°N) site in northeastern Siberia (Sikora et al. 2019). The rare case of unquestionable evidence for human involvement without artifact association was an unearthing of a mammoth kill site on the Taymyr Peninsula, dated to 45,000 years BP (Pitulko et al. 2016). This discovery expanded the populated area to almost 72°N. The advancement of mammoth hunting probably allowed people to survive and spread widely across northernmost Arctic Siberia (Pitulko et al. 2016). Local groups indisputably could not survive in isolation as they needed to access crucial resources clustered irregularly over landscapes. Benefiting from information preserved over generations and from contacts with neighboring groups, they likely resorted to foraging-area exploration (Rolland 2014).

These considerations, along with the emerging picture of similarities in culture, environmental conditions, and lifeways among Paleolithic Siberians suggest that "a continental culture of sedentary Arctic Paleolithic hunters" existed in the north of Siberia and Eastern Europe by the end of the Last Glacial Maximum (LGM; Bogoraz 1929; Okladnikov 1964). This hypothesis was supported by recent studies demonstrating that wide-ranging networks of mate exchange and gene flow were present among Upper Paleolithic small foraging bands at high latitudes on the pre-LGM Eurasian landscape (Sikora et al. 2017, 2019). The Russian ethnographer Simchenko (1968) developed this hypothesis further and suggested that by the end of the Glacial period, all ancient Arctic populations from the Kola Peninsula up to the Bering Strait represented an ethnically related (or homogeneous) ancient aboriginal population. He posited that modern northern Samoyeds (Nganasans, Entsi, and Nentsi), Yukagirs, Evenks, Evens, Chukchi, and Koryaks originated from the same or closely related ancestral tribe(s). The Nganasan language and material culture suggest prehistoric contacts with Tungus and Yukagir populations (Anikin and Helimsky 2007; Simchenko 1968).

Our results are compatible with this hypothesis. The MDS plot and maximum likelihood tree (with and without migration events; Figures 7.2 and 7.4) cluster northern Siberian populations (Nganasans, Yukagirs, Evenks, and Koryaks) despite the large geographic distances separating these populations. ADMIXTURE analysis demonstrated that the "Siberian" component that is highly pronounced in Nganasans is also frequent in northern Siberian populations (Karafet et al. 2018). Positive f3 values do not support a history of admixture between these Arctic and sub-Arctic populations. At the same time, this analysis implicated Nganasans and/or Evenks as the mixing source population for the other Siberian groups. These results are consistent with our finding that the Nganasans and Evenks exhibit the highest level of short-IBD-segment sharing with nearly all other Siberian populations (Karafet et al. 2018). Indirect support for the existence of prehistoric Arctic populations also derives from Y-chromosome data. The oldest modern human genome, Ust'-Ishim, belongs to the NO clade (Wong et al. 2016), the predecessor of the most frequent North Eurasian haplogroup N.

There is no consensus among linguists and ethnographers about the original homeland of Samoyeds. The assumption that all extant Samoyed-speaking peoples were related to Sayan tribes was put forward back in the 18th century by Finnish ethnologist M. A. Castren (Prokof'yeva 1964). It was hypothesized that the ancient Samoyeds and Finno-Ugric peoples lived on the slope of the Altai and Sayan Mountains and from there they moved to the north and to the west. The Hungarian linguist Péter Hajdú (1963) claimed that the ancient Uralic home was not in southern Siberia but on the western side of the Ural Mountains, around the bend of the Volga. According to Hajdú (1963), the ancestors of the Samoyeds left the Uralic community in the fourth millennium BC and moved to the forest regions of

Western Siberia. Based on his linguistic historical reconstruction, Helimsky (1990) suggested a wide area between Middle Ob and Yenisey as the proto-Samoyed original home, matching the Kulay archaeological culture of the 500 BC to 500 AD. Finnish linguist J. Janhunen (2009) placed the homeland of Proto-Samoyed in the area of Minusinsk basin on the Upper Yenisey, which is bounded on the east by the Sayan Mountains and on the west by Kuznetsk Alatau.

The Minusinsk basin region has an exceptionally well-documented direct sequence of archaeological cultures, extending from the Eneolithic Afanasievo culture (3500–2500 BC) through the Bronze and Iron Age Okunevo (2500–2000 BC), to the Andronovo (2000–1500 BC), Karasuk (1500–800 BC), Tagar (800–100 BC), and Tashtyk (100 BC–400 AD) cultures, and continuing up to the historical Yenisei Kirghiz (from 400 AD) and Mongols (from 1300 AD). Yeniseian-speaking Kets were suggested to be connected to Karasuk culture (Chlenova 1972; van Driem 2007). Samoyeds were hypothesized to be associated with the Tagar archaeological culture (Janhunen 2009). There is evidence of likely migration among settled groups of Kulay and Tagar cultures (Bobrov 2011; Makarov 2013).

Autosomal genome data from 18 ancient samples from the Afanasievo, Okunevo, Andronovo, and Karasuk archeological sites have recently become available (Allentoft et al. 2015). They have been used to investigate Siberian populations (Karafet et al. 2018). Samples from the Eneolithic and Bronze Age Afanasievo, Okunevo, and Andronovo cultures in the Altai-Sayan region formed a rather tight cluster, with closer genetic relationships with present-day Iranians and Germans (Karafet et al. 2018). On the other hand, ancient samples from Karasuk Bronze Age culture were broadly dispersed among Europeans, Central Asians, and Siberians. The origin of the Karasuk culture is very controversial among archaeologists. This period of transition from the Bronze to the Iron Age (from the second to the first millennium BC) coincided with the time of climatic shifts to more humid conditions, which resulted in the spread of forests southward and expansion of the habitat for hunters and fishermen (Schneeweiss et al. 2018). It is commonly accepted that Karasuk culture represents the continuation and transformation of the Okunevo and Andronovo cultures, along with the incorporation of new migrants. Yet, there is no consensus about the origin of migrants in the Karasuk culture. Many suggestions of the southern source population have been put forward, including Central Asia, the steppes of Kazakhstan, Mongolia, and China (Chlenova 1972; Jettmar 1950) or northern forest source population (Martynov 1991). Contradicting suggestions for migrations from East Asia to the Karasuk (Karafet et al. 2018), our ADMIXTURE, PCA, f3, and D-statistic analyses provided several lines of evidence that show closer affinity of the Karasuk culture to modern Siberian populations. The Tagar culture is commonly believed to have developed out of the traditions of the previous Bronze Age Karasuk culture without significant immigration (Bokovenko 2006). Autosomal data on ancient Tagar samples are not yet available. We cannot determine whether Karasuk and Tagar populations were genetically close. However, the proposed descendant populations of Kets and Selkups from Karasuk and Tagar cultures, respectively, show very close connections; clustering together in the MDS plot and tree, sharing a substantial amount of long and short IBD blocks, and returning a divergence that is estimated at ~ 1.7 kya. This divergence time does coincide with the suggested time when ancestors of Selkups migrated to the north during the early part of the first millennium AD (Prokof'yeva 1964).

Summary

Our data are consistent with several hypotheses about the origin of modern Samoyedic populations. The Nganasans are very likely direct descendants of the ancient sedentary Arctic Paleolithic hunters of northern Asia and Neolithic hunters of wild deer who were culturally assimilated by various Samoyedic peoples. The Taymyr Peninsula might serve as the last refuge for local and migrant hunters due to its remoteness, low population density, and the harshness of the ecosystem.

Our data are compatible with the hypothesis that Tundra Nentsi and Forest Nentsi are descended from the same aboriginal tribe that gave rise to the Nganasans (Prokof'ev 1940) and who were assimilated by ancient Samoyeds genetically and culturally. PCA, MDS plot, TreeMix tree, ADMIXTURE, and IBD analyses reveal close genetic ties between the Tundra and Forest Nentsi (Karafet et al. 2018). Structure analysis demonstrates that the Paleo-Asiatic ("Nganasan") component is present at a higher frequency in the Tundra Nentsi (51%) than in the Forest Nentsi (46%; Karafet et al. 2018). Both populations also share the common Y-chromosome P63 variant with Nganasans (Tundra Nentsi, 63%; Forest Nentsi, 49%). Evidence of higher Samoyedic than autochthonous proportion in the Forest Nentsi also comes from information on their patrilineal clan affiliation. The patrilineal clan system was universal for almost all Siberian populations. Based on the names of clans, toponymics, and legends, ethnographers were able to trace clan origins either to ancient Samoyeds or to aboriginal tribes in Forest Nentsi population (Vasil'ev 1977). Among 1,552 Forest Nentsi studied in 1978 (Abanina and Sukernik 1980), 58% traced back to Samoyedic clans, while 42% belonged to the clans of aboriginal origin.

The composition of Selkup ancestry was suggested to be different than that of the Nganasans or of the Nentsi. Prokof'ef (1940) proposed that the Selkups shared the same Paleo-Asiatic components with Ugric-speaking Khants and Mansi. Pelikh (1972) argued that in addition to Paleo-Asiatic people from western Siberia and Samoyeds from southern Siberia, some other ancestral components can be identified in the Selkups. Vasil'ev (1983) hypothesized that the Selkups did not assimilate the aboriginal people of western Siberia but rather represent the direct descendants of southern Samoyeds. We found some support for Vasil'ev's hypothesis. The Selkups differ significantly from the Nganasans and Forest and Tundra Nentsi, sharing common genetic features not with the Khants, as proposed by Prokof'ev (1940), but with the Kets, who speak the only surviving member of the

Yeneseian linguistic family. Y-chromosome data also demonstrate the close genetic affinity of the Selkups and the Kets. IBD analysis revealed that the Selkups exhibit the highest IBD sharing with the Kets in all segment classes. This may suggest not only recent gene flow between populations, but also their common ancestry or ancient admixture.

Conclusions

Genetic relationship is observed only between Forest and Tundra Nentsi. Selkups are affiliated with the Kets from the Yenisey River, while the Nganasans are separated from their linguistic neighbors, showing closer affinities with the Evenks and Yukagirs.

Acknowledgments

This project was supported by a National Science Foundation grant to T.M.K. and M.F.H. (PLR-1203874), and by the State Research Project (No 0324-2019-0041) to L.P.O. The authors thank Olga Savina for assistance in analyses.

References

Abanina, Tatiana A., and Rem I. Sukernik. 1980. "Population structure of the Forest Nentsi: Results of genealogical research." *Genetika* no. 16: 156–164.

Allentoft, Morten E., Sikora Martin, Karl G. Sjögren, Simon Rasmussen, Morten Rasmussen, Jesper Stenderup, Peter B. Damgaard, Hannes Schroeder, Torbjörn Ahlström, Lasse Vinner, et al. 2015. "Population genomics of Bronze Age Eurasia." *Nature* 522, no. 7555: 167–172.

Anikin, Alexandr E., and Eugene A. Helimsky. 2007. *Samoyedic-Tungus-Manchu Lexical Relations*. Moscow: Nauka.

Bobrov, Vladimir V. 2011. "Tagar Culture in North Forest-Steppe." Paper presented at the Terra Scythica, Novosibirsk.

Bogoraz, Waldemar G. 1929. "Elements of the culture of circumpolar zone." *American Anthropologist* no. 31: 579–601.

Bokovenko, Nikolay. 2006. "The emergence of the Tagar culture." *Antiquity* no. 80: 860–879.

Chard, Chester S. 1974. *Northeast Asia in Prehistory*. Madison: University of Wisconsin.

Chlenova, Natalya L. 1972. *The Chronology of the Monuments in Karasuk Epoch*. Moscow: Nauka.

Fu, Qiaomei, Heng Li, Priya Moorjani, Flora Jay, Sergey M. Slepchenko, Alexsei A. Bondarev, Philip L. F. Johnson, Aynuer Aximu-Petri, Kay Prüfer, Cesare de Filippo, et al. 2014. "Genome sequence of a 45,000-year-old modern human from western Siberia." *Nature* no. 514: 445–449.

Goebel, Ted, Anatoli P. Derevianko, and Valerii T. Petrin. 1993. "Dating the Middle-to-Upper-Paleolithic transition at Kara-Bom." *Current Anthropology* no. 34: 452–458.

Goltsova, T. V., Ludmila. P. Osipova, Sergey I. Zhadanov, and Richard Villems. 2005. "The effect of marriage migration on the genetic structure of the Taimyr Nganasan population: Genealogical analysis inferred from mtDNA markers." *Russian Journal of Genetics* no. 41: 779–788.

Hajdú, Peter. 1963. *The Samoyed Peoples and Languages*. The Hague: Indiana University.

Hammer, M. F., Tatiana M. Karafet, Alan J. Redd, Hamdi Jarjanazi, Silvana Santachiara-Benerecetti, Himla Soodyall, and Stephen L. Zegura. 2001. "Hierarchical patterns of global human Y-chromosome diversity." *Molecular Biology and Evolution* no. 18: 1189–1203.

Helimsky, E. A. 1990. "The external connections and early contacts of the Uralic languages." *Problems of Uralistics* no. 1: 19–43.

Jettmar, Karl. 1950. "The Karasuk culture and its south-eastern affinities." *Bulletin of the Museum of Far Eastern Antiquities* no. 22: 83–126.

Janhunen, Juha. 2009. "Proto-Uralic—What, where, and when?" In *The Quasquicentennial of the Finno-Ugrian Society*, edited by J. Ylikoski, 79–94. Helsinki: Suomalais-Ugrilainen Seura.

Karafet, Tatiana M., Ludmila P. Osipova, Marina A. Gubina, Olga L. Posukh, Stephen L. Zegura, and Michael F. Hammer. 2002. "High levels of Y-chromosome differentiation among native Siberian populations and the genetic signature of a boreal hunter-gatherer way of life." *Human Biology* no. 74: 761–789.

Karafet, Tatiana M., Ludmila P. Osipova, and Michael F. Hammer. 2008. "The Effect of History and Life-style on Genetic Structure of North Asian Populations." In *Past Human Migrations in East Asia: Matching Archaeology, Linguistics and Genetics*, edited by A Sanchez-Mazas, R Blench, M. D Ross, I. Peiros, and M. Lin, 395–415. New York: Routledge.

Karafet, Tatiana M., Ludmila P. Osipova, Olga V. Savina, Brian Hallmark, and Michael F. Hammer. 2018. "Siberian genetic diversity reveals complex origins of the Samoyedic-speaking populations." *American Journal of Human Biology* no. 30: e23194.

Khlobystin, Leonid P. 2005. "Taymyr: The archaeology of northernmost Eurasia." In *Contributions to Circumpolar Anthropology*, Vol. 5, edited by W. W. Fitzhugh and V. V. Pitulko. Washington DC: Smithsonian Institution.

Kuzmin, Yaroslav V., A. J. Timothy Jull, Lubov A. Orlova, and Leopold D. Sulerzhitsky. 1998. "C-14 chronology of Stone Age cultures in the Russian Far East." *Radiocarbon* 40: 675–686.

Kuzmin, Yaroslav V., Pavel A. Kosintsev, Dmitry I. Razhev, and Gregory W. L. Hodgins. 2009. "The oldest directly-dated human remains in Siberia: AMS 14C age of talus bone from the Baigara locality, West Siberian Plain." *Journal of Human Evolution* no. 57: 91–95.

Makarov, Nikolai P. 2013. "The ancient stages of the culture genesis of the Krasnoyarsk northern indigenous peoples." *Journal of Siberian Federal University, Humanities and Social Sciences* no. 6: 816–841.

Martynov, Anatoly I. 1991. *The Ancient Art of Northern Asia*. Urbana: University of Illinois Press.

McEvoy, Brian P., Joseph E. Powell, Michael E. Goddard, and Peter M. Visscher. 2011. "Human population dispersal 'out of Africa' estimated from linkage disequilibrium and allele frequencies of SNPs." *Genome Research* no. 21: 821–829.

Molodin, Vyacheslav L., and Alexandr S. Pilipenko. 2016. "Hunters for ancient genes: Genetic chronicle of the West Siberian population in the Paleometal age." *Science First Hand* no. 43: 60–73.

Napolskikh, Vladimir V. 1995. *Uralic Original Home: History of Studies*. Izhevsk: Udmurt Institute for History, Language and Literature.

Okladnikov, Alexei P. 1964. "Ancient population of Siberia and its culture." In *The Peoples of Siberia*, edited by M. G. Levin and P. Potapov, 13–98. Chicago: University of Chicago Press.

Pelikh, Galina I. 1972. *The Origin of Selkups*. Tomsk: Tomsk University.

Pickrell, Joseph K., and Jonathan K. Pritchard. 2012. "Inference of population splits and mixtures from genome-wide allele frequency data." *PLOS Genetics* 8: e1002967.

Pitulko, Vladimir V., Alexei N. Tikhonov, Elena Y. Pavlova, Pavel A. Nikolskiy, Konstantin E. Kuper, and Roman N. Polozov. 2016. "Paleoanthropology—Early human presence in the Arctic: Evidence from 45,000-year-old mammoth remains." *Science* no. 351: 260–263.

Prokof'ev, Georgiy N. 1940. "Ethnogenesis of the ethnic groups on the Ob-Yenisei basin." *Soviet Ethnography* no. 3: 67–76.

Prokof'yeva, Ekaterina D. 1964. "The Nentsy." In *The Peoples of Siberia*, edited by M. G. Levin and P. Potapov, 547–570. Chicago: University of Chicago Press.

Pugach, Irina, Rostislav Matveev, Victor Spitsyn, Sergey Makarov, Innokentiy Novgorodov, Vladimir Osakovsky, Mark Stoneking, and Brigitte Pakendorf. 2016. "The complex admixture history and recent southern origins of Siberian populations." *Molecular Biology and Evolution* 33: 1777–1795.

Rolland, Nicolas. 2014. "The Pleistocene peopling of the North: Paleolithic milestones and thresholds horizons in Northern Eurasia: Part I: Lower Paleolithic antecedents." *Archaeology Ethnology and Anthropology of Eurasia* no. 42: 2–17.

Sikora, Martin, Vladimir V. Pitulko, Victor C. Sousa, Morten E. Allentoft, Lasse Vinner, Simon Rasmussen, Ashot Margaryan, Peter de Barros Damgaard, Constanza de la Fuente, Gabriel Renaud, Melinda Yang, et al. 2019. "The population history of northeastern Siberia since the Pleistocene." *Nature* no. 570: 182–188.

Sikora, Martin, Andaine Seguin-Orlando, Victor C. Sousa, Anders Albrechtsen, Thorfinn Korneliussen, Amy Ko, Simon Rasmussen, Isabelle Dupanloup, Philip R. Nigst, Marjolein D. Bosch, et al. 2017. "Ancient genomes show social and reproductive behavior of early Upper Paleolithic foragers." *Science* no. 358: 659–662.

Simchenko, Yuri B. 1968. "Some data on ancient ethnic substrate in the composition of North Eurasia peoples." In *The Problems in the Anthropology and Ethnic History of the Peoples of the Asia*, edited by V. P. Alekseev and I. S. Gurvich, 194–213. Moscow: Nauka.

Schneeweiss, Jens, Fabian Becker, Vyacheslav I. Molodin, Hermann Parzinger, Zhanna V. Marchenko, and Svetlana V. Svyatko. 2018. "Radiocarbon chronology of occupation of the site Chicha and Bayesian statistics for the assessment of a discontinuous transition from Late Bronze to Early Iron Age (West Siberia)." *Russian Geology and Geophysics* no. 59: 635–651.

van Driem, George. 2007. "Endangered Languages of South Asia." In *Language Diversity Endangered*, edited by M. Brenzinger, 303–308. Berlin: Mouton de Gruyter.

Vasil'ev, Vladimir I. 1977. "Problems in the ethnogenesis and ethnic history of the peoples of the North (based upon Samodiyan data)." *Soviet Ethnography* no. 4: 3–17.

Vasil'ev, Vladimir I. 1983. "Main Problems of Ethnic History of Samoyedic Peoples." In *Uralo-Altaistika. Archeology, Ethnography, Language*, edited by E. I. Ubryatova, 119–123. Novosibirsk: Nauka.

Wong, Emily H. M., Andrey Khrunin, Larissa Nichols, Dmitry Pushkarev, Denis Khokhrin, Dmitry Verbenko, Oleg Evgrafov, James Knowles, John Novembre, Svetlana Limborska, et al. 2016. "Reconstructing genetic history of Siberian and Northeastern European populations." *Genome Research* no. 27: 1–14.

Zenin, Vasiliy N. 2002. "Major stages in the human occupation of the West Siberian Plain during the Paleolithic." *Archaeology, Ethnology and Anthropology of Eurasia* no. 4: 22–44.

8

Linguistic Diversity and Human Migrations in Gabon

Franz Manni and John Nerbonne

Introduction

Bantu Languages and Classifications

Bantu languages belong to the Niger-Congo phylum and include about 600 varieties spoken in almost all of sub-Saharan Africa. Their geographic continuity is nearly perfect, interrupted only by the Khoisan languages spoken in South Africa. Wilhelm Heinrich Immanuel Bleek (1862) was the first to hypothesize the genetic unity of Bantu languages. The most important classification of Bantu languages, still used as a practical taxonomic reference, is that done by Guthrie (1967), but more recent classifications are available (Grimes 2000; Mann and Dalby 1987).

Guthrie defined the geographical boundary of the Bantu linguistic domain and divided it into an eastern and a western zone. The western region includes Cameroon, Gabon, Congo, the west of the Democratic Republic of Congo (DRC), Angola, and part of Zambia. The eastern region includes the eastern part of DRC and all the eastern countries of sub-Saharan Africa. This split, although debated, has been consensual for a long time. Using a lexicostatistics approach, Heine (1973) corroborated the division of Bantu languages into two clusters, a western and an eastern one: the eastern group derives from the western one, while the languages of Gabon and Cameroon are independent lineages. Later, Ehret (1999) suggested that eastern, central, and southern Bantu languages should be merged into a single group called Savannah Bantu. Other studies (Bastin et al. 1999; Nurse and Philippson 2003; Rexova et al. 2006) confirm a western/eastern division of the entire Bantu linguistic domain, suggesting that the western part is older because the languages spoken in zones A, B, and C emerged first from proto-Bantu varieties initially spoken in the middle Benue River valley, located between Cameroon and Nigeria about 5000 BP (Greenberg 1955). The identification of a homeland is essential to describe how Bantu languages later disseminated, and the middle Benue River valley is a good candidate because it includes speakers of the only linguistic varieties close enough to another branch (Benue-Congo, spoken in Nigeria) of the Niger-Congo linguistic family to which Bantu languages belong as well.

Franz Manni and John Nerbonne, *Linguistic Diversity and Human Migrations in Gabon* In: *Human Migration*. Edited by: Maria de Lourdes Muñoz-Moreno and Michael H. Crawford, Oxford University Press. © Oxford University Press 2021. DOI: 10.1093/oso/9780190945961.003.0009

Migration Waves

Concerning the diffusion of Bantu languages, Bastin et al. (1979) suggested two migration waves: first, a western wave from the south of Cameroon, then to the equatorial forest along the rivers and, finally, progressing further southward along the Atlantic coast; and second, an eastward wave avoiding the equatorial forest. Of course, the general picture is blurred by secondary contact between the languages, by a different migration speed according to the route (to the south or to the east), by secondary migrations, and by continuous population displacements until today.

The reader's attention should turn to the extremely fast spread of Bantu languages, which have disseminated throughout half of Africa and have replaced almost all preexisting languages, similarly to what Latin did in Europe. A possible explanation is related to the lifestyle of early Bantu-speaking populations: proto-Bantu lexical roots[1] show that they generally were agriculturalists and farmers. In contrast to hunting and gathering, agriculture requires a considerable workforce before it can be sustained viably, which then leads to large societies. Once populations begin growing, migration processes and population diffusion follow, and they are further promoted when the soil rapidly (but only temporarily) depletes due to its exploitation. This is how the Bantu expansion could have progressed.

Archaeology and Linguistics

A first attempt to combine archaeological and linguistic evidence to understand the dispersal process can be traced back to Oliver (1966). Later, Phillipson (1976, 1977a, 1977b, 2002) provided an ambitious reconstruction by arguing for the development of early Bantu language varieties in Cameroon from about 3000 BP on. According to his scenario, when Bantu-speakers dispersed eastward along the northern fringe of the equatorial forest, they met other farmers, probably speaking Central Sudanic languages. After a long phase of contact, they started herding domestic cattle and sheep, learned about the cultivation of certain cereal crops and acquired metal-working techniques. Vansina (1984; 1990, 49–57; 1995) criticized Phillipson about the need for a reliable dispersal route and stressed that the major driving force for the Bantu linguistic divergence was a phenomenon of linguistic fission between varieties that had diverged in an earlier phase, with outermost dialects developing into languages after each fission (Heggert 2004, 315). According to Vansina, if indeed a first migration happened eastward to the Great Lakes Region, a simultaneous movement took place to the south of Cameroon and Gabon, and more southward. Later on, the continuous pattern of habitat resulting from the first major migration was disrupted by the presence of both a dense forest and large stretches of

---

[1] For a reference database including about 10,000 entries proposed for Proto-Bantu reconstructions, see www.africamuseum.be/collections/browsecollections/humansciences/blr

marshland. New computational analyses (Holden 2002; Holden and Gray 2006) on 75 Bantu languages extracted from the 542 Bantu languages published by Bastin et al. (1999) agreed with the archaeological hypothesis about the large migration eastward of Bantu-speaking agriculturalists, meaning that Bantu languages diffused together with their speakers.

Population Genetics and Linguistics

An interesting study bringing together genetic and linguistic evidence over the whole Bantu linguistic domain (de Filippo et al. 2013) was aimed at testing two models of Bantu population-language dispersal: an Early Split, north of the rainforest, of the eastern and western groups, about 4000 BP; a Late Split, south of the rainforest, of the eastern group from the western group, about 2000 BP. The authors measured DNA genetic diversity as a function of the geographic distance from the Bantu homeland, along possible inferred itineraries of migration, finding a progressive reduction of the genetic diversity from the homeland. This pattern supports the demic diffusion[2] of the Bantu expansion and better correlates with the Late Split model. Li et al. (2014) estimated the first expansion of the Bantu-speaking groups to have occurred at around 5600 years ago but found that a migration to the east and then to the south is statistically as likely as other models.

While the DNA genetic diversity of Gabon populations is very low and detectable only by genome sequencing (Patin et al. 2017), the linguistic differences are not negligible when compared to those in other Bantu-speaking areas, suggesting that the peopling—by populations having a similar genetic background—happened early in the dispersal from Cameroon, leaving time for later differentiation in situ.

Current Linguistic Diversity in Gabon

All the ethnic groups living in Gabon speak Bantu languages (with the exception of some Pygmy groups), but the use of French (the official language) is widespread in the increasingly multiethnic towns and many indigenous language varieties are now threatened.

The Bantu varieties of Gabon include languages from Guthrie zones A (Benga A34; Fang A75; Shiva A83; Bekwil A85b) and B (B10/20/30/40/50/60/70; see Figure 8.1). According to grammatical and lexical traits, the languages of the group B10 (MYENE) and B30 (TSOGO) are related and distinct from other languages in the region, but it is not clear if this is the consequence of a common genealogical

---

[2] *Demic diffusion* is a demographic term referring to a migratory model of population diffusion into and across an area that had been previously uninhabited by that group, with the group possibly, but not necessarily, displacing, replacing, or intermixing with preexisting populations.

origin or the result of linguistic convergence due to contact (Mouguiama-Daouda and Van der Veen, 2005; Nurse and Philippson, 2003). The languages B20 (KOTA-KELE) have an ambiguous status, too, because their genetic unity is unclear. While the languages belonging to the three groups B50 (NJABI), B60 (MBETE), and B70 (TEKE) are close to those of zone C and might be classified into a single cluster B50-B60-B70, languages of the group B40 (SIRA) are related to those spoken in zone H (south of Gabon).

According to Clist (2005, 490), Gabon has been progressively peopled by waves of Bantu-speaking populations coming from the northeast, but also from the south and the east, starting about 2600 BP. This peopling scenario is more complex than the simple southward movement from Cameroon proposed by Vansina (1995), in which Gabon would have been crossed only by the western Bantu expansion wave moving to the south. Clist (2005) suggests that, in reality, the main migration wave from the Benue River valley to the south might not have passed through Gabon, but

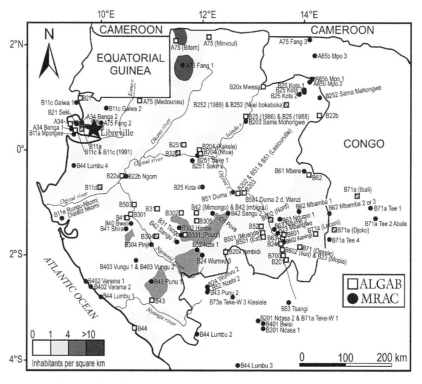

Figure 8.1 The Republic of Gabon (see map) is an equatorial African country largely covered by rainforest, with a total population of about a million and half inhabitants. More than half of the population live in the bigger cities, and population density outside urban areas is low (see shades of gray). The map shows the location of the 53 varieties reported in the ALGAB database (empty squares; when a diagonal appears inside ta square, the position is approximated) and the location of the 64 varieties of the MRAC database (see the section "ALGAB and MRAC Data Sets (https://research.rug.nl/en/publications/linguistic-probes-into-human-history)" for details).

rather further to the east, in a savannah corridor created by dry climatic conditions, in what previously was a dense equatorial forest. This corridor is believed to have lasted from 2800 BP to 2100 BP, that is, about seven centuries (Maley 2001), a time-span long enough to enable continued human migrations. Whatever the general migration scenario, it is known that the Fang languages (A75) correspond to a rather recent migration wave from Cameroon that started 500 BP and continued until the 1930s (Hombert et al. 1989).

The Aims of This Study

By processing word lists accounting for the linguistic diversity of Gabon with a computational linguistics method measuring the phonetic difference between two words, the Levenshtein distance (Heeringa 2004), we computed distance matrices accounting for the aggregate lexical difference. Then, we analyzed the distance matrices using both bootstrap phylogenetic trees and multidimensional scaling (MDS) to identify major linguistic groups that we related to specific migration waves.

We processed two independently collected and largely overlapping data sets: the ALGAB (*Atlas Linguistique du Gabon,* Hombert 1990) and the MRAC (*Musée Royal de l'Afrique Centrale*, Tervuren, Belgium).[3] The ALGAB data set lists 158 words for 53 linguistic varieties, while the MRAC is based on 88 words and accounts for 64 varieties (Figure 8.1).

The Levenshtein method is different from the cognate-sharing approach used so far in studying the diversity of Bantu languages. The earlier work (Bastin et al. 1999; Grollemund et al. 2015; Holden 2002; Holden and Gray 2006) proceeded by establishing 0/1 matrices, where a 0 pairwise difference is attributed to a pair of words having the same meaning and sharing a common ancestor term (cognates), while a difference of 1 is attributed to a pairwise comparison where the words originated independently (not cognates). This method relies on having expert judgments of cognacy and is less sensitive than Levenshtein distances, which measure the difference in pronunciation according to string alignment.

Results of the Levenshtein Classification

ALGAB and MRAC Data Sets

Concerning the ALGAB (wordlists of 158 items), the consensus tree (Figure 8.2A) shows that there are five main clusters: {A75}; {B10, B30}; {B20}; {B40}; {B50, B60, B70}. When we represent these groups on a geographic map (Figure 8.2B), we note a striking degree of geographic coherence, suggesting a significant

---

[3] The MRAC is a subset of the database processed by Bastin et al. (1999).

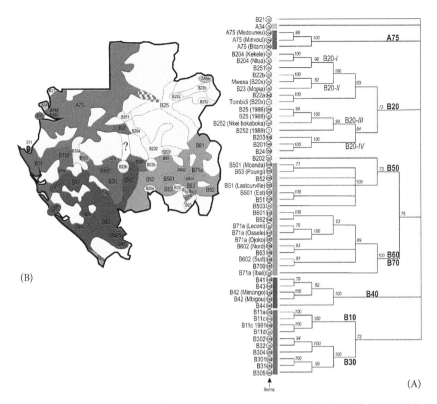

Figure 8.2 A: UPGMA bootstrap consensus tree concerning the classification of the 53 varieties listed in the ALGAB database. Nodes supported by fewer than 70% of the bootstrap subreplicates have been collapsed. The number of lexical items available for each language is reported after the labels. B: Mapping of the major clusters on the consensus map obtained using several references (Grimes 2000; Maho 2009; Simons 2016).

correlation between geographic and linguistic distances ($r = 0.461^{**}$; Mantel test). When MRAC data (wordlists of 88 items) are processed in the same way (figure not shown), the five major clusters are not exactly the same ones: {A75}; {B10, B30}; {B20*a*}; {B20*b*}; {B20*c*, B40, B50, B60, B70}. While the varieties labeled as A75 and B10/B30 are classified in a similar way, the latter data set shows that the linguistic group B20 is split in three clusters and that heterogeneous varieties belonging to the groups B20, B40, B50, B60, and B70 are clustered together as in Bastin et al. (1999), who called it North-central Bantu. It is reasonable that we found similar results because we were processing the same data set used by Bastin.

Are the two clusterings different because the two databases, the ALGAB and the MRAC, account for word lists of a different length? A closer look shows that the wordlists of the MRAC are a subset of those used in the ALGAB, the latter including 65 additional items (50 nouns + 15 verbs), meaning that ALGAB data are likely to provide a better classification, as more information is available.

Merging Data Sets

To verify whether identical varieties, documented independently in the two databases, would cluster together, we merged the two data sets to compute a unique consensus bootstrap tree (not shown) based on only shared lexical items. It points to:

1. The unity of the cluster of languages B10-B30 (bootstrap score = 99)
2. The close association of the languages B60-B70 (bootstrap score = 82)
3. The coherence of the languages B50 that form a cluster (bootstrap score = 79) distinct from B60/B70
4. The looseness of the cluster B40 (bootstrap score = 60)
5. The "explosion" of the group B20, split in five different and independent clusters.

Two other aspects of the classification of the merged data set are interesting:

6. Identical varieties, documented by two independent databases, are generally clustered next to each other in the tree, thus suggesting that the discrepancies between the classifications are related to the different lengths of the wordlists. This leads us to trust the ALGAB data (158 words) more than the MRAC (88 words).
7. The bootstrap support for the North-central western Bantu cluster advocated by Bastin et al. (1999) becomes weak (bootstrap score = 54).

To compare the outcomes of our analysis based on a new method, the Levenshtein distance, with the classical cognate-sharing approach, we aligned our classification of the ALGAB data set to the classification of Grollemund et al. (2015) that processed a subset of the same varieties. We found that clusters and subclusters did correspond (not shown), meaning that the Levenshtein algorithm captures the same signal of linguistic relatedness (or difference) that a method based on cognate-sharing does, but with the immense advantage of not requiring the aid of experts to provide judgments about cognacy. The only discrepancy between the two approaches concerns, again, the clustering of the group B20.

Discussion and Conclusions

The Peopling of Gabon: Savannah Corridors Versus the Rainforest

The question of the Bantu dispersal has vigorously resurfaced thanks to the work of the team of Koen Bostoen (University of Ghent). The routes of Bantu expansion they suggest rely on the geographical plot of consensus Bayesian phylogenetic trees (for a review, see Bostoen et al. 2015; Grollemund et al. 2015). These authors code linguistic

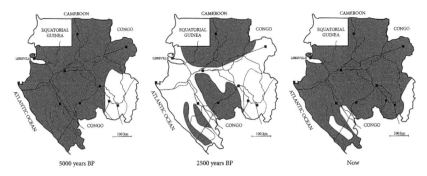

Figure 8.3 Gabon: the progressive appearance of savannah corridors (white) in the rainforest (gray) according to paleoenvironmental data (adapted from Grollemund et al. 2015). Equatorial Guinea is not shaded. BP means "before the present."

diversity according to cognacy and provide a temporal frame for the different splits calibrated according to archaeological dates. By interpreting the topology of trees as reliable migration routes,[4] they highlight that savannah corridors were migration routes preferable to rainforest crossings, the rationale being that Bantu-speaking societies were adapted to this kind of environment, since their earlier homeland was savannah. According to palynological evidence, a progressive formation of savannah corridors took place through the rainforest during the Middle and Late Holocene, from 4000 BP to 2500 BP, when the surface of savannah was at its maximum extension (Bostoen et al. 2015; Lézine et al. 2013). Concerning Gabon, the formation of savannah corridors can be inferred as shown in Figure 8.3; if the Bantu migration proceeded through them, the peopling would have been possible from the east and the south (the west being the Atlantic seashore), but not from the north.

But, did the Bantu-speaking populations enter Gabon following the progressive formation of savannah corridors starting 4000 BP or were their migrations independent of them? We cannot answer this question directly, because the linguistic analyses we conducted do not explicitly address temporal issues related to peopling phases. To set a timeframe according to archaeological sites excavated in Gabon, we referred to the calibrated radiocarbon (^{14}C) dates compiled by Oslisly et al. (2013) and to their classification in four main stages of technological knowledge (Figure 8.4). While some dating uncertainty cannot be excluded, the temporal sequence of occupation addressed by Oslisly et al. (2013) overlaps with the time frame of the Bantu expansion and, interestingly, points to a population decline starting about 2400 BP and lasting until recent centuries (Oslisly 2001; Wozka 2006), finally reaching a new maximum five centuries ago, after which it declined again until the colonial period (Figure 8.4). The Neolithic stage corresponds to the transition between the Late Stone Age and the Early Iron Age, that is, when people started to become sedentary, made polished stone tools and pottery, and used stone hoes and

---

[4] By hypothesizing that the present-day location of languages corresponds to the location they had in the past.

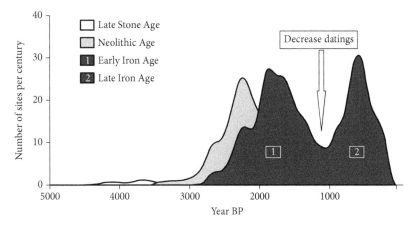

Figure 8.4 Survey of radiocarbon dates over the past 4,000 years in central Africa, including Gabon (from Oslisly et al. 2013). Late Stone Age (5500–3500 years BP)—14 sites, shown between 4400 and 3500 years BP; Neolithic Stage (3500–1900 years BP)—33 sites; Early Iron Age (2800–1000 years BP) —79 sites; Late Iron Age (1000–100 years BP) —40 sites. The periods sometimes overlap because two technological phases can coexist at the same time, like typewriters and computers in the late 1990s.

axes to practice slash-and-burn agriculture. This phase is related to the arrival of Bantu migrations, with a demographic explosion in the subsequent period, the Iron Age. Because the Neolithic stage started 3500 BP, we date the first arrival of Bantu speakers to Gabon at this point. This time frame fits well with the scenario and theory of savannah corridors (Grollemund et al. 2015), but is also compatible with earlier Bantu migrations southward, directly though the rainforest. Interestingly, the maximum extension of the savannah corridors (Figure 8.3) corresponds to a possible population maximum (Figure 8.4).

Was the rainforest a real impediment to migrations from Cameroon? While a migration route from Cameroon southward along the generally sandy seashore advocated by Bastin et al. (1979) is possible, we suggest that the rivers crossing the rainforest could have been potential paths of displacement and that the practice of slash-and-burn agriculture (characterizing many Bantu-speaking groups today) would also have been practiced in the past, in a forest environment. For all of the above reasons, we will not assume that the Bantu peopling of Gabon was necessarily dependent on the climatic change that led the rainforest to shrink and the savannah habitat to increase its surface.

The Current Challenge, Going Beyond Geography

In linguistics, it has long been admitted that the geographic distance between varieties has an effect on their evolution—namely, that closer varieties are generally more similar than distant ones. Lyle Campbell (1995) summarized this succinctly, stating, "neighboring languages often turn out to be related." The first model about the spread of linguistic innovations, a form of contact, was the wave theory of Johannes Schmidt (1872). While Seguy (1971) presented linguistic distances as function of the square root of geographic distance, Trudgill (1974) suggested that the spread of innovations declines quadratically. Nerbonne and Heeringa (2007) and Nerbonne (2010) found a logarithmic model to function better, similarly to the models of population genetics concerning the biological differences of neighboring populations that are a function of migration processes (Wright 1943, Malécot 1948).

When a matrix of Levenshtein aggregate linguistic distances is found to be significantly correlated with the corresponding matrix of geographic distances, as is the case for the ALGAB data set, it is possible to compute a linear regression (linguistic distance vs. log [geographic distance]) in order to compute, from it and for each pairwise comparison, the linguistic distance that is expected between two linguistic varieties according to the geographic distance separating them. This procedure leads to a matrix of *expected* pairwise linguistic distances that can be subtracted from the linguistic distances obtained from the original data. The matrix that results after the subtraction consists of residual distances that can be positive, negative, or null. They will be positive when the linguistic distance computed on original data is higher than the one expected from the regression; they will be negative when two varieties at a given distance exhibit a linguistic distance that is lower than what is expected according to the regression. The idea is that residual distances represent the fraction of the linguistic variability that is not explained by "normal" linguistic contact between neighbors (geography).

We can then analyze these residual distances via multidimensional scaling (MDS). Our goal in this is to detect latent influences on linguistic similarity beyond simple proximity. If varieties are assigned nearby positions in the multidimensional space, then those varieties are similar for reasons other than the contact promoted by nearness. One obvious candidate influence is, of course, the history of the varieties, that is, their genealogy. The MDS plot of the residuals of the geographic analysis is likely to reflect the relations among the varieties before they drifted apart due to migration or were subject to convergence/divergence influences related to contact. Residual distances (Figure 8.5) convey clues about the possible historical scenario of linguistic diversification of Gabon. They point to linguistic diversity between different languages that long-lasting linguistic contact and convergence have progressively defaced (see Table 8.1).

Two features of the analysis deserve particular note. First, the MDS plot of the residuals confirms the historical division proposed by Guthrie, in that his groups are generally close to one another. Given the status of Guthrie's work, we hasten to add that it is perhaps better to regard the correspondence between the two as a confirmation of the step of applying MDS to the residuals of the geographic analysis. Second, despite the overall excellent correspondence between Guthrie's classification and the MDS-residuals analysis, the B20 group stands out in reflecting the Guthrie classification less faithfully. This leads us to hypothesize that the B20 group

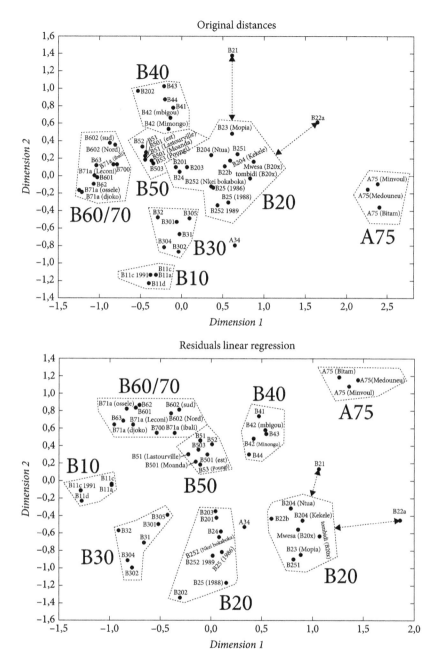

Figure 8.5 Multidimensional scaling projections concerning the 53 linguistic varieties listed in the ALGAB database. Top: Original Levenshtein distances. Stress values in 1 dimension = 0.3247, in 2 dimensions = 0.1641 (plot reported), and in 3 dimensions = 0.1215. Correlation between geographic and linguistic distances = 0.478**. Bottom: Residual distances after computing the regression (R^2 = 0.216; the logarithmic transformation of geographical distances makes almost no difference: R^2 = 0.222) between the kilometric distances and the corresponding Levenshtein distances. Stress values in 1 dimension = 0.399, in 2 dimensions = 0.249 (plot reported), and in 3 dimensions = 0.171. Residuals are normally distributed. See Table 9.1.

Table 8.1 Summary of the possible effects on the Bantu varieties in Gabon that linguistic contact has determined (ALGAB data)

Varieties	Scenario inferred from residual distances (inferred by comparing the two plots of Fig. 9.5)
B10	Initially separate, later converged with B30
B30	Initially separate, later converged with B10
B20	Two initially separate groups converged
B40	Initially separate, later converged with B50
B50	Initially part of same group including B60 and B70, later slightly diverged from those, becoming closer to B40 and to some varieties B20
B60/B70	Initially part of a single group, together with B50, later diverged from B50
A75	A75 arrived in Gabon recently (~5 centuries ago), when B40 was already spoken. Their closeness might correspond to a similar geographic origin in Cameroon: today, they are very different.

has a more complex history, and that it includes two groups that were distinct in earlier history.

Gabon: Immigration Scenarios

Palaeoenvironmental and archaeological studies show that the opening of savannah plains on the coastal region of Gabon started about 4000 BP (Bostoen et al. 2015), with a Neolithization process dated at around 3500 BP (Oslisly 2001) and a detectable sedentarization starting at 2700 BP in northern Gabon (Bostoen et al. 2015). According to linguistic cartography, B20 varieties correspond to an early migration south of Cameroon, through the rainforest or along savannah corridors, to the north of Gabon in at least two independent and early waves: the DNA genetic diversity of B20-speakers is the highest among all the Bantu speakers of Gabon (Patin et al. 2017), meaning that these populations and the languages they spoke had more time to evolve than varieties brought by more recent migration waves. Based on the genetic diversity reported by Patin et al. (2017), we propose that other early migrations took place by following the Atlantic coast from Cameroon to Gabon. These migrations might encompass two different and separate waves corresponding to B10 and B30. The other two groups emerged or arrived later (i.e., B40 and the single group B50/B60/B70).

Conclusions

We do not know whether the present-day location of each group of languages corresponds to the position they had in the past, and we consider extensive migrations over millennia more than likely, meaning that ancestral languages might have been spoken elsewhere, not necessarily where they are today. We also judge it possible that, after an initial stage of peopling, some Bantu languages diffused from one Bantu group to another, in the absence of population movements. Finally, we recognize that the majority of African populations today are multilingual, and there is no reason to think that the past situation was different. Multilingualism, in itself, is a source of language diversification, and this is not a recent phenomenon. We also find it reasonable to admit that borrowing and secondary contact between differentiated languages might have been a major force in the process of linguistic differentiation. These phenomena might well explain the high degree of correlation between linguistic and geographic distances that we partially took into account by dealing with residual distances, which we consider to be closer to the historical scenario of the peopling of Gabon.

The Levenshtein distance measures the signal of historical relatedness and the contact between the languages, and its ability to match classifications based on shared cognates identified by experts is much higher than past criticisms would suggest (for a review, see Manni 2017, 278–286). The very good match between the clusters identified by Grollemund et al. (2015) and the corresponding Levenshtein classifications has been reported. This result might be explained by the fact that Bantu languages are linguistically quite close, often forming dialect chains. This is a scenario closer to the initial application of the Levenshtein method to dialectology; nevertheless, the Levenshtein classification of more distantly related languages not forming dialect chains (Mennecier et al. 2016) is equally convincing. The Levenshtein distance captures the same historical signal that a cognate-based approach does, without the need to seek assistance of an expert in assessing the historical (genetic) relatedness of lexical items (shared vocabulary). This is a remarkable advantage because it allows the phylogenetic classification of linguistic corpora that might otherwise be neglected, which can lead to straightforward hypotheses about past migrations and peopling stages that are poorly documented.

References

Bastin, Y., A. Coupez, and B. de Halleux. 1979. "Statistique lexicale et grammaticale pour la classification historique des langues bantoues." [Lexical and grammatical statistics for historically classifying the Bantu languages] *Bulletin des séances de l'Académie royale de Sciences d'Outre Mer* 3: 375–387.

Bastin, Y., A. Coupez, and M. Mann. 1999. *Continuity and Divergence in the Bantu Languages: Perspectives from a Lexicostatistic Study.* Tervuren: MRAC.

Batini, C., G. Ferri, G. Destro-Bisol, F. Brisighelli, D. Luiselli, P. Sánchez-Diz, J. Rocha, T. Simonson, A. Brehm, V. Montano, N. E. Elwali, G. Spedini, M. E. D'Amato, N. Myres, P. Ebbesen, D. Comas, and C. Capelli. 2011. "Signatures of the preagricultural peopling processes in sub-Saharan Africa as revealed by the phylogeography of early Y chromosome lineages." *Molecular Biology and Evolution* 28: 2603–2613.

Bleek, W. H. I. 1862. *A Comparative Grammar of South African Languages,* Vol. 1. Gregg International. London, Trübner & Co.

Bostoen, K., B. Clist, C. Doumenge, R. Grollemund, J. M. Hombert, J. K. Muluwa, and Jean Maley. 2015. "Middle to late Holocene paleoclimatic change and the early Bantu expansion in the rain forests of Western Central Africa." *Current Anthropology* 56: 354–384.

Campbell, L. 1995. "The Quechumaran hypothesis and lessons for distant genetic comparison." *Diachronica* XII, no. 2: 157–200.

Clist, B. 2005. Des premiers villages aux premiers européens autour de l'estuaire du Gabon: quatre millénaires d'interaction entre l'homme et son milieux.[From the first villages to the first Europeans around the Gabon estuary: four millennia of human interaction with the environment] PhD diss., Université Libre de Bruxelles.

de Filippo, C., C. Barbieri, M. Whitten, S.W. Mpoloka, E. D. Gunnarsdóttir, K. Bostoen, T. Nyambe, K. Beyer, H. Schreiber, P. de Knijff, D. Luiselli, M. Stoneking, and B. Pakendorf. 2011. "Y-chromosomal variation in sub-Saharan Africa: Insights into the history of Niger-Congo groups." *Molecular Biology and Evolution* 28: 1255–1269.

de Filippo, C., K. Bostoen, M. Stoneking, and B. Pakendorf. 2012. "Bringing together linguistic and genetic evidence to test the Bantu expansion." *Proceedings of the Royal Society B* 279 (1741): 3256–3263 doi:10.1098/rspb.2012.0318

Ehret, C. 2002. "Language Family Expansion: Broadening Our Understandings of Cause from an African Perspective." In *Examining the Farming/Language Dispersal Hypothesis*, edited by P. Bellwood and C. Renfrew, 163–176. Cambridge: McDonald Institute Monographs.

Greenberg, J. H. 1955. *Studies in African Linguistics Classification*. New Haven: The Compass Publishing Company.

Grimes, B. F., ed. 2000. *Ethnologue*. 2 vols., 14th ed. Dallas: SIL International.

Grollemund, R., S. Branford, K. Bostoen, A. Meade, C. Venditti, and M. Pagel. 2015. "Bantu expansion shows that habitat alters the route and pace of human dispersals." *PNAS* 112: 13296–13301.

Guthrie, M. 1967. *Comparative Bantu*. Vols. 1–4. Farnborough: Gregg International Publishers Ltd.

Heeringa, W. 2004. *Measuring Dialect Pronunciation Differences Using Levenshtein Distance*. PhD diss., Rijksuniversiteit Groningen.

Heeringa, W., and B. Joseph. 2007. "The Relative Divergence of Dutch Dialect Pronunciations from Their Common Source: An Exploratory Study." In *SigMorPhon 07 ACL 2007, Computing and Historical Phonology, Proceedings of the Ninth Meeting of the ACL Special Interest Group in Computational Morphology and Phonology*, edited by J. Nerbonne, T. Mark Ellison, and G. Kondrak, 31–39. Stroudsburg: The Association for Computational Linguistics (ACL). Heggert, M. 2004. "The Bantu Problem and African Archaeology." In *African Archaeology: A Critical Introduction*, edited by A. B. Stahl, 301–326. Blackwell Publishing, Oxford.

Heine, B. 1973. "Zur genetischen Gliederung der Bantu-sprachen. [On the genealogical classification of the Bantu languages]" *Afrika und Ürbersee* 56: 164–195.

Holden, C. J. 2002. "Bantu language trees reflect the spread of farming across sub-Saharan Africa: A maximum-parsimony analysis." *Proceedings of the Royal Society B. Biological Sciences* 22: 793–799.

Holden, C. J., and Gray R. D. 2006. "Rapid Radiation, Borrowing and Dialect Continua in the Bantu Languages." In *Phylogenetic Methods and the Prehistory of Languages,* edited by P. Forster and C. Renfrew, 19–32. Cambridge: McDonald Institute Monographs.

Hombert, J. M. 1990a. "Atlas linguistique du Gabon [Linguistic atlas of Gabon]." *Revue Gabonaise des Sciences de L'homme* 2: 37–42.

Hombert, J. M., P. Medjo Mvé, and R. Nguéma. 1989. "Les Fangs sont-ils Bantu? [Are the Fang people Bantu?]" *Pholia* 4: 133–147.

Lézine, A-C., C. Assi-Khaudjis, E. Roche, A. Vincens, and G. Achoundong. 2013. "Towards an understanding of West African montane forest response to climate change." *Journal of Biogeography* 40: 183–196.

Li, S., C. Schlebusch, and M. Jakobsson. 2014. "Genetic variation reveals large-scale population expansion and migration during the expansion of Bantu-speaking peoples." *Proceedings of the Royal Society B* 281(1793): 20141448. doi:10.1098/rspb.2014.1448

Maho, J. F. 2009. NUGL online: *"The online version of the New Updated Guthrie List, a referential classification of the Bantu languages."* goto.glocalnet.net/mahopapers/nuglonline.pdf

Malécot, G. 1948. *Les Mathématiques de L'hérédité.* Paris: Masson.

Maley, J. 2001. "La destruction catastrophique des forêts d'Afrique centrale survenue il y a environ 2500 ans exerce encore une influence majeure sur la répartition actuelle des formations végétales [The catastrophic decline of Central African forests 2500 years ago strongly influences the current geographic distribution of vegetation]." *Systematic and Geography of Plants* 71: 777–796.

Mann, M., and D. Dalby. 1987. *A Thesaurus of African Languages.* London: Hans Zell Publishers.

Manni, F. 2017. *Linguistic Probes into Human History.* Groningen: University of Groningen. 320 pp. Groningen dissertations in linguistics n° 162. ISBN: 978-90-367-9871-6 (print version); ISBN: 978-90-367-9872-3 (electronic version).

Mennecier, P., J. Nerbonne, E. Heyer, and F. Manni. 2016. "A Central-Asian survey." *Language Dynamics and Change* 6: 57–98.

Mouguiama-Daouda, P., and L. J. Van der Veen. 2005. "B10-B30: Conglomérat Phylogénétique ou Produit d'une Hybridation [B10-B30: A phylogenetic grouping or the product of hybridization?]" In *Studies in African Comparative Linguistics, with Special Focus on Bantu and Mande,* edited by K. Bostoen and J. Maniacky, 1781–1857. Tervuren: Royal Museum for Central Africa (RMCA/MRAC), Sciences Humaines.

Nerbonne, J. 2010. "Measuring the diffusion of linguistic change." *Philosophical Transactions of the Royal Society B* 365: 3821–3828.

Nurse, D., and G. Philippson. 2003. "Towards a Historical Classification of the Bantu." In *The Bantu Languages,* edited by D. Nurse and G. Philippson, 164–179. London: Routledge.

Oliver, R. 1966. "The problem of the Bantu expansion." *Journal of African History* 7: 361–376.

Oslisly, R. 2001. "The History of Human Settlement in the Middle Ogooué Valley (Gabon): Implications for the Environment." In *African Rain Forest Ecology and Conservation,* edited by W. Weber, L. J. T. White, A. Vedder, and L. Naughton-Treves, 101–118. New Haven: Yale University Press.

Oslisly, R., I. Bentaleb, C. Favier, M. Fontugne, and J. F. Gillet. 2013. "West Central African peoples: Survey of radiocarbon dates over the past 4000 years." In *Proceedings of the 21st International Radiocarbon Conference,* edited by A. J. Jull & C. Hatté. *Radiocarbon* 55: 1377–1382.

Patin, E., G. Laval, L. B. Barreiro, A. Salas, O. Semino, S. Santachiara-Benerecetti, K. K. Kidd, J. R. Kidd, L. Van der Veen, Hombert J.M., A. Gessain, A. Froment, S. Bahuchet, E. Heyer, and L. Quintana-Murci. 2009. "Inferring the demographic history of African farmers and pygmy hunter-gatherers using a multilocus resequencing data set." *PLOS Genetics* 5: e1000448.

Patin E, Lopez M, Grollemund R, Verdu P, Harmant C, Quach H, Laval G, Perry GH, Barreiro LB, Froment A, Heyer E, Massougbodji A, Fortes-Lima C, Migot-Nabias F, Bellis G, Dugoujon JM, Pereira JB, Fernandes V, Pereira L, Van der Veen L, Mouguiama-Daouda P, Bustamante CD, Hombert JM, Quintana-Murci L. 2017. "Dispersals and genetic adaptation of Bantu-speaking populations in Africa and North America." *Science* 356: 543–546.

Phillipson, D. W. 1976. "Archaeology and Bantu linguistics." *World Archaeology* 8: 65–82.

Phillipson, D. W. 1977a. *Later Prehistory of Eastern and Southern Africa.* London: Heinemann.

Phillipson, D. W. 1977b. "The spread of the Bantu languages." *Scientific American* 236: 106–114.

Phillipson, D. W. 2002. "Language and Farming Dispersals in Sub-Saharan Africa." In *Examining the Farming/Language Dispersal Hypothesis,* edited by P. Bellwood and C. Renfrew. Cambridge: McDonald Institute Monographs.

Rexova, K., Y. Bastin, and D. Frynta. 2006. "Cladistics analysis of Bantu languages: A new tree based on combined lexical and grammatical data." *Naturwissenschaften* 93: 189–194.

Schmidt, J. 1872. *Die Verwandtschaftsverhältnisse der indogermanischen Sprachen [The relationships among the Indoeuropean languages].* Weimar: H. Böhlau.

Séguy, J. 1971. "La relation entre la distance spatiale et la distance lexicale [The relation between spatial distance and lexical distance]." *Revue de Linguistique Romane* 35: 335–357.

Simons, G., ed. 2016. *Ethnologue: Languages of the World.* Dallas: SIL International. Internet publication accessible at www.ethnologue.com

Trudgill, P. 1974. "Linguistic change and diffusion: Description and explanation in sociolinguistic dialect geography." *Language in Society* 2: 215–246.

Vansina, J. 1984. "Western Bantu expansion." *Journal of African History* 25: 129–145.

Vansina, J. 1990. *Paths in the Rainforest.* London: Currey.

Vansina, J. 1995. "New linguistic evidence of the Bantu expansion." *Journal of African History* 36: 173–195.

Verdu, P., F. Austerlitz, A. Estoup, R. Vitalis, M. Georges, S. Thery, A. Froment, S. Le Bomin, A. Gessain, J. M. Hombert, L. Van der Veen, L. Quintana-Murci, S. Bahuchet, and E. Heyer. 2009. "Origins and genetic diversity of pygmy hunter-gatherers from western Central Africa." *Current Biology* 19: 312–318.

Whiteley, W. H. 1971. "Introduction." In *Language Use and Social Change: Problems of Multilingualism with Special Reference to Eastern Africa,* edited by Wilfred H. Whiteley, 1–23. Oxford: Oxford University Press.

Wotzka, H-P. 2006. "Records of Activity: Radiocarbon and the Structure of Iron Age Settlement in Central Africa." In *Grundlegungen. Beiträge zur europäischen und afrikanischen Archäologie für Manfred K.H. Eggert,* edited by H-P. Wotzka, 271–289. Tübingen: Francke Attempto Verlag and Co.

Wright, S. 1943. "Isolation by distance." *Genetics* 28: 114–138.

9

Migration Patterns in the Republic
of Sakha (Yakutia)

*Larissa Tarskaia, A. G. Egorova, A. S. Barashkova,
S. A. Sukneva,and W. Leonard*

The sources of population formation within a region are natural reproduction and migration. Migration is the movement of people across the borders of certain territories with a change of place of residence, either permanently or over a period of time. Along with birth rates and mortality rates (deaths), patterns of migration have a large role in shaping long-term demographic trends and genetic diversity within and among populations. In Russia, there is a long history of migration in Siberia (Forsyth 1992, 10). The long-term expansion and decline of the many indigenous Siberian populations are largely the product of migration patterns over the last two centuries (Sukneva 2017). This chapter examines migration patterns and other demographic changes in the Republic of Sakha (Yakutia) in Siberia.

The Republic of Sakha (Yakutia) is one of the federal subjects of the Russian Federation and is located in the northeastern part of Siberia. As depicted in Figure 9.1, the territory of the Republic of Sakha (Yakutia) covers more than three million km^2, or one fifth of Russia's territory. The Republic of Sakha (Yakutia) is the largest administrative subdivision of Russia, comprised of 35 districts (*uluses*), 13 cities, 57 worker settlements, and 354 village administrative offices. As of 2010, its population was 958,528 (2010 Census), consisting mainly of ethnic Yakuts (also known as the Sakha) and Russians, 49.9% and 37.8%, respectively. Over time, the republic has undergone significant demographic changes that have resulted in fluctuations in its population size due to the economic and political changes that have occurred in Russia. The economic and political changes were dramatic in the northeastern regions of Russia, especially during the post-Soviet period. In the period of 1991 to 2010, the Siberian and Far Eastern regions lost more than 2.2 million people in total (Rybakovsky and Kozhevnikova 2015). Yet, the Republic of Sakha (Yakutia) has a long history of population growth due to natural increase and as well as a positive net migration.

The history of Yakutia can be divided into several periods: prehistoric, pre-Soviet, Soviet, and post-Soviet. During the prehistoric period, the territory was sparsely populated by different indigenous ethnic groups, such as the Yakut, Evenks, Evens, Dolgans, Yukagirs, and Chukchi.

Larissa Tarskaia, A. G. Egorova, A. S. Barashkova, S. A. Sukneva, and W. Leonard, *Migration Patterns in the Republic of Sakha (Yakutia)* In: *Human Migration*. Edited by: Maria de Lourdes Muñoz-Moreno and Michael H. Crawford, Oxford University Press.

Figure 9.1 Map showing the location of the Sakha Republic within the Russian Federation. Adapted from Snodgrass et al. 2005.

Genetic and archaeological research suggests that the Yakut people originated from Turkic populations in south Siberia and migrated northward to the Sakha Republic ~1,000 years BP (Zlojutro et al. 2008, 2009). The population size of these peoples is unknown due to the lack of written information. The first written record of the population took place during the pre-Soviet period, in the 17thth century, after the Russian conquest of Siberia. For the first time, the native people were recorded by the Russian annual tribute bookkeepers for the collection of *yasak* (fur tribute) that was required of every adult male. For this reason, the data did not reflect the total population of the region. However, based on the population record for 1763, the number of residents was 81,802, with Russians constituting only 4.0% (Ignat'eva 1994, 15).

During the prehistoric and pre-Soviet periods, growth of the Russian population was mainly due to resettlement from Russia. Thus, according to I. Naumov (2006), from the total number of settlers from Russia of the prehistoric era, 85% were volunteers, who received government loans to set up new homes and were excused from taxes and work/service duties (except military service). The involuntary settlers included soldiers, assigned peasants (*pripisnye*, who were involved in agriculture), exiled criminals, religious dissidents (such as the Old Believers, *Staroobriaddtsy*), national liberal activists (Poles, Lithuanians, Ukrainians, people from the Caucasus, and others), prisoners of war, and political opponents/exiles. In the 1880s, gold deposits were discovered in Yakutia, and development of mining resulted in the arrival of immigrants from Russia and other Slavic regions. In 1897, the Yakuts and other indigenous groups constituted 87.3% of the population, while the Russians increased to 7.9% (Fedorova and Zheleznova 2003).

The Soviet period was characterized by collectivization of agriculture, industrialization, and the formation of industrial centers for the extraction of natural resources. According to the 1926 Census, the total population was 285,471, including 235,926 (82.6%) Yakuts and 30,156 (10.6%) Russians. The development of the mining regions has greatly altered the republic's ethnographic, demographic, and social structure (Fedorova and Zheleznova 2003, 92–93). Migrants were attracted by higher wages (known as "northern wage-increments"), bonuses, and some other privileges. From 1985 to 1989, the number of immigrants reached 304,938, while 262,112 persons left the republic. The percentage of Yakuts in the overall population of the republic declined from 90% in 1920 to 43% in 1970, then to 36.6% in 1979, and to 33.4% in 1989 (Khazanov 1995, 44).

As shown in Table 9.1, from 1763 to 1989, the republic added more than one million people, mainly Russians, Ukrainians, Tatars, and other ethnic groups from the former Soviet Union. However, this demographic trend shifted after the fall of the Soviet Union in 1991; by 2002, the population of the Republic of Sakha (Yakutia) declined to 949,300 people, of which 610,000 (64.3%) lived in urban areas. Compared to the previous census in 1989, the total population had declined by 144,800 people, while the continuing natural increase share was 100,300 people, and therefore total population decreased, mainly due to the large migration outflow.

Table 9.1 Dynamics of correlation of natural and migration population growth, Republic of Sakha (Yakutia)

Years	Growth or Decline (%)		
	Total	Natural	Migratory
1961–1965	100	66.7	33.3
1966–1970	100	62.4	27.6
1971–1975	100	52.3	47.7
1976–1980	100	38.4	61.6
1981–1985	100	57.0	43.0
1986–1990	100	89.7	10.3
1991–1995	−100	47.1	−147.1
1996–2000	−100	43.7	−143.7

Source: Sukneva and Mostakhova (2002).

Thus, the population size declined by 13.3%, with a 16.7% decline in urban areas and a 6.3% decrease in rural areas.

Substantial fluctuations in the population were recorded in the developing industrial regions between 1979 and 2002. For example, during the Soviet period, two of the main industrial areas, the Neryunginsky and Mirninsky Districts, were founded. In the 1950s, diamond-bearing deposits were discovered in Mirny, and in the 1970s Neryungi was founded as a coal-mining town. From 1979 to 1989, the population of Neryungi increased by 109% (62,704 people). By 2002, its population had decreased by 30,400 people. In the Mirninsky region, the population increase between 1979 and 1989 was estimated as 47.7%, or 29,819 persons, and by 2002, its decline was less than that in Neryungri and amounted 6,300 people (or 6.9%). These two major industrial centers together lost almost one fourth of their population by 2002. Although the natural increase in both these industrial cities was 10,700 people, the migration outflow was significant, 36,700 people.

In contrast, the population of the city of Yakutsk has been continually increasing. Between 1979 and 1989, the city gained 41,800 people (23.5%), and from 1989 to 2002, it increased by 26,800 people (12.3%), with 14,000 due to natural increases and 12,800 due to migration. Similar demographic trends have been observed among many of the *uluses* of central Yakutia.

According to the 2002 Census, the ethnic composition of the population of the Sakha Republic was as follows: Yakuts (45.5%), Russians (41.2%), Ukrainians (3.6%), Evenks (1.9%), and Evens (1.2%). Yakuts have been a growing population throughout their history, increasing from 3,212 individuals in 1763 to 432,290 in 2002. As shown in Figure 9.2, the Yakut population was dominant in the region before industrial development in the republic, especially before 1959, when the first diamond kimberlite pipe was discovered in the area that became Mirny. From 1979 to 1989, the number of Russians in the Mirninsky District increased by 120,675 and then it declined by 159,592 people in 2002.

As shown in Figure 9.3, the two largest ethnic groups in the republic are Russians and Yakuts, comprising about 90% of the total population. In 1979, 1989, 2002, and 2010, ethnic Russians composed 50.4%, 50.3%, 41.2%, and 37.8% of the population,

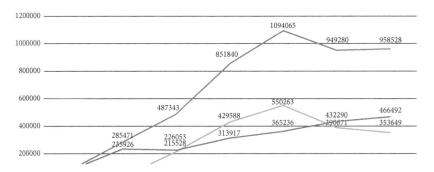

Figure 9.2 Population dynamics in Yakutia between 1763 and 2010.

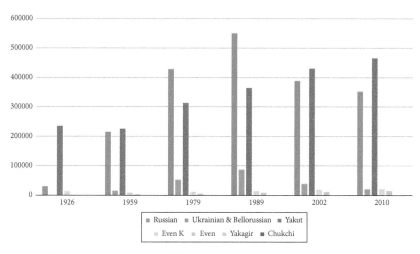

Figure 9.3 Ethnic composition in Yakutia (1926–2010).

respectively, and ethnic Yakuts composed 36.9%, 33.4%, 45.5%, and 49.9%, respectively. A majority of urban population in Yakutsk, the Aldansky, Neryunginsky and Mirninsky Districts presented by ethnic Russians, while the majority of ethnic Yakuts live in rural districts such as Amginsky, Namsky, Viluysky, Suntasky, Verkhne-Viluysky and Khangalassky Districts (Ignat'eva 1994, 15).

The two components of population change (natural changes and migration) for the Sakha Republic between 1961 and 2000 are presented in Table 9.1. Reproduction has contributed a positive trend of natural increase despite the high migratory outflow in 1995 (Sukneva and Mostakhova 2002, 15).

Between 1961 and 1980, the share of migratory growth almost doubled. The share of natural growth was higher than migratory growth in the following 10 years. Since 1991, natural growth has not been sufficient to compensate for the continuing migratory outflow from the republic, and there has been a net decline in overall population size. Migratory outflow reached its peak in 1994 (Table 9.1), and it played a significant role in the population decline. Natural growth contributed 100,300 people, but the republic lost 245,100 people in the large migratory outflow (Sukneva and Mostakhova 2002, 15).

The dynamics of the net migration of urban and rural populations have also been changing. From 1979 to 2002, the highest in-migrations occurred in urban areas (in 1979 there were 81 out-migrants per 100 immigrants), but this was followed by a gradual increase in the number of people moving out of urban areas. However, out-migration from rural areas stayed almost the same from 1979 to 1989 (107 and 111, respectively) and then almost doubled by 2002 (186 out-migrants per 100 immigrants). From the mid-1970s to the beginning of the 1980s, high migration was recorded in southern Yakutia, home to the region's mining industry. In the Mirninsky District, the center of the Siberian diamond industry, there were 38 out-migrants per 100 in-migrants. In the Aldansky District, where the gold and iron ore reserves are located, there were 46 out-migrants per 100 in-migrants, and in the Neriungrinsky District, the center of coal mining, there were 59 out-migrants per 1000 in-migrants (Fedorova and Zheleznova 2003, 91).

Since 1970s, the republic's population distribution has shifted from primarily rural to increasingly urban. The peak was recorded in 1989, when 66.9% of the total population were urban dwellers, due to immigrants from outside the republic and migration from the rural areas within the republic. The urban population included residents of cities, towns, and urban-type settlements or workers' settlements, in Yakutia. This shift occurred because rural areas had been affected by an economic crisis that led to harsh living conditions, and urban areas became more attractive due to their production facilities and infrastructure, media, and better job opportunities. Hwever, the urban growth process declined to 64.3% by 2002 as the result of an out-migration process in 2002. Consequently, in 2002, 36% of the republic's population lived in rural areas, down from 38.7% in 1979. Therefore, the urbanization that had grown intensively during the industrialization period showed a decline between the censuses of 1989 and 2002.

Conclusion

During the period of the industrialization of northeastern Siberia from 1960 to 1990, the population of the Sakha Republic grew due to migration. Additionally, over this time, the population of the republic shifted from being majority rural to majority urban. However, since 1991, migratory outflow from the republic has exceeded natural growth, contributing to a net decline in population size. Therefore, population dynamics in the Sakha Republic are closely connected to key transitions in economic development and political history. Future trends in the republic's demography will depend on the state's management of migration by improving the socioeconomic situation.

References

Fedorova, E. N., and G. A. Zheleznova. 2003. *Migration of Population of Yakutia: Past and Present.* Novosibirsk: Nauka.

Forsyth, James. 1992. *A History of the Peoples of Siberia: Russia's North Asian Colony, 1581–1990.* Cambridge: Cambridge University Press.

Ignat'eva, V. B. 1994. *Natsional'nyi sostav naselenia Yakutii.* Ethnic composition of the population of Yakutia. Yakutsk: Yakutskii nauchny tsentr SO RAN.

Khazanov, A. 1995. *After the USSR: Ethnicity, Nationalism, and Politics in the Commonwealth of Independent States.* Madison: The University of Wisconsin Press.

Naumov, I. 2006. *The History of Siberia,* edited by D. Collins. New York: Routledge.

2002 Census. Russian Federal State Statistics Service (2004-05-21). *Territoria, chislo rayonov, naselennykh punktov I sel'skikh administratsiy po sub'ektam Rossiyskoi Federatsii* [Territory, Number of Districts, Inhabited Localities, and Rural Administration by Federal Subjects of the Russian Federation]. *Vserossiyskaia perepis' naselenia 2002 goda* [All-Russia Population Census of 2002]. Federal State Statistics Service. Retrieved from perepis2002.ru/ct/html/ TOM_01_03.htm

2010 Census. Russian Federal State Statistics Service. 2011. "Vserossiyskaia perepis' naselenia 2010 goda. Tom1" [2010 All-Russian Population Census, vol. 1]. *Всероссийская перепись населения 2010 года* [2010 All-Russia Population Census]. Federal State Statistics Service. Retrieved from http://www.gks.ru/free_doc/new_site/perepis2010/croc/perepis_itogi1612. htm

Rybakovsky, L. L., and N. I. Kozhevnikova. 2015. "The eastern vector of demographic development of Russia." *Narodonaselenie V.*1: 4–16.

Snodgrass, J. J., W. R. Leonard, L. A. Tarskaia, V. P. Alekseev, and V. G. Krivoshapkin. 2005. "Basal metabolic rate in the Yakut (Sakha) of Siberia." *American Journal of Human Biology V.* 17(2): 155–172.

Sukneva, S. A., and T. S. Mostakhova. 2002. *Demographic Development of the Region.* Novosibirsk: Nauka.

Zlojutro, M., L. A. Tarskaia, M. Sorensen, J. J. Snodgrass, W. R. Leonard, and M. H. Crawford. 2008. "The Origins of the Yakut People: Evidence from Mitochondrial DNA Diversity." *International Journal of Human Genetics V.* 8: 119–130.

Zlojutro, M., L. A. Tarskaia, M. Sorensen, J. J. Snodgrass, W. R. Leonard, and M. H. Crawford. 2009. "Coalescent Simulations of Yakut mtDNA Variation Suggest Small Founding Population." *American Journal of Physical Anthropology V.* 139: 474–482.

10

Rapid, Adaptive Human Evolution Facilitated by Admixture in the Americas

Emily T. Norris, Lavanya Rishishwar, and I. King Jordan

Human Migration, Genetic Divergence, and Admixture

The story of human evolution is one of nearly constant migration. The impulse to leave one's home, to explore, and to settle new territories is a seemingly universal hominid trait, manifest across multiple species and subspecies of the genus *Homo*, and one that ultimately allowed for humans to populate nearly every corner of the globe. Our hominid ancestors and their earliest human descendants embarked on numerous long-distance migrations around the world since their origins on the African continent. Fossil evidence suggests that *Homo erectus* migrated out of Africa to Eurasia just over 2 million years ago (ya), and *H. heidelbergensis*, a putative ancestor to both archaic and modern humans, left Africa ~ 800 thousand ya (kya; Fleagle et al. 2010; Zhu et al. 2018). The modern human subspecies—*Homo sapiens sapiens*—is thought to have originated from an African lineage of *H. heidelbergensis* ~ 300 kya (Schlebusch et al. 2017) and began to migrate out of Africa and around the world starting ~ 75 kya (Henn et al. 2012; Nielsen et al. 2017). The phenomenon of migration has had a profound impact on the genetic composition of human populations worldwide, simultaneously driving the joint processes of genetic divergence and admixture. We are particularly interested in how admixture, and the resulting introgression of alleles from ancestral source populations, may have accelerated adaptive evolution in human populations.

The long and steady march of *H. sapiens* out of Africa, through Asia, Oceania, and Europe, and finally throughout the Americas, entailed repeated episodes of population divergence followed by admixture, whereby previously separated populations came together and mixed (Figure 10.1A). Indeed, we can consider human evolution to be characterized by a recurrent pattern of (e)migration, isolation, divergence, (im)migration, and admixture (Figure 10.1B). Human populations constantly migrate to new lands, which often results in physical isolation, which in turns leads to genetic diversification of the isolated populations. Genetic divergence of isolated populations can occur via genetic drift, owing to small population sizes, and/or natural selection based on local adaptations. However, population isolation does not last forever; eventually, additional waves of migration bring previously isolated

Emily T. Norris, Lavanya Rishishwar, and I. King Jordan, *Rapid, Adaptive Human Evolution Facilitated by Admixture in the Americas*
In: *Human Migration*. Edited by: Maria de Lourdes Muñoz-Moreno and Michael H. Crawford, Oxford University Press.
© Oxford University Press 2021. DOI: 10.1093/oso/9780190945961.003.0011

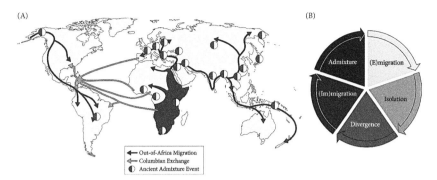

Figure 10.1 Human migration, genetic divergence, and admixture. A: Out-of-Africa migration routes around the world are indicated by gray lines/arrows. Ancient admixture events are indicated by circles (locations taken from Reich, 2018), and the modern admixture event that brought together African, European, and Indigenous American populations, i.e., the Columbian Exchange (Crosby, 2003; Mann, 2011), is shown by arrows. B: The recurrent and joint processes of migration-driven genetic divergence and admixture are illustrated.

populations together again. Any time previously isolated human populations encounter one another, even when those populations are from distinct subspecies, as was the case when modern humans encountered Neandertals and Denisovans, they interbreed, exchanging genes and yielding new hybrid lineages with genetic contributions from multiple ancestral source populations. As David Reich detailed in his book-length treatment of the ancient DNA revolution, this pattern of migration-driven divergence and admixture has been repeated countless times in archaic and modern human populations around the world (Reich 2018). Admixture is not a bug of human evolution—it is in fact a ubiquitous feature of our species (Hellenthal et al. 2014).

We posit that the process of genetic admixture has had a major impact on accelerating human adaptive evolution, in particular by stimulating the rapid evolution of introgressed alleles in recently admixed populations (Jordan 2016). This model of human evolution has much in common with Sewall Wright's shifting balance theory, which emphasized the importance of population subdivision and subsequent migration in facilitating adaptive evolution (Wright 1932). In this chapter, we briefly review studies related to the pace of human adaptive evolution, explaining how admixture can speed up this process, followed by a more detailed treatment of adaptive introgression in both archaic and modern human populations. We emphasize rapid adaptive introgression in the Americas, the region of the world that has experienced perhaps the greatest single admixture event in human history, whereby African, European, and Indigenous American populations that were previously isolated for tens of thousands of years were suddenly brought together again following Columbus's arrival in the New World (Crosby, 2003; Mann, 2011).

Admixture and the Pace of Adaptive Evolution in Human Populations

In their book *The 10,000 Year Explosion*, Cochran and Harpending took aim at the anthropological doctrine that holds that human biological evolution came to a halt around 50,000 ya, thereafter being superseded by a far more dynamic cultural evolution (Cochran and Harpending 2009). In distilling this so-called conventional wisdom regarding human evolution, they quote Stephen Jay Gould as saying "There's been no biological change in humans in 40,000 or 50,000 years. Everything we call culture and civilization we've built with the same body and brain." The basic idea underlying this assertion is that the explosion of human culture and behavioral modernity that marked the Upper Paleolithic essentially liberated humans from the strictures of biological evolution. This happened because rapid cultural evolution, in the form of tool and technology development, obviated the need to respond to environmental pressures by the slower process of natural selection. The authors convincingly dismiss this (perhaps slightly straw man) argument and stress instead that technological developments, the invention of agriculture in particular, actually accelerated human adaptive evolution by allowing for larger population sizes and consequently more adaptive mutations (Hawks et al. 2007).

The recent acceleration of human adaptive evolution covered in *The 10,000 Year Explosion* was based on the authors' own research, along with the work of many other scientists who have taken advantage of the accumulation of human genome sequence variation data to detect signals of adaptive evolution genome-wide in multiple populations around the world (Fan et al. 2016; Oleksyk et al. 2010; Sabeti et al. 2006; Vitti et al. 2013). This impressive body of research has leveraged the ongoing growth of human population genomic data sets, along with the development of increasingly sensitive methods for detecting adaptive evolution, to steadily decrease the amount of elapsed time needed to observe adaptation events. For instance, the agricultural revolution 10 kya led to adaptive evolution for calcium absorption in European populations (Akey et al. 2004). Adaptive mutations that conferred lactose tolerance in Europeans (Bersaglieri et al. 2004) and increased energy metabolism in East Asia (Helgason et al. 2007) emerged independently 8 kya. Lighter skin pigmentation and increased height were selected for in Europeans 6 kya and 5 kya, respectively (Field et al. 2016; Gibbons, 2007). Sickle cell mutations for protection against malaria were initially proposed to have arisen multiple times in African populations, with estimates around 3 kya (Currat et al. 2002; Ohashi et al. 2004), but a recent study has proposed that these haplotypes are derived from a common ancestral haplotype that emerged 7.3 kya (Shriner and Rotimi 2018). Perhaps the most recent sequence-based evidence for human adaptive evolution, at the lactase and major histocompatibility loci, dates to 2 kya (Field et al. 2016). Here, we present evidence in support of our thesis that adaptive human evolution has occurred in the Americas within the last 500 years, an exceedingly short amount of time with respect to human evolution, via introgression of beneficial haplotypes from ancestral source populations.

Despite the findings on selection outlined in the previous paragraph, human adaptive evolution is still largely regarded as a slow process, which is constrained by the introduction of new adaptive alleles via mutation. This can be illustrated by the classic population genetic model showing the rate at which the frequency of an adaptive allele will increase in a population (Figure 10.2). A new mutant allele will be introduced at the low population frequency of $1/(2 \times N_2)$, where N_2 is the effective population size. For example, a relatively small N_2 of 5,000 will yield an initial mutant allele frequency of 0.01%. If the new allele is adaptive, selection will act to increase its population frequency (p) over time proportional to the selection (s) and dominance (h) coefficients, according to the recursion equation $p_{i,t+1} = p_{i,t} w_i / \bar{w}$, where i is the allele, t is the current generation, p is the allele frequency, w_i is the marginal fitness of the allele i, and $\bar{w}$ is the population mean fitness. The increase in adaptive allele frequency happens extremely slowly at the end of the low end of the allele frequency spectrum. Under an additive dominance model ($h = 0.75$) with a strong selection coefficient of $s = 0.1$, it will take more than 100 generations to see a 20% increase in the initial frequency of the adaptive allele. However, as can be seen in Figure 10.2, the rate of adaptive allele frequency increase speeds up tremendously at intermediate allele frequencies. Under the same dominance and selection parameters, but starting from an intermediate

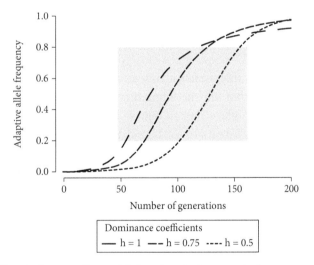

Figure 10.2 Population genetic model showing the increase in frequency of an adaptive allele over time. Adaptive allele frequencies are shown on the y axis, and time in generations is shown on the x axis. The model corresponds to a selection coefficient (s) of 0.1, and three different allele increase trajectories are shown according to different dominance coefficients (h). The gray shading corresponds to the zone at intermediate allele frequencies where the rate of adaptive allele frequency increase is most rapid.

allele frequency of 35%, a doubling of the adaptive allele frequency can occur in less than 20 generations. This feature of adaptive evolution is what leads us to believe that admixture can facilitate extremely rapid human adaptation. When two or more previously isolated populations converge and mix, they introduce alleles to the newly formed admixed population at intermediate frequencies proportional to the percent contributions of each ancestral population. Since admixture introduces new alleles at intermediate frequencies via introgression in this way, it has the potential to allow for substantial increases in the frequency of adaptive alleles over a relatively small number of generations.

While our notion that admixture can dramatically speed up adaptive human evolution is based on a theoretical conjecture, there are numerous empirically observed cases of adaptive introgression in human populations that support this view of evolution. A majority of the studies on adaptive introgression focus on the impacts of admixture between modern humans and archaic hominids, such as Neandertals and Denisovans. In the sections that follow, we first discuss a few examples of ancient adaptive introgression, because they are by now more widely accepted. Then we review an example of more recent adaptive introgression in Africans. Finally, we discuss adaptive introgression in the Americas, where there are fewer studies and more contention regarding the results. Findings from admixed American populations are particularly provocative in the sense that they point to adaptive evolution occurring over a span of 500 years, or approximately 20 to 25 generations, an exceptionally short period of time for human evolution.

Ancient Adaptive Introgression

Evidence of adaptive introgression acting as a means of rapid human adaptive evolution has been shown via the admixture of modern humans with archaic hominids. Before the modern human out-of-Africa migration, Europe and Asia were populated by other *Homo* species: Neandertals and Denisovans. These archaic populations had isolated and genetically diverged from the ancestors of modern humans hundreds of thousands of years before and were adapted to their respective local environments in Europe and Asia. As modern humans emerged from Africa, they encountered not only new environments, but also new humanlike populations. Admixture of archaic and modern humans created new genomes that combined intermediate-frequency adaptive alleles from the archaic populations into the genomic background of the modern humans. This is thought to have helped modern human populations to more quickly adapt to new environments as they settled Europe and Asia, including fighting off novel pathogens. Here we review a few examples of ancient adaptive introgression; for a more comprehensive review, see Racimo et al. (2015).

Immune System

A key factor in the response of the immune system is the HLA class I genes of the major histocompatibility complex (MHC). These genes are involved in antigen presentation for immune cell recognition and are very diverse across human populations. Abi-Rached et al. were interested in the evolutionary source of the deeply divergent human HLA-B allele, *HLA-B*73:01* (Abi-Rached et al. 2011). This allele is found mainly in west Asia and is in linkage with *HLA-C*15:05*, found in west and southeast Asia. The authors suggested that this allele combination was not present in Africa prior to the out-of-Africa migration. To test this hypothesis, they characterized the archaic HLA class I genes from Denisovans and Neandertals, finding that two HLA-A and two HLA-C alleles in Denisovans were most similar to modern sequences, including the *HLA-C*15* allele. Geographic locations of these similar modern alleles show a very low presence in Africa, with higher presence in Asia and Oceania. Divergence estimates of these alleles show that they were formed before the out-of-Africa migration, but since the alleles are not present in Africa, they are likely to have come from an archaic source populations, namely Denisovans. The authors completed the same analysis with similar Neandertal HLA-A and HLA-C alleles and found them to be present in Eurasians and absent in Africans. These findings suggest that the immune systems of archaic *Homo* populations were better adapted to local pathogens in the Eurasian environments, and, upon migrating out of Africa, modern humans rapidly adapted the immune system through introgression of these sequences.

After the study on HLA adaptive introgression, Mendez et al. interrogated *STAT2*, an immune system gene with a role in interferon-mediated responses, for signals of adaptive introgression with Neandertals in Europeans (Mendez et al. 2012). Initial sequencing of *STAT2* revealed the N haplotype, a deeply divergent haplotype present only in the non-African populations used in this study. When compared with the draft Neandertal reference sequence, the N haplotype most closely matched the Neandertal sequence. In addition to this, the haplotype linkage disequilibrium (LD) including *STAT2* is very long in both West Eurasians (~ 130 kb) and East Asians and Melanesians (~ 260 kb), a finding expected if the haplotype was introduced via introgression. The upper divergence estimate between the Neandertal and N haplotype of ~ 160 kya is more recent than the estimates of divergence between the Neandertal and human reference sequence of ~ 600 kya. The N haplotype, particularly the long form of the haplotype, has a 10-fold higher frequency in Melanesians than in the rest of the Eurasian populations. This suggests that *STAT2*, or some other gene in the haplotype, was the target of positive selection in Melanesians. In either case, the introgression of *STAT2* into the modern human genome affected interferon signaling in the immune system.

Genome-wide studies on Neandertal and Denisovan introgression identified a number of putative Neandertal introgressed regions in modern humans (Sankararaman et al. 2014; Vernot and Akey 2014). Dannemann et al. utilized these genomic maps to characterize the adaptive introgression potential of a region containing a haplotype of three Toll-like receptors (TLRs), *TLR6-TLR1-TLR10*, which show some of the highest probabilities of Neandertal introgression (Dannemann et al. 2016). These genes play an important role in the innate immune system because they are the first line of defense against pathogens and help to activate the adaptive immune response. To characterize this region of chromosome 4 as resulting from an adaptive introgression event, the authors identified evidence of introgression by showing that there are seven main haplotypes for the TLRs and three of these are more similar to archaic sequences than to the rest of modern humans as determined by sequence comparisons. In addition to this, the geographic distribution of the archaic-like sequences provides evidence for introgression, because they are mostly found outside of Africa, which is to be expected if introgression occurred when modern humans moved into Neandertal and Denisovan environments. A high differentiation, as determined by Fixation Index (F_{ST}), and previous studies reporting signatures of positive selection on single-nucleotide polymorphisms (SNPs) in *TLR10* provided the authors with evidence that the haplotype was under positive selection. Expression quantitative trait loci (eQTL) analysis showed tissue-specific regulatory effects increasing the expression of these genes in white blood cells. Overall, it was posited that this TLR haplotype could have been adaptively introgressed into the modern human population to affect the innate immune system response to potential pathogens, including *Helicobacter pylori*. The authors mentioned that a diversity of haplotypes for TLRs could have increased the adaptation of modern humans in novel environments, such as those encountered after migration out of Africa.

Integumentary System

As discussed in two studies published in 2014, adaptive introgression of Neandertal sequences has also affected the hair and skin phenotypes of Europeans and East Asians. Vernot and Akey, as mentioned previously, created a catalog of putative introgressed Neandertal sequences in 379 Europeans and 298 East Asians, part of the Phase 1 data release of the 1000 Genomes Project, using a modified S^* summary statistic to identify signals of introgression followed by sequence comparison to a Neandertal reference (Vernot and Akey 2014). The authors then interrogated these sequences to find significantly differentiated introgressed regions between Europe and East Asia as well as shared regions in the two populations with relatively high allele frequency. Using F_{ST} to identify differentiated variants between the two populations, the authors identified two regions with genes that are part of the integumentary system, in addition to other regions and genes. *BCN2* on chromosome 9 was found to be at ~ 70% frequency in Europeans but absent in East Asians; this gene has been related to skin pigmentation in Europeans. *POU2F3* on chromosome 11 was found to be at ~ 66% frequency in East Asians and < 1% in Europeans;

this gene is expressed in the epidermis and mediates keratinocyte proliferation and differentiation. Both populations shared a total of six regions with > 40% allele frequency and with signals of adaptive introgression. One of these regions on chromosome 12q13 contains a type II cluster of keratin genes, providing evidence for adaptive variation of skin phenotypes.

A second study to find evidence of adaptive introgression in the integumentary system used a conditional random field approach to identify putative Neandertal introgressed regions in 1,004 modern humans from the Phase I data release of the 1000 Genomes Project (Sankararaman et al. 2014). They also created a map of Neandertal introgression events in the modern human genome and analyzed the top 5% of genes with the highest inferred Neandertal ancestry. They found that there was a significant enrichment of genes involved in keratin filament formation and posit that Neandertal variants may have helped modern humans adapt to the novel European and Asian environments by affecting their skin and hair.

Altitude Adaptation

A well-studied example of how archaic introgression can provide an adaptive benefit to a modern human population is that of altitude adaptation in Tibetans. Early studies elucidated the molecular underpinnings of adaptation to high-altitude living in Tibetans and the high-altitude Sherpa population and noted that *EPAS1* and *EGLN1* are crucial to controlling signals of hemoglobin concentration in Tibetans, compared with other lowland East Asian populations (Beall et al. 2010; Simonson et al. 2010; Yi et al. 2010). Jeong et al. (2014) determined that the Tibetan population was a result of an ancestral "high-altitude" population admixing with a "low-altitude" population. Because *EPAS1* had the strongest signal of selection for Tibetans, Huerta-Sanchez et al. (2014) moved forward to identify the source of the adaptive introgression for this gene. After resequencing 40 Tibetan individuals (high-altitude) and 40 Han Chinese individuals (low-altitude), the authors found that F_{ST} for this gene was highly differentiated, as expected if there was selection on the high-altitude haplotype in Tibetans. When comparing the Tibetan-specific haplotype to potential donor sequences, it was determined that the haplotype shared more sequence similarity with the Denisovan haplotype than with any other extant or archaic population. The authors concluded that introgression from a Denisovan population allowed the modern Tibetan population to rapidly adapt to the high altitude of the Tibetan plateau.

Adaptive Introgression in Modern Humans

The Bantu-speaking populations (BSPs) of Africa experienced multiple periods of adaptive introgression during their migration throughout Africa over the last

1,500 years (Patin et al. 2017). As the BSPs migrated, they encountered other African populations already adapted to their local environments. Genetic ancestry characterization of the BSPs showed admixture with other African populations, including western rainforest hunter-gatherers, eastern African farmers, and San populations. In each of the BSP populations—western, eastern, and southeastern—the authors searched for evidence of excess ancestry and compared their findings with signals of positive selection to identify putative adaptively introgressed regions. In western BSPs, there was an overlap of excess western rainforest hunter-gatherer ancestry with a strong signal of positive selection for the HLA region and a moderately strong signal for CD36. These regions are both related to the immune system response, with CD36 being associated with susceptibility to malaria caused by *Plasmodium falciparum*. In eastern BSPs, excess eastern African ancestry overlapped a moderately strong signal of positive selection for the lactase gene (*LCT*), providing the lactase persistence phenotype for eastern BSPs. Thus, both the immune-system and diet-related phenotypes of BSPs have been influenced by recent admixture and adaptive introgression.

Adaptive Introgression in the Americas

We are particularly interested in studying adaptive introgression in admixed American populations (Jordan 2016). Modern (cosmopolitan) human populations in the Americas were formed primarily by admixture among ancestral source populations from Africa, Europe, and the Americas (Figure 10.1A). This is considered to be one of the largest and most abrupt admixture events in all of human evolution and one that has had a profound effect on world history (Crosby 2003; Mann 2011). Admixed American genomes can be considered evolutionarily novel, in the sense that they contain combinations of ancestry-specific alleles that never previously existed together on the same genomic background. The creation of such novel admixed genome sequences has important implications for health and fitness in modern American populations (Norris et al. 2018). The possibility of adaptive introgression in the Americas can be considered controversial in light of the fact that the ~ 20 to 25 generations that have elapsed since the process of admixture in the Americas began represent a very short amount of time in terms of human evolution, less than 1% of the time that has elapsed since humans migrated out of Africa. In principle, it should be very difficult to observe adaptive human evolution over such a short time. Nevertheless, we contend that the Americas represent an ideal laboratory for studying adaptive introgression and for exploring the possibility of extremely rapid adaptation in human populations.

Our working hypothesis is that numerous alleles were "pre-selected" in ancestral source populations over thousands of years based on their utility in the local environments of Africa, Europe, and the Americas. Subsequently, when the ancestral source populations were suddenly brought back together, some of the pre-selected alleles could have also provided an adaptive benefit in the New World environment. For example, alleles that served to protect their human hosts against infectious disease may have been particularly important in the pathogen-rich environment of the Americas. In addition, neutral alleles that diverged in frequency among ancestral populations based on genetic drift could later become adaptively beneficial in the new environment. In either case, adaptively beneficial alleles introduced at intermediate frequencies via introgression could quickly increase in frequency owing to their utility in the novel admixed populations (see Figure 10.2 and the previous discussion on how introgression can accelerate adaptive evolution).

The analytical approach used to test this hypothesis is based on the delineation of genome-wide "local ancestry" patterns in admixed populations. Local ancestry refers to the specific ancestral origins—African, European, or Native American—for specific chromosomal regions (i.e., haplotypes). Local ancestry is assigned by comparing individual chromosomal segments with the corresponding genomic regions from panels of ancestral population reference genomes (Geza et al. 2018; Maples et al. 2013). Once this is done for an admixed American population, the population's local ancestry fractions for any given region of the genome can be compared to the genome-wide population ancestry averages to search for "ancestry-enriched" regions (Figure 10.3). Ancestry-enriched regions are genomic segments that show anomalously low or high ancestry fractions for any given ancestry component, compared to the genome-wide population averages. Statistically significant local ancestry deviations are taken to represent evidence of adaptive introgression. The presence of independent signals of previous positive selection on these same regions in ancestral source populations can be used to provide additional evidence in support of adaptive introgression (Patin et al. 2017). Below, we review all currently known cases of adaptive evolution in admixed American populations.

Puerto Rico

One of the first studies on adaptive introgression in the Americas uncovered signals of recent selection and introgression in Puerto Ricans (Tang et al. 2007). Genetic ancestry characterization of the study population—192 Puerto Ricans as part of the Genetics of Asthma in Latino Americans study—showed that the population was mainly of European descent, with relatively equal fractions of African and Native American ancestries. Local ancestry estimates, when compared with global averages, revealed excess African ancestry on chromosome 6 and excess Native American ancestry on chromosomes 8 and 11. The region of chromosome 6 with African enrichment harbors the MHC, the first response to invading pathogens for the adaptive immune system. Chromosome 11 shows Native American enrichment of an olfactory gene cluster, which, as the authors mentioned, has been shown to be under positive selection (Gilad et al. 2003; Voight et al. 2006). The final Native

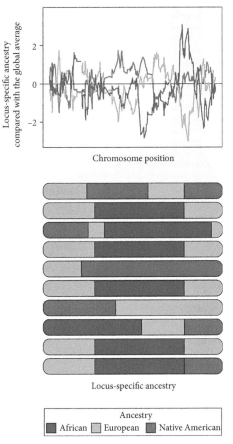

Figure 10.3 Ancestry-enrichment analysis for identifying adaptive introgression events. Locus-specific ancestry patterns—African (blue), European (orange), and Native American (red)—are characterized across all chromosomes in the population. Locus-specific ancestry fractions are compared with the global ancestry fractions along the genome to identify ancestry-enriched (or depleted) segments. An example is shown (gray shading) for a region that is enriched for African ancestry and depleted for European ancestry.

American enrichment on chromosome 8 did not have any clear candidates for adaptive introgression, but the one gene in the region, *CSMD3*, did have some tissue-specific expression. The authors suggested that the enrichment of African alleles of the immune system could be due to their being more advantageous in the new environment, because they are genetically more diverse than those of Europeans or Native Americans and are better able to confront the numerous pathogens that are both imported and endemic to the area. Another reason for the ancestry enrichment could be that the genes played a role in a phenotype that offered a fitness advantage and thus was acted upon by natural selection upon admixture in the New World. For either reason, the immune system of Puerto Rican individuals seems to have been influenced by adaptive introgression of ancestry-specific alleles.

Colombia

A 2015 study to characterize the genetic ancestry of the Colombian population from the 1000 Genomes Project found that, like Puerto Ricans, the Colombian population has mostly European ancestry, followed by Native American and finally African ancestry fractions (Rishishwar et al. 2015). In addition to understanding the impacts of sex-biased admixture, the authors evaluated local ancestry patterns, looking for locus-specific patterns using a trinomial probability metric to signal areas with excess ancestry. The regions with anomalous ancestry patterns were then interrogated for previously identified signs of positive selection and their phenotypic associations, specifically health-related phenotypes. The authors found signals of African ancestry enrichment on chromosome 6 for HLA-B, part of the MHC. There was also evidence of enrichment of European ancestry on chromosome 15 for *SLC24A5*, a gene that influences skin pigmentation with decreased melanin. In addition to single genes with anomalous ancestry patterns, the authors performed a gene set enrichment analysis to identify pathways containing an abundance of ancestry-enriched genes. They found both the innate and adaptive immune response contained signs of ancestry enrichment. These lines of evidence provide support for the hypothesis that the Colombian genomes retained adaptive ancestry-specific loci that were better suited to combat the wide variety of pathogens in the environment.

As a caveat to the findings, it should be noted that the program used for local-ancestry assignment generated very short ancestry tracts, compared with the tract lengths generated from current local-ancestry assignment programs, such as RFMix (Maples et al. 2013). It is unlikely that, in the ~ 20 to 25 generations since the Columbian Exchange, there would be sufficient genomic recombination to generate the size tracts seen in this analysis, and the findings should be validated using contemporary programs.

Mexico

In 2014, a novel two-layer hidden Markov model was proposed for inferring local ancestry of admixed individuals based on the detection of haplotype structure (Guan 2014). This method was used to characterize the genetic ancestry and to highlight regions with excess ancestry-specific loci in Mexican individuals from the HapMap3 Project and 1000 Genomes Project. For both projects, the author found excess African ancestry on chromosome 6 in the MHC region, as well as on chromosome 8p23.1 at a known chromosomal inversion.

A follow-up study on recent selection in Mexican individuals found a significant excess of African ancestry of the MHC on chromosome 6 (Zhou et al. 2016). The authors analyzed genetic ancestry data from two cohorts of Mexican individuals and found excess ancestry in both cohorts. By being able to replicate the results in a second cohort, there was more confidence in the findings of African ancestry enrichment. In addition to this, the authors developed a technique to infer local ancestry without the concern of inaccurate Native American samples and estimated the amount of selection necessary for this locus to remain significantly enriched with African ancestry. The results suggest that selection was at similar strength to those of lactase selection in Europeans and the sickle-cell trait in Africans. The authors suggested that the challenging conditions of the New World, with the existing pathogens and those brought over by the Spaniards and Africans, could have provided the necessary selection pressure for the African MHC alleles to increase in frequency due to the greater diversity of African MHC. Infectious diseases and epidemics, as well as harsh living conditions, caused many of the Native American populations to perish, leading to a lack of Native American ancestry at this important immune system locus.

African Americans

A study on ancestry-specific selection in 1,890 African Americans characterized the genetic ancestry of the population to be ~ 72% African and ~ 28% European (Jin et al. 2012). The authors then identified loci with European (or African) ancestry 3 standard deviations above or below the genome-wide average as those potentially under natural selection. Four regions were found to have excess European ancestry and two were found to have excess African ancestry. The excess European ancestry was found to be related to response to influenza infection and the African ancestry was found to be related to general immune system signaling. Using F_{ST} between African Americans and putative ancestral African populations, the authors identified signals of positive selection and found four regions that carried differentiated SNPs, including those related to malaria response and the MHC, suggesting that the immune system of African Americans in the New World was under pressure with response to the novel pathogen environment. It is important to note, however, that the regions with excess African ancestry do not overlap with those putatively under positive selection.

In response to these findings, another group performed a genetic ancestry analysis on 29,141 African American individuals and found there to be no evidence of directional selection (Bhatia et al. 2014). Instead, the authors stated that the earlier findings could have been due to chance or systematic biases in data handling. When directly comparing the 2012 results to the authors' calculations, the authors showed that they got much lower ancestry deviation estimates than previously reported and the 2012 enrichment findings were not replicated. The same results were found when focusing on the positive selection estimates using F_{ST}; the previous results were not replicated and there were no overlaps between the two findings.

Conclusions and Future Prospects

Adaptive human evolution was postulated to have stopped ~ 40 to 50 kya, coincident with the emergence of cultural evolution (Gould 2000; Mayr 1963). Not only has this anthropological doctrine been shown to be false, but there is abundant evidence that adaptive evolution has actually accelerated over the last 10,000 years (Cochran and Harpending 2009). Throughout this chapter, we provide examples of adaptive introgression acting as a means for facilitating rapid human evolution, through admixture with Neandertals and Denisovans thousands of years ago, via admixture in modern Africans 1,500 years ago, and via admixture in the Americas 500 years ago. Admixture and introgression have allowed modern humans to colonize new environments, such as the Tibetan plateau, resist novel pathogens and combinations of pathogens throughout Europe, Africa, and the Americas, and generally be better suited to their local environments. The immune system has been shown to be a hotspot of adaptive introgression throughout time, as admixture among previously adapted populations allowed modern humans to quickly adapt and thrive in their new environments.

As seen in the contradictory studies of African Americans, signals of adaptive introgression can be due to biases in the data or small sample sizes. To increase confidence in ancestry enrichment as a signal of adaptive introgression, future studies can ensure that regions with ancestry enrichment also have multiple lines of evidence of positive selection from ancestral populations on the same genomic regions in the admixed populations. In addition, if there are multiple admixed populations with similar ancestral source populations, finding signals of ancestry enrichment and positive selection shared among the multiple populations could provide more confidence that a finding is true, because there is likely shared selection pressure in the same area of the world. Selection is also likely to occur on phenotypes that are caused by multiple genes, not just one genomic locus. If the same ancestry enrichment and positive selection signals are seen in multiple genes encoding a polygenic phenotype, then it is likely that adaptive introgression could have led to rapid, adaptive human evolution.

Finally, we propose that admixture, now recognized as a ubiquitous feature of human evolution (Hellenthal et al. 2014; Reich 2018), has had a far greater impact on human adaptive evolution than formerly recognized. The linked processes of genetic divergence followed by admixture (Figure 10.1B) have provided abundant raw material for adaptive introgression throughout human evolutionary history, which in turn can provide for extremely rapid adaptive evolution unconstrained by mutation. It may well be the case that, owing to admixture, adaptive evolution of human populations is far more common and much more dynamic than previously imagined.

References

Abi-Rached, L., Jobin, M. J., Kulkarni, S., McWhinnie, A., Dalva, K., Gragert, L., Babrzadeh, F., Gharizadeh, B., Luo, M., Plummer, F. A., Kimani, J., Carrington, M., Middleton, D., Rajalingam, R., Beksac, M., Marsh, S. G., Maiers, M., Guethlein, L. A., Tavoularis, S., Little, A. M., Green, R. E., Norman, P. J., Parham, P. 2011. "The shaping of modern human immune systems by multiregional admixture with archaic humans." *Science* 334, no. 6052: 89–94.

Akey, J. M., Eberle, M. A., Rieder, M. J., Carlson, C. S., Shriver, M. D., Nickerson, D. A., Kruglyak, L.. 2004. "Population history and natural selection shape patterns of genetic variation in 132 genes." *PLOS Biology* 2, no. 10: e286.

Beall, C. M., Cavalleri, G. L., Deng, L., Elston, R. C., Gao, Y., Knight, J., Li, C., Li, J. C., Liang, Y., McCormack, M., Montgomery, H. E., Pan, H., Robbins, P. A., Shianna, K. V., Tam, S. C., Tsering, N., Veeramah, K. R., Wang, W., Wangdui, P., Weale, M. E., Xu, Y., Xu, Z., Yang, L., Zaman, M. J., Zeng, C., Zhang, L., Zhang, X., Zhaxi, P., Zheng, Y. T.. 2010. "Natural selection on EPAS1 (HIF2alpha) associated with low hemoglobin concentration in Tibetan highlanders." *PNAS* 107, no. 25: 11459–11464.

Bersaglieri, T., Sabeti, P. C., Patterson, N., Vanderploeg, T., Schaffner, S. F., Drake, J. A., Rhodes, M., Reich, D. E., Hirschhorn, J. N.. 2004. "Genetic signatures of strong recent positive selection at the lactase gene." *American Journal of Human Genetics* 74, no. 6: 1111–1120.

Bhatia, G., et al. 2014. "Genome-wide scan of 29,141 African Americans finds no evidence of directional selection since admixture." *American Journal of Human Genetics* 95, no. 4: 437–444.

Cochran, G., and H. Harpending. 2009. *The 10,000 Year Explosion: How Civilization Accelerated Human Evolution*. Basic Books.

Crosby, A. W. 2003. *The Columbian Exchange: Biological and Cultural Consequences of 1492*. Westport: Greenwood Publishing Group.

Currat, M., Trabuchet, G., Rees, D., Perrin, P., Harding, R. M., Clegg, J. B., Langaney, A., Excoffier, L.. 2002. "Molecular analysis of the beta-globin gene cluster in the Niokholo Mandenka population reveals a recent origin of the beta(S) Senegal mutation." *American Journal of Human Genetics* 70, no. 1: 207–223.

Dannemann, M., A. M. Andres, and J. Kelso. 2016. "Introgression of Neandertal- and Denisovan-like haplotypes contributes to adaptive variation in human toll-like receptors." *American Journal of Human Genetics* 98, no. 1: 22–33.

Fan, S., Hansen, M. E., Lo, Y., Tishkoff, S. A. 2016. "Going global by adapting local: A review of recent human adaptation." *Science* 354, no. 6308: 54–59.

Field, Y., Boyle, E. A., Telis, N., Gao, Z., Gaulton, K. J., Golan, D., Yengo, L., Rocheleau, G., Froguel, P., McCarthy, M. I., Pritchard, J. K. . 2016. "Detection of human adaptation during the past 2000 years." *Science* 354, no. 6313: 760–764.

Fleagle, J. G., Fleagle, J.G., Shea, J.J., Grine, F.E., Baden, A.L., Leakey, R.E. (Eds.) 2010. *Out of Africa I: The First Hominin Colonization of Eurasia*. Springer Science & Business Media. Berlin/Heidelberg, Germany. .

Geza, E., Mugo, J., Mulder, N. J., Wonkam, A., Chimusa, E. R., Mazandu, G. K.. 2019. "A comprehensive survey of models for dissecting local ancestry deconvolution in human genome." *Briefings in Bioinformatics*. 20(5):1709-1724.

Gibbons, A. 2007. "European skin turned pale only recently, gene suggests." *Science* 316, no. 5823: 364.

Gilad, Y., Bustamante CD, Lancet D, Pääbo S. 2003. "Natural selection on the olfactory receptor gene family in humans and chimpanzees." *American Journal of Human Genetics* 73, no. 3: 489–501.

Gould, S. J. 2000. "The spice of life." *Leader to Leader* 15: 14–19.

Guan, Y. 2014. "Detecting structure of haplotypes and local ancestry." *Genetics* 196, no. 3: 625–642.

Hawks, J., Wang, E. T., Cochran, G. M., Harpending, H. C., Moyzis, R. K. 2007. "Recent acceleration of human adaptive evolution." *PNAS* 104, no. 52: 20753–20758.

Helgason, A., Pálsson, S., Thorleifsson, G., Grant, S. F., Emilsson, V., Gunnarsdottir, S., Adeyemo, A., Chen, Y., Chen, G., Reynisdottir, I., Benediktsson, R., Hinney, A., Hansen, T., Andersen, G., Borch-Johnsen, K., Jorgensen, T., Schäfer, H., Faruque, M., Doumatey, A., Zhou, J., Wilensky, R. L., Reilly, M. P., Rader, D. J., Bagger, Y., Christiansen, C., Sigurdsson, G., Hebebrand, J., Pedersen, O., Thorsteinsdottir, U., Gulcher, J. R., Kong, A., Rotimi, C., Stefánsson, K.. 2007. "Refining the impact of TCF7L2 gene variants on type 2 diabetes and adaptive evolution." *Nature Genetics* 39(2):218–225.

Hellenthal, G., Busby, G. B. J., Band, G., Wilson, J. F., Capelli, C., Falush, D., Myers, S. 2014. "A genetic atlas of human admixture history." *Science* 343, no. 6172: 747–751.

Henn, B. M., L. L. Cavalli-Sforza, and M. W. Feldman. 2012. "The great human expansion." *PNAS* 109, no. 44: 17758–17764.

Huerta-Sanchez, E., Jin, X., Asan, Bianba, Z., Peter, B. M., Vinckenbosch, N., Liang, Y., Yi, X., He, M., Somel, M., Ni, P., Wang, B., Ou, X., Huasang, Luosang, J., Cuo, Z. X., Li, K., Gao, G., Yin, Y., Wang, W., Zhang, X., Xu, X., Yang, H., Li, Y., Wang, J., Wang, J., Nielsen, R.. 2014. "Altitude adaptation in Tibetans caused by introgression of Denisovan-like DNA." *Nature* 512, no. 7513: 194–197.

Jeong, C., Alkorta-Aranburu, G., Basnyat, B., Neupane, M., Witonsky, D. B., Pritchard, J. K., Beall, C. M., Di Rienzo, A. 2014. "Admixture facilitates genetic adaptations to high altitude in Tibet." *Nature Communications* 5: 3281.

Jin, W., Wang, H., Yu, Y., Shen, Y., Wu, B., Jin, L.. 2012. "Genome-wide detection of natural selection in African Americans pre- and post-admixture." *Genome Research* 22, no. 3: 519–527.

Jordan, I. K. 2016. "The Columbian exchange as a source of adaptive introgression in human populations." *Biology Direct* 11, no. 1: 17.

Mann, C. C. 2011. *1493: Uncovering the New World Columbus Created*. Vintage. New York: Knopf.

Maples, B. K., Gravel, S., Kenny, E. E., Bustamante, C. D.. 2013. "RFMix: a discriminative modeling approach for rapid and robust local-ancestry inference." *American Journal of Human Genetics* 93, no. 2: 278–288.

Mayr, E. 1963. *Animal Species and Evolution*. Cambridge: Belknap Press of Harvard University Press.

Mendez, F. L., J. C. Watkins, and M. F. Hammer. 2012. "A haplotype at STAT2 introgressed from Neanderthals and serves as a candidate of positive selection in Papua New Guinea." *American Journal of Human Genetics* 91, no. 2: 265–274.

Nielsen, R., Akey, J. M., Jakobsson, M., Pritchard, J. K., Tishkoff, S., Willerslev, E. 2017. "Tracing the peopling of the world through genomics. *Nature* 541, no. 7637: 302–310.

Norris, E. T., Wang, L., Conley, A. B., Rishishwar, L., Mariño-Ramírez, L., Valderrama-Aguirre, A., Jordan, I.K . 2018. "Genetic ancestry, admixture and health determinants in Latin America." *BMC Genomics* 19 (Suppl 8): 861.

Ohashi, J., Naka, I., Patarapotikul, J., Hananantachai, H., Brittenham, G., Looareesuwan, S., Clark, A. G., Tokunaga, K.. 2004. "Extended linkage disequilibrium surrounding the hemoglobin E variant due to malarial selection." *American Journal of Human Genetics* 74, no. 6: 1198–1208.

Oleksyk, T. K., M. W. Smith, and S. J. O'Brien. 2010. "Genome-wide scans for footprints of natural selection." Philosophical Transactions of the Royal Society B: Biological Sciences 365, no. 1537: 185–205.

Patin, E., Lopez, M., Grollemund, R., Verdu, P., Harmant, C., Quach, H., Laval, G., Perry, G. H., Barreiro, L. B., Froment, A., Heyer, E., Massougbodji, A., Fortes-Lima, C., Migot-Nabias, F., Bellis, G., Dugoujon, J. M., Pereira, J. B., Fernandes, V., Pereira, L., Van der, V.een, L., Mouguiama-Daouda, P., Bustamante, C. D., Hombert, J. M., Quintana-Murci, L.. 2017. "Dispersals and genetic adaptation of Bantu-speaking populations in Africa and North America." *Science* 356, no. 6337: 543–546.

Racimo, F., Sankararaman S, Nielsen R, Huerta-Sánchez E. 2015. "Evidence for archaic adaptive introgression in humans." *Nature Reviews Genetics* 16, no. 6: 359–371.

Reich, D. 2018. *Who We Are and How We Got Here: Ancient DNA and the New Science of the Human Past*. Oxford: Oxford University Press.

Rishishwar, L., Conley, A. B., Wigington, C. H., Wang, L., Valderrama-Aguirre, A., Jordan, I.K. 2015. "Ancestry, admixture and fitness in Colombian genomes." *Scientific Reports* 5: 12376.

Sabeti, P. C., Schaffner, S. F., Fry, B., Lohmueller, J., Varilly, P., Shamovsky, O., Palma, A., Mikkelsen, T. S., Altshuler, D., Lander, E.S . 2006. "Positive natural selection in the human lineage." *Science* 312, no. 5780: 1614–1620.

Sankararaman, S., Mallick S, Dannemann M, Prüfer K, Kelso J, Pääbo S. 2014. "The genomic landscape of Neanderthal ancestry in present-day humans." *Nature* 507, no. 7492: 354–357.

Schlebusch, C. M., Malmström, H., Günther, T., Sjödin, P., Coutinho, A., Edlund, H., Munters, A. R., Vicente, M., Steyn, M., Soodyall, H., Lombard, M., Jakobsson, M.. 2017. "Southern African ancient genomes estimate modern human divergence to 350,000 to 260,000 years ago." *Science* 358, no. 6363: 652–655.

Shriner, D., and C. N. Rotimi. 2018. "Whole-genome-sequence-based haplotypes reveal single origin of the sickle allele during the Holocene wet phase." *American Journal of Human Genetics* 102, no. 4: 547–556.

Simonson, T. S., Yang, Y., Huff, C. D., Yun, H., Qin, G., Witherspoon, D. J., Bai, Z., Lorenzo, F. R., Xing, J., Jorde, L. B., Prchal, J. T., Ge, R... 2010. "Genetic evidence for high-altitude adaptation in Tibet." *Science* 329, no. 5987: 72–75.

Tang, H., Choudhry, S., Mei, R., Morgan, M., Rodriguez-Cintron, W., Burchard, E. G., Risch, N.J.. 2007. "Recent genetic selection in the ancestral admixture of Puerto Ricans." *American Journal of Human Genetics* 81, no. 3: 626–633.

Vernot, B., and J. M. Akey. 2014. "Resurrecting surviving Neandertal lineages from modern human genomes." *Science* 343, no. 6174: 1017–1021.

Vitti, J. J., S. R. Grossman, and P. C. Sabeti. 2013. "Detecting natural selection in genomic data. *Annual Review of Genetics* 47: 97–120.

Voight, B. F., Kudaravalli, S., Wen, X., Pritchard, J. K. 2006. "A map of recent positive selection in the human genome." *PLOS Biology* 4, no. 3: e72.

Wright, S. W. 1932. "The roles of mutation, inbreeding, crossbreeding and selection in evolution." In *Proceedings of the 6th International Congress of Genetics* (Jones, D. F., ed.), pp. 356–366. Austin: Genetics Society of America.

Yi, X., Y. Liang, E. Huerta-Sanchez, et al. 2010. "Sequencing of 50 human exomes reveals adaptation to high altitude." *Science* 329, no. 5987: 75–78.

Zhou, Q., L. Zhao, and Y. Guan. 2016. "Strong selection at MHC in Mexicans since admixture." *PLOS Genetics* 12, no. 2: e1005847.

Zhu, Z., Dennell, R., Huang, W., Wu, Y., Qiu, S., Yang, S., Rao, Z., Hou, Y., Xie, J., Han, J., Ouyang, T.. 2018. "Hominin occupation of the Chinese Loess Plateau since about 2.1 million years ago." *Nature* 559, no. 7715: 608–612.

11

Diversity of Mexican Paternal Lineages Reflects Evidence of Migration and 500 Years of Admixture

R. Gómez, T. G. Schurr, and M. A. Meraz-Ríos

Introduction

By all accounts, Mexico was the cradle of the major Mesoamerican cultures—Olmec, Teotihuacan, Aztec, and Mayan (Beezley 2011). Colonization by Europeans and the importation of enslaved Africans subsequently influenced the genetic composition of Mexican populations over the past several centuries (Beezley 2011; Santana et al. 2014). During this period, most Native Mexican populations experienced some degree of admixture, with gene flow happening largely between indigenous women and European and African men, leading to the formation of the mestizo gene pool (Bryc et al. 2010). As a result of this process, more than 90% of Mexicans are carriers of indigenous maternal lineages (Gonzalez-Sobrino et al. 2016; Martinez-Cortes et al. 2013), whereas they exhibit a great diversity of nonindigenous paternal lineages due to historical events (Beezley 2011).

The Y chromosome is one of the most useful markers in anthropological genetic and genealogical studies (Calafell and Larmuseau 2017). It is acrocentric and represents at most 3% of the haploid genome (Mangs and Morris 2007; Quintana-Murci and Fellous 2001). The Y chromosome has two small pseudoautosomal regions (PAR) that exhibit homology with regions in the X chromosome, and a nonrecombining region (NRY) of ~ 50 Mb between them (Mangs and Morris 2007). The NRY is uniparentally inherited, being passed down from father to son without genetic recombination (Calafell and Larmuseau 2017; Rubicz, Melton, and Crawford 2007).

Thee NRY region also presents a wide range of polymorphisms, such as single-nucleotide polymorphisms (SNPs; ~ 60,555) and short tandem repeats (STRs; ~ 4,500; Calafell and Larmuseau 2017; Poznik et al. 2016; Willems et al. 2016). The set of alleles obtained from Y-chromosome STR (Y-STR) or SNP genotyping in a single individual is referred to as a haplotype. Such haplotypes can be used to predict the paternal lineage or haplogroup to which a Y chromosome belongs. Nowadays, it is more common to use SNPs to delineate the haplogroup status of a sample, and therefore the geographic origin of a Y chromosome, because of their relative evolutionary uniqueness (Calafell and Larmuseau 2017; ISOGG 2018; Wells 2006). Individuals sharing such haplogroups and related haplotypes have a common ancestral origin. With information about the geographic source of Y chromosomes, it becomes possible to reconstruct the admixture process; therefore, the Y-chromosome composition of Mexican mestizos reflects the contributions of paternal lineages from diverse source populations.

In an effort to illuminate the genetic diversity of Mexican males, we evaluated a compendium of NRY data from previous studies (Ramos-Gonzalez et al. 2017; Salazar-Flores et al. 2010; Santana et al. 2014). This data set included a total of 1,614 unrelated Y-STR haplotypes defined by 17 different markers. These haplotypes were obtained from populations inhabiting different geographic regions of Mexico, including the Central Valley of Mexico (CVM), Chiapas (CHIS), Guanajuato (GTO), Jalisco (JAL), Nuevo Leon (NL; Monterrey City specifically), and Yucatan (YUC), with the CVM including GTO, Puebla (PUE), and Queretaro (QRO) states. Haplogroup assignment was made using Bayesian methods, with haplogroup probability assignments ≥ 80% being included in the analyses (Athey 2005).

Haplotype Diversity

Mexican males exhibited a great diversity of haplotypes. The study populations had a mean number of pairwise differences (MPD) of 9.682 ± 4.439 and a mean population variance (Vp) of 0.881. YUC and CHIS had the highest MPD values (9.946 ± 4.575, $Vp = 0.884$, and 9.808 ± 4.519, $Vp = 0.899$, respectively) of all Mexican populations, reflecting their remarkable diversity. Interestingly, these two populations shared very few haplotypes between them, a trend observed across all populations. Overall, YUC shared the greatest number of haplotypes with NL (4%), and CVM (2%); , while CHIS shared the most haplotypes with NL (2.3%) and CMV (1.15%); and YUC and CHIS shared 0.67% of the haplotypes.

Urban zones with the greatest economic development and notable internal migration, such as JAL (9.682 ± 4.467, $Vp = 0.975$) and CVM (9.652 ± 4.441, $Vp = 0.915$), presented the highest Vp values (Varela-Llamas, Ocegueda-Hernández, and Castillo-Ponce 2015). In this setting, CVM shared haplotypes with NL (7.58%), YUC (0.95%), GTO, JAL, and CHIS (0.47% for each). The highly economically developed state of NL exhibited the lowest values (9.534 ± 4.377, $Vp = 0.844$), possibly due to the number of haplotypes (> 6%) that it shared with other states (Ramos-Gonzalez et al. 2017; Varela-Llamas et al. 2015). Of these, 3.14% were within the same region, followed by CVM (1.74%), YUC (0.67%), CHIS (0.33%), and GTO (0.12%). These findings likely reflect the large-scale movements of people seeking employment in the largest cities within the Mexican national territory.

An analysis of molecular variance (AMOVA) suggested that the patterns of genetic variation were not strongly related to geography (1.71%). This finding may reflect the migratory movements that contributed to the genetic architecture of each state (precontact and colonial; 98.29%, $P \leq 0.0001$). However, certain geographic regions showed genetic similarities to each other. For example, the southeastern states of CHIS and YUC were closely related ($R_{st} = 0.00048$, $P = 0.324$), while urban zones (CVM, JAL, and NL) presented similar diversity patterns (average $R_{st} = 0.00738$, $P > 0.05$). By contrast, GTO exhibited major differences with both the urban zones and the southeastern states, being closest to JAL in terms of its patterns of diversity.

The similar patterns of genetic diversity in GTO and JAL could be explained by the geographic closeness of these regions, and the possible sharing of ancestral indigenous roots in pre- or postcolonial times. Perhaps more importantly, GTO and JAL, together with the states of Aguascalientes, Colima, Nayarit, and Zacatecas (ZAC) and the northwest corner of San Luis Potosí (SLP), once constituted the Nueva Galicia kingdom (Velázquez 1959). The mining industry in GTO, ZAC, and SLP resulted in significant gene flow in Nueva Galicia (Velázquez 1959). It also attracted immigrants from France and England and employed enslaved Africans in the labor force, which explain the presence of haplogroups R1b, E1b1a, E1b1b, and T in the region (see below). The large-scale movement of Michoacan natives (i.e., Tarascos) to serve as laborers in mines, as well as population dispersals from ZAC associated with a devastating famine (1749–1750), could have further contributed to the genetic similarities between GTO and JAL (Zavala and Castelo 1980).

NRY Phylogeography

Overall, Mexican males possessed 21 different haplogroups (Figure 11.1). Most prominent of these paternal lineages were R1b (0.344), Q-M3 (0.341), and E1b1b (0.077). The European-derived lineages (i.e., R1b) appeared at the highest frequency in the biggest cities in CVM (0.445) and NL (0.379), whereas Native American lineages were more prominent in the southeast Mexican region.

Native American Founder Lineages

As mentioned earlier, Native American haplogroup Q is commonly seen in mestizo populations of Mexico. In our study, it was most frequent in the southeast states of CHIS (0.623) and YUC (0.428), being nearly double the mean frequency (0.390) in the entire country. These findings highlight the Native American contribution to mestizo genomes in different geographical regions, as ascertained with ancestral informative markers (AIMs; Cahua-Pablo et al. 2017; Risch et al. 2009).

Analysis of the ancestral origins of the first Americans constitutes an enormous challenge due to the extent to which European colonizers genetically transformed Mesoamerican populations (Beezley 2011). Both pre- and

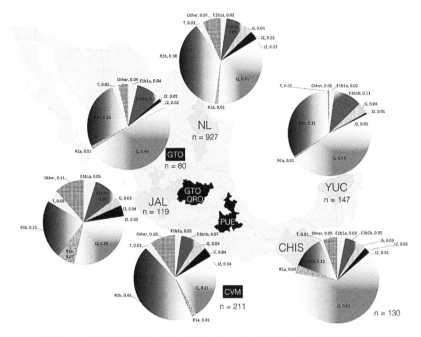

Figure 11.1 A map of the Mexican Republic showing the localities included in the present study and their lineage contributions. The comparison data were obtained from Santana et al. 2014; Salazar-Flores et al. 2010; and Ramos-Gonzalez et al. 2017. Abbreviations: CHIS = Chiapas; CVM = Central Valley of Mexico; GTO = Guanajuato; JAL = Jalisco; NL = Nuevo Leon; PUE = Puebla; QRO = Queretaro; YUC = Yucatan. CVM includes the states of GTO, PUE, and QRO.

postcolonization bottlenecks seriously decreased the genetic diversity of indigenous groups. In particular, smallpox, pneumonia, and measles carried by the Spaniards greater diminished Native Mexican diversity (Beezley 2011). Nonetheless, a remarkable diversity of haplotypes within haplogroup Q has been maintained in contemporary Native Mexicans, and even among mestizos (Battaglia et al. 2013; Gomez et al. 2019; Gonzalez-Sobrino et al. 2016; Sandoval et al. 2012; Santana et al. 2014).

Haplogroup Q is the most common and ancient lineage in the Americas and has evolved into at least four sublineages, including Q-MEH2 (Q1a), Q-M346 (Q1b), Q-L54 (Q1b1a), and Q-M3 (Q1b1a1a; Battaglia et al. 2013; Dulik et al. 2012a). Of these, Q-MEH2 is a haplogroup restricted to circumpolar populations, with its possible presence in Mesoamerican populations being poorly documented (Gomez et al. 2019; Olofsson et al. 2015). Q-L54 connects the Native Mexican populations with the Athabaskan-speaking groups in North America and with Altaians and Tuvinians indigenous to Siberia (ancestral to Q-M3; Dulik et al. 2012a; Gomez et al. 2021). The most common lineage in Mexico and the rest of the Americas, haplogroup Q-M3, appears in an increasing clinal frequency from North to South America (Roewer et al. 2013). All three of these haplogroups were brought to the

Americas through Beringia, and, as a result, connect Mexican populations with ancient Asian ancestors. Furthermore, both Q-L54 and Q-M3 have significantly diversified since arriving in the Americas, each having different sublineages within it (Battaglia et al. 2013; Belbin et al. 2018; Sandoval et al. 2012).

Owing to limited high-resolution studies of haplogroup Q, the exact sublineages carried by Mexican males are yet unknown. However, it is likely that the majority of Native Mexican males belong to Q-M3, followed by Q-L54, both of which having been reported by prior studies in Native and mestizo Mexican populations (Battaglia et al. 2013; Dulik et al. 2012b; Gomez et al. 2019; Jota et al. 2011; Santana et al. 2014).

Genetic Legacy from the Iberian Peninsula

Once the New World was "discovered," migrations from the Iberian Peninsula arrived in Mexico via Spanish conquest around 1519 (Beezley 2011). Previous studies have pointed out that the first Spanish conquerors came from the southern region of Spain (Andalusia and Extremadura), likely bringing paternal lineages like R1b with them (Beezley 2011; Salazar-Flores et al. 2010; Santana et al. 2014). The Andalusian region has a R1b frequency close to 63%, being more prominent in the east (EAND, 72.6%) compared with the west (WAND, 53.4%), while the R1b frequency in Extremadura (EXT) is 50% (Adams et al. 2008). Geographic zones like the Spanish Pyrenees (SPY, 77.5%), the Basque Country (BAC, 86.5%), and Catalonia (CAT, 67%) also exhibit higher R1b frequencies than the regions described before (Garcia et al. 2016; Lopez-Parra et al. 2009; Sole-Morata et al. 2017). By comparison, haplogroup R1b is the most prominent lineage among central and west European populations, being present in more than 50% of men from these regions (Myres et al. 2011; Villaescusa et al. 2018).

All these Spanish regions, as well as Aragon (ARA) and Castilla La Mancha (CLM), showed strong similarities to Mexican haplotypes (Figure 11.2A). As seen in this network, several haplotypes in Mexican mestizos are shared with populations from the Iberian Peninsula. This heterogeneous contribution from different peninsular regions of Spain to different Mexican states has clearly increased the haplotype diversity within R1b in Mexican populations.

On a finer scale, sublineages of R1b that arose in distinct regions of the Iberian Peninsula, such as R1b3d and R1b3f, are observed in Mexican males (Adams et al. 2008). It is also likely that other sublineages, such as R1b-DF27, are present in Mexican men, given its substantial frequency in BAC (Adams et al. 2008; Sole-Morata et al. 2017; Villaescusa et al. 2017). However, the paucity of high-resolution haplotype studies of this and other paternal lineages makes their identification difficult. Nevertheless, a remarkable number of these haplotypes appeared as singletons, suggesting a significant diversity within this haplogroup (data not shown).

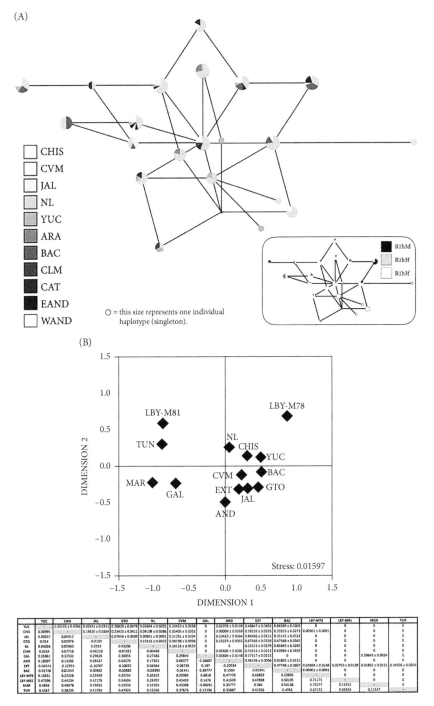

Figure 11.2 Comparison among Mexican, Spanish, and North African populations using median-joining networks based on 10 Y-STRs (R1b), and 15 Y-STRs (E1b1b). A: Haplogroup R1b, with the dotted box showing sublineages within R1b. B: Haplogroup E1b1b multidimensional scale plot based on genetic distances using R_{ST} values. The comparative data were obtained from Adams et al. 2008; Lopez-Parra et al. 2009; Garcia et al. 2016; Sole-Morata et al. 2017; Villaescusa et al. 2017;

Ambrosio, Dugoujon et al. 2010; Ambrosio, Hernandez et al. 2010; Regueiro et al. 2015. Abbreviations: CHIS = Chiapas; CVM = Central Valley of Mexico; JAL = Jalisco; NL = Nuevo Leon; YUC = Yucatan; AND = Andalusia; ARA = Aragon; BAC = Basque Autonomous Community; CLM = Castilla La Mancha; CAT = Catalonia; EAND = East Andalusia; EXT = Extremadura; LBY-M78 = Libya subclade M78; LBY-M81 = Libya subclade M81; MAR = Morocco; TUN = Tunisia; WAND = West Andalusia. The loci were weighted using the variance inverse; DYS385a/b loci were excluded from the network analysis.

Circum-Mediterranean Influences

Like the current Mexican population, populations from the Iberian Peninsula exhibit a complex genetic background that is influenced by the different historical, economic, and religious events contributing to Spanish biological diversity (Lafuente et al. 1887). Because of the peninsula's geographic position, it was involved in trade routes extending from southeastern Europe (Greeks), the Middle East (Phoenicians), and North Africa (Carthaginians, modern Tunisia; Lafuente et al. 1887). The genetic contribution of these cultures has been delineated through phylogeographic studies of haplogroups I, J1, and J2 (Ambrosio et al. 2010b; Rey-Gonzalez et al. 2017).

Since the 1st century AD, Jewish populations were dispersed across the Mediterranean and subsequently occupied the north of Spain (Ayton-Shenker 1993). The Roman expansion brought with it communities of Jews (some being slaves) that expanded into the rest of the peninsula and then were forcibly converted to Christianity (*conversos*) during the 13th and 14th centuries (Adams et al. 2008; Ayton-Shenker 1993). Moreover, the Barbarian invasions and subsequent occupation by Germanic Visigoths also contributed to Spain's genetic diversity (Adams et al. 2008; Lafuente et al. 1887). During the Muslim domination in the 12th to 15th centuries AD, Arabs and Berbers invaded different regions of Spain (Ambrosio et al. 2010a). Genetic vestiges of these population movements are reflected in the presence of lineages like E, G, I, J (M-172 and M-267), and K, which are quite common in the parental population from which the Mexican mestizo arose, i.e., Spaniards (Adams et al. 2008; Alvarez et al. 2014; Ambrosio et al. 2010a; Ambrosio et al. 2010b; Sutton et al. 2006).

Like the rest of the peninsula, the Andalusian region was influenced by Sephardic Jewish, North African, and Middle East populations, as evidenced by the presence of haplogroups E (M81 and V13), G, I, and K (Adams et al. 2008; Alvarez et al. 2014; Ambrosio et al. 2010a; Ambrosio et al. 2010b; Rey-Gonzalez et al. 2017). The Berber component of the Muslim invasion and the North African gene flow are evidenced by the presence of haplogroups E-M78, E-M81, and E-V12 (Adams et al. 2008; Alvarez et al. 2014; Ambrosio, et al. 2010a; Ambrosio, et al. 2010b; Rey-Gonzalez et al. 2017; Sutton et al. 2006). Other sublineages likely contributed by Middle East populations (E-V65), as well as Sephardic Jews and Tartessians (protohistorical Andalusian civilization), such as E-M34, and E-M123, E-M78, and E-V13, respectively, are also distributed throughout the peninsula (Adams et al. 2008; Alvarez et al. 2014; Ambrosio, Doujon et al. 2010; Ambrosio, Hernandez et al. 2010; Rey-Gonzalez et al. 2017; Sutton et al. 2006).

This complex genetic legacy was observed in all Mexican populations (Figure 11.1), where different sublineages within E1b1b appeared in these Mexican states. Of note, the Mexican haplotypes revealed some genetic influences from Moroccan (MAR), Libyan (LBY), and Tunisian (TUN) populations. The closeness of these populations and their genetic similarity to those from EXT, AND, BAC, and Galicia (GAL) could also reflect the Moorish influence in Spain (Figure 11.2B) (Ambrosio, Doujon et al. 2010; Ambrosio, Hernandez et al. 2010; Regueiro et al. 2015; Sole-Morata et al. 2017; Triki-Fendri et al. 2015). Urban zones like NL and CVM had the most populations bearing the E1b1b lineage and were also fairly close to Spanish regions like AND, BAC, and EXT. CHIS and NL also showed connections with LBY, MAR, and TUN, possibly reflecting recent migration from urban zones to rural regions of these states. Nonetheless, only CHIS exhibited a slight genetic closeness ($R_{ST} = 0.222$; $P = 0.009$) with the sublineage M78 of LBY. YUC appeared genetically distinctive from the rest of the states, showing a genetic connection with Sephardic Jews (data not shown; Adams et al. 2008).

Within haplogroup I, the Mexican population showed genetic similarities to populations from different regions of Spain, such as AND, BAC, EXT, and GAL (Figure 11.3A), although Aragon and Asturias (data not shown) also exhibited the presence of this haplogroup (Adams et al. 2008). These connections could reflect the geographic origins from which colonial and even postcolonial migrants came (Adams et al. 2008; Garcia et al. 2016; Lopez-Parra et al. 2009; Sole-Morata et al. 2017). Mexicans with haplogroup J were found mainly in the northern state of NL (Figure 11.1, Figure 11.3B), where similarities with haplotypes from LBY, MAR, Oman, Qatar, United Arab Emirates, and Yemen were also seen (data not shown; Adams et al. 2008; Alvarez et al. 2014). The relationship between NL and Middle Eastern populations could possibly be related to *conversos* Jews arriving during the colonial period, when NL played a critical role in their settlement there (Beezley 2011; Chacon-Duque et al. 2018). Carriers of this lineage were further observed in present-day large cities, such as CVM and JAL, possibly as a consequence of significant Lebanese migrations during the last century (Beezley 2011).

The frequencies of all these Y-chromosome lineages were reinforced by a prior study (Gonzalez-Sobrino et al. 2016). However, the present analysis included a greater representation of mestizo populations (i.e., in CHIS, CVM, JAL, and NL) that had not previously been analyzed (Gonzalez-Sobrino et al. 2016).

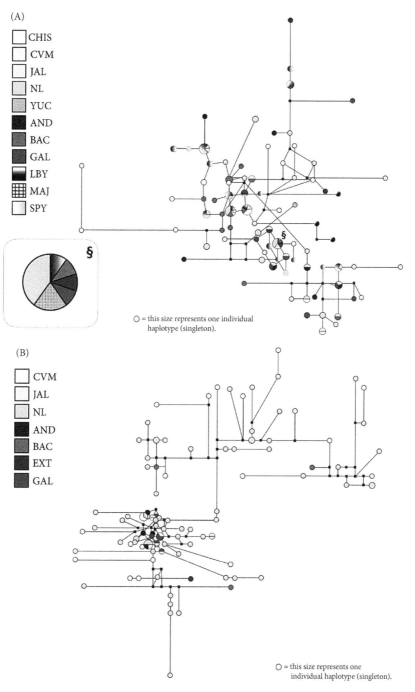

(A)

CHIS
CVM
JAL
NL
YUC
AND
BAC
GAL
LBY
MAJ
SPY

§

○ = this size represents one individual
haplotype (singleton).

(B)

CVM
JAL
NL
AND
BAC
EXT
GAL

○ = this size represents one
individual haplotype (singleton).

Figure 11.3 Comparison among Mexican, Spanish, and North African populations using median-joining networks based on 10 Y-STRs (haplogroup I), and 15 Y-STRs (haplogroup J). A: Haplogroup I. B: Haplogroup J. The loci were weighted using the variance inverse, with the DYS385a/b loci being excluded from the network analysis. §The box depicts a zoom-in of the shared haplotype among the different populations. The comparative data were obtained from Adams et al. 2008; Ambrosio, Dugoujon et al. 2010; Ambrosio, Hernandez et al. 2010; Rey-Gonzalez et al. 2017; Sutton et al. 2006. Abbreviations: CHIS = Chiapas; CVM = Central Valley of Mexico; JAL = Jalisco; NL = Nuevo Leon; YUC = Yucatan; AND = Andalusia; BAC = Basque Autonomous Community; EXT = Extremadura; GAL = Galicia; MAJ = Majorca; LBY = Libya; SPY = Spaniard Pyrenees.

African Imprint on Mestizo Genomes

Enslaved Africans from Angola, Congo, Gambia, Guinea, Mozambique, and Senegal were brought to Mexico with the Spaniards as replacements for enslaved Native Mexicans, due to their drastic population reduction because of the diseases carried by the European colonizers (Beezley 2011; Velázquez and Nieto 2016). Thus, Africans contributed to the genetic diversity of some Mexican genomes. Even so, Y-chromosome lineages linked to African ancestry have not yet been broadly documented in the Mexican population. Previous studies using autosomal AIMs revealed that the African component varied between 6.1% and 8.9%, being more prominent in the coastal regions (Cahua-Pablo et al. 2017; Risch et al. 2009). The coastal states of JAL and YUC presented the highest frequencies of haplogroup T, which has been found in MAR, Ethiopia, Kenya, and Egypt (Mendez et al. 2011). Haplogroup T also appears in Middle East populations, such as in Iran, Iraq, and Oman, as well as in Spain (in Cadiz and Ibiza; Adams et al. 2008; Mendez et al. 2011).

Anthropological genetic studies in other Latino populations (i.e., in Colombia and Venezuela) have further documented the presence of African paternal lineages, such as B*, B2*, E1a, E1b1, and E2b1 (Gómez-Camargo et al. 2015; Guerra et al. 2011; Noguera et al. 2014). Given that enslaved Africans were brought to the Colombian Caribbean for their subsequent distribution to the rest of Latin America, it is likely that some of these haplogroups could also be present in Mexican males. Nonetheless, the main legacy from Africa likely came with the Spanish conquerors more so than enslaved Africans, whose contribution to Afro-mestizo populations should be thoroughly analyzed.

Conclusions

Given the indigenous roots of Mexican populations and the range of paternal lineages brought to Mexico by colonizing populations from the Iberian Peninsula, it is not surprising to find a great diversity of Y chromosomes within the Mexican population. It is worth mentioning here that each state presented a distinctive distribution of these haplogroups, possibly due to the demographic events that shaped their genetic make-up. Overall, haplogroups R1b, Q, and E1b1b were the most frequent paternal lineages in Mexican mestizo males. During the historical period, R1b and E1b1b lineages arrived first with the Spanish conquest, then with the subsequent emigrations during the colonial period, and more recently with the exodus provoked by the Spanish Civil War, further enriching the Mexican genetic diversity. Nonetheless, the geographic regions where these lineages came from and the subclades to which they belong are partially known. Thus, high-resolution studies should be conducted to more completely reconstruct the historical past of Mexico 500 years after Spanish conquest. These studies could help to elucidate immigrants' geographic origins and better describe the patrilineal genetic architecture of contemporary Mexican men.

References

Adams, S. M., E. Bosch, P. L. Balaresque, S. J. Ballereau, A. C. Lee, E. Arroyo, A. M. Lopez-Parra, M. Aler, M. S. Grifo, M. Brion, et al. 2008. "The genetic legacy of religious diversity and intolerance: paternal lineages of Christians, Jews, and Muslims in the Iberian Peninsula." *American Journal of Human Genetics* 83, no. 6: 725–736.

Alvarez, L., E. Ciria, S. L. Marques, C. Santos, and M. P. Aluja. 2014. "Y-chromosome analysis in a Northwest Iberian population: Unraveling the impact of Northern African lineages." *American Journal of Human Biology* 26: 740–746.

Ambrosio, B., J. M. Doujon, C. Hernandez, D. De La Fuente, A. Gonzalez-Martin, C. A. Fortes-Lima, A. Novelletto, J. N. Rodriguez, and R. Calderon. 2010. "The Andalusian population from Huelva reveals a high diversification of Y-DNA paternal lineages from haplogroup E: Identifying human male movements within the Mediterranean space." *Annals of Human Biology* 37: 86–107.

Ambrosio, B., C. Hernandez, A. Novelletto, J. M. Dugoujon, J. N. Rodriguez, P. Cuesta, C. Fortes-Lima, and R. Calderon. 2010. "Searching the peopling of the Iberian Peninsula from the perspective of two Andalusian subpopulations: A study based on Y-chromosome haplogroups." *Journal of the Croatian Anthropological Society* 34: 1215–1228.

Athey, W. 2005. "Haplogroup prediction from Y-STR values using an allele-frequency approach." *Journal of Genetic Genealogy* 1: 1–7.

Ayton-Shenker, D. 1993. "The Jewish community of Spain." *Journal Storage* 5, no. 3/4: 159–208.

Battaglia, V., V. Grugni, U. A. Perego, N. Angerhofer, J. E. Gomez-Palmieri, S. R. Woodward, A. Achilli, N. Myres, A. Torroni, and O. Semino. 2013. "The first peopling of South America: New evidence from Y-chromosome haplogroup Q." *PLOS One* 8: e71390. doi:10.1371/journal.pone.0071390.

Beezley, W. H. 2011. *Mexico in World History*. New York: Oxford University Press.

Belbin, G. M., M. A. Nieves-Colon, E. E. Kenny, A. Moreno-Estrada, and C. R. Gignoux. 2018. "Genetic diversity in populations across Latin America: implications for population and medical genetic studies." *Current Opinion in Genetics and Development* 53: 98–104.

Bryc, K., C. Velez, T. Karafet, A. Moreno-Estrada, A. Reynolds, A. Auton, M. Hammer, C. D. Bustamante, and H. Ostrer. 2010. "Colloquium paper: Genome-wide patterns of population structure and admixture among Hispanic/Latino populations." *PNAS* 107 (Suppl 2): 8954–8961.

Cahua-Pablo, J. A., M. Cruz, P. V. Tello-Almaguer, L. C. Del Alarcon-Romero, E. J. Parra, S. Villerias-Salinas, A. Valladares-Salgado, V. A. Tello-Flores, A. Mendez-Palacios, C. P. Perez-Macedonio, et al. 2017. "Analysis of admixture proportions in seven geographical regions of the state of Guerrero, Mexico." *American Journal of Human Biology* 29, no. 6: 1–10.

Calafell, F., and M. H. D. Larmuseau. 2017. "The Y chromosome as the most popular marker in genetic genealogy benefits interdisciplinary research." *Human Genetics* 136: 559–573.

Chacon-Duque, J. C., K. Adhikari, M. Fuentes-Guajardo, J. Mendoza-Revilla, V. Acuna-Alonzo, R. Barquera, M. Quinto-Sanchez, J. Gomez-Valdes, P. Everardo Martinez, H. Villamil-Ramirez, et al. 2018. "Latin Americans show wide-spread Converso ancestry and imprint of local Native ancestry on physical appearance." *Nature Communications* 9: 5388.

Dulik, M. C., A. C. Owings, J. B. Gaieski, M. G. Vilar, A. Andre, C. Lennie, M. A. Mackenzie, I. Kritsch, S. Snowshoe, R. Wright, et al. 2012a. "Y-chromosome analysis reveals genetic divergence and new founding native lineages in Athapaskan- and Eskimoan-speaking populations." *PNAS* 109: 8471–8476.

Dulik, M. C., S. I. Zhadanov, L. P. Osipova, A. Askapuli, L. Gau, O. Gokcumen, S. Rubinstein, and T. G. Schurr. 2012b. "Mitochondrial DNA and Y chromosome variation provides evidence for a recent common ancestry between Native Americans and Indigenous Altaians." *American Journal of Human Genetics* 90: 229–246.

Garcia, O., I. Yurrebaso, I. D. Mancisidor, S. Lopez, S. Alonso, and L. Gusmao. 2016. "Data for 27 Y-chromosome STR loci in the Basque Country autochthonous population." *Forensic Science International: Genetics* 20: e10–e12. doi:10.1016/j.fsigen.2015.09.010

Gomez, R., M. Villar, M. A. Meraz-Rios, D. Veliz, G. Zuñiga, E. Hernandez-Tobias, P. Figueroa-Corona, A. Owings, J. Gaieski, T. G. Schurr, and The Genographic Consortium. 2020. "Y-chromosome diversity in Aztlan descendants and its implications for the history of Central Mexico." *iScience*; 24(5):102487. doi: 10.1016/j.isci.2021.102487.

Gomez-Camargo, D., R. Camacho-Mejorado, C. Gómez-Alegría, A. Alario, E. Henández-Tobías, G. Mora-García, M. Meraz-Ríos, and R. Gómez. 2015. "Genetic structure of Cartagena de Indias population using hypervariable markers of Y chromosome." *Open Journal of Genetics* 5: 27–41. doi:10.4236/ojgen.2015.51003

Gonzalez-Sobrino, B. Z., A. P. Pintado-Cortina, L. Sebastian-Medina, F. Morales-Mandujano, A. V. Contreras, Y. E. Aguilar, J. Chavez-Benavides, A. Carrillo-Rodriguez, I. Silva-Zolezzi, and L. Medrano-Gonzalez. 2016. "Genetic diversity and differentiation in urban and indigenous populations of Mexico: Patterns of mitochondrial DNA and Y-chromosome lineages." *Biodemography and Social Biology* 62: 53–72.

Guerra, D. C., C. F. Perez, M. H. Izaguirre, E. A. Barahona, A. R. Larralde, and M. V. Lugo. 2011. "Gender differences in ancestral contribution and admixture in Venezuelan populations." *Human Biology* 83: 345–361.

Mangs, A. H., and B. J. Morris. 2007. "The human pseudoautosomal region (PAR): Origin, function and future." *Current Genomics* 8: 129–136.

ISOGG. 2018, August 21. *International Society of Genetic Genealogy*. https://isogg.org

Jota, M. S., D. R. Lacerda, J. R. Sandoval, P. P. Vieira, S. S. Santos-Lopes, R. Bisso-Machado, V. R. Paixao-Cortes, S. Revollo, Y. M. C. Paz, R. Fujita, et al. 2011. "A new subhaplogroup of native American Y-chromosomes from the Andes." *American Journal of Physical Anthropology* 146: 553–559.

Lafuente, M., J. Valera, A. Pirala, and A. S. Borrego. 1887. *Historia general de España desde los tiempos primitivos hasta la muerte de Fernando VII [General history of Spain from primitive times to the Ferdinand VII death]*. Barcelona: Montaner y Simón.

Lopez-Parra, A. M., L. Gusmao, L. Tavares, C. Baeza, A. Amorim, M. S. Mesa, M. J. Prata, and E. Arroyo-Pardo. 2009. "In search of the pre- and post-Neolithic genetic substrates in Iberia: Evidence from Y-chromosome in Pyrenean populations." *Annals of Human Biology* 73: 42–53.

Martinez-Cortes, G., J. Salazar-Flores, J. Haro-Guerrero, R. Rubi-Castellanos, J. S. Velarde-Felix, J. F. Munoz-Valle, M. Lopez-Casamichana, E. Carrillo-Tapia, L. M. Canseco-Avila, C. M. Bravi, et al. 2013. "Maternal admixture and population structure in Mexican-Mestizos based on mtDNA haplogroups." *American Journal of Physical Anthropology* 151: 526–537.

Mendez, F. L., T. M. Karafet, T. Krahn, H. Ostrer, H. Soodyall, and M. F. Hammer. 2011. "Increased resolution of Y chromosome haplogroup T defines relationships among populations of the Near East, Europe, and Africa." *Human Biology* 83: 39–53.

Myres, N. M., S. Rootsi, A. A. Lin, M. Jarve, R. J. King, I. Kutuev, V. M. Cabrera, E. K. Khusnutdinova, A. Pshenichnov, B. Yunusbayev, et al. 2011. "A major Y-chromosome haplogroup R1b Holocene era founder effect in Central and Western Europe." *European Journal of Human Genetics* 19: 95–101.

Noguera, M. C., A. Schwegler, V. Gomes, I. Briceno, D. Alvarez, D. Uricoechea, A. Amorim, E. Benavides, C. Silvera, M. Charris, et al. 2014. "Colombia's racial crucible: Y chromosome evidence from six admixed communities in the Department of Bolivar." *Annals of Human Biology* 41: 453–459.

Olofsson, J. K., V. Pereira, C. Borsting, and N. Morling. 2015. "Peopling of the North circumpolar region—Insights from Y chromosome STR and SNP typing of Greenlanders." *PLOS One* 10: e0116573. doi:10.1371/journal.pone.0116573

Poznik, G. D., Y. Xue, F. L. Mendez, T. F. Willems, A. Massaia, M. A. Wilson Sayres, Q. Ayub, S. A. McCarthy, A. Narechania, S. Kashin, et al. 2016. "Punctuated bursts in human male demography inferred from 1,244 worldwide Y-chromosome sequences." *Nature Genetics* 48: 593–599.

Quintana-Murci, L., and M. Fellous. 2001. "The human Y chromosome: The biological role of a 'functional wasteland.'" *Journal of Biomedicine and Biotechnology* 1: 18–24.

Ramos-Gonzalez, B., J. A. Aguilar-Velazquez, M. de Lourdes Chavez-Briones, M. Del Rocio Escareno-Hernandez, E. Alfaro-Lopez, and H. Rangel-Villalobos. 2017. "Genetic population data of three Y-STR genetic systems in Mexican-Mestizos from Monterrey, Nuevo Leon (Northeast, Mexico)." *Forensic Science International: Genetics* 29: e21–e22. doi:10.1016/j.fsigen.2017.04.016

Regueiro, M., R. Garcia-Bertrand, K. Fadhlaoui-Zid, J. Alvarez, and R. J. Herrera. 2015. "From Arabia to Iberia: A Y chromosome perspective." *Gene* 564: 141–152.

Rey-Gonzalez, D., M. Gelabert-Besada, R. Cruz, F. Brisighelli, M. Lopez-Soto, M. Rasool, M. I. Naseer, P. Sanchez-Diz, and A. Carracedo. 2017. "Micro and macro geographical analysis of Y-chromosome lineages in South Iberia." *Forensic Science International: Genetics* 29: e9–e15. doi:10.1016/j.fsigen.2017.04.021

Risch, N., S. Choudhry, M. Via, A. Basu, R. Sebro, C. Eng, K. Beckman, S. Thyne, R. Chapela, J. R. Rodriguez-Santana, et al. 2009. "Ancestry-related assortative mating in Latino populations." *Genome Biology* 10: R132.

Roewer, L., M. Nothnagel, L. Gusmao, V. Gomes, M. Gonzalez, D. Corach, A. Sala, E. Alechine, T. Palha, N. Santos, et al. 2013. "Continent-wide decoupling of Y-chromosomal genetic variation from language and geography in native South Americans." *PLOS Genetics* 9: e1003460. doi:10.1371/journal.pgen.1003460

Rubicz, R., P. Melton, and M. Crawford. 2007. "Molecular Markers in Anthropological Genetic Studies." In *Anthropological Genetics: Theory, Methods and Applications,* edited by M. Crawford, 141–186. New York: Cambridge University Press.

Salazar-Flores, J., R. Dondiego-Aldape, R. Rubi-Castellanos, M. Anaya-Palafox, I. Nuno-Arana, L. M. Canseco-Avila, G. Flores-Flores, M. E. Morales-Vallejo, N. Barojas-Perez, J. F. Munoz-Valle, et al. 2010. "Population structure and paternal admixture landscape on present-day Mexican-Mestizos revealed by Y-STR haplotypes." *American Journal of Human Biology* 22: 401–409.

Sandoval, K., A. Moreno-Estrada, I. Mendizabal, P. A. Underhill, M. Lopez-Valenzuela, R. Penaloza-Espinosa, M. Lopez-Lopez, L. Buentello-Malo, H. Avelino, F. Calafell, et al. 2012. "Y-chromosome diversity in Native Mexicans reveals continental transition of genetic structure in the Americas." *American Journal of Physical Anthropology* 148: 395–405.

Santana, C., G. Noris, M. A. Meraz-Rios, J. J. Magana, E. S. Calderon-Aranda, L. Munoz, and R. Gomez. 2014. "Genetic analysis of 17 Y-STRs in a Mestizo population from the Central Valley of Mexico." *Human Biology* 86: 289–312.

Sole-Morata, N., P. Villaescusa, C. Garcia-Fernandez, N. Font-Porterias, M. J. Illescas, L. Valverde, F. Tassi, S. Ghirotto, C. Ferec, K. Rouault, et al. 2017. "Analysis of the R1b-DF27 haplogroup shows that a large fraction of Iberian Y-chromosome lineages originated recently in situ." *Scientific Reports* 7: 7341. doi.10.1038/s41598-017-07710-x

Sutton, W. K., A. Knight, P. A. Underhill, J. S. Neulander, T. R. Disotell, and J. L. Mountain. 2006. "Toward resolution of the debate regarding purported crypto-Jews in a Spanish-American population: Evidence from the Y chromosome." *Annals of Human Biology* 33: 100–111.

Triki-Fendri, S., P. Sanchez-Diz, D. Rey-Gonzalez, I. Ayadi, A. Carracedo, and A. Rebai. 2015. "Paternal lineages in Libya inferred from Y-chromosome haplogroups." *American Journal of Physical Anthropology* 157: 242–251.

Varela Llamas, R., J. M. Ocegueda Hernandez, and R. A. Castillo Ponce. 2015. "Migración interna en México y causas de su movilidad [Internal migration in Mexico and its mobility causes]." *Perfies Latinoamericanos* 25: 141–167.

Velázquez, M. E., and G. I. Nieto. 2016. *Afrodescendientes en México [Afrodescendants in Mexico].* Instituto Nacional de Antropologia e Historia. Mexico City, Mexico.

Villaescusa, P., M. J. Illescas, L. Valverde, M. Baeta, C. Nunez, B. Martinez-Jarreta, M. T. Zarrabeitia, F. Calafell, and M. M. de Pancorbo. 2017. "Characterization of the Iberian Y chromosome haplogroup R-DF27 in Northern Spain." *Forensic Science International: Genetics* 27: 142–148.

Villaescusa, P., L. Palencia-Madrid, M. A. Campaner, J. Jauregui-Rada, M. Guerra-Rodriguez, A. M. Rocandio, and M. M. de Pancorbo. 2018. "Effective resolution of the Y chromosome sublineages of the Iberian haplogroup R1b-DF27 with forensic purposes." *International Journal of Legal Medicine* 133: 17–23.

Wells, S. 2006. *Deep Ancestry: Inside the Genographic Project*. Washington DC: National Geographic.

Willems, T., M. Gymrek, G. D. Poznik, C. Tyler-Smith, 1000 Genomes Project Chromosome Y Group, and Y. Erlich. 2016. "Population-scale sequencing data enable precise estimates of Y-STR mutation rates." *American Journal of Human Genetics* 98: 919–933.

Zavala, S., and M. A. Castelo. 1980. *Fuentes para la historia del trabajo en Nueva España* [Sources for the history of work in the New Spain]. Centro de Estudios Históricos del Movimiento Obrero Mexicano. Mexico City, Mexico.

12

Migration of Garifuna

Evolutionary Success Story

Michael H. Crawford, Christine Phillips-Krawczak,
Kristine G. Beaty, and Noel Boaz

Introduction

The Garifuna (also known as the Black Caribs) are a unique mélange of African, Native American, and European ancestry originally established on the island of St. Vincent but forcibly transplanted by the British to the Bay Islands and by the Spanish to the coast of Central America. Although some archaeological evidence exists for the initial settlement of the Lesser Antilles approximately 8,000 years ago, it is likely that St. Vincent was settled 2,000 years BP (Fitzpatrick 2015). The earliest inhabitants of the Lesser Antilles were Arawak-speakers who expanded from South America, which explains the Arawak language spoken by the Garifuna today. Their settlement was followed by Carib movement through the Lesser Antilles for trade and possible invasion of the Caribbean Islands (Taylor 2012). In the 17th century, African slaves, including those who escaped from the Barbados slave trade center, plus two shipwrecks off the coast of St. Vincent, resulted in a mixed Native American/African population. The Africans were accepted into the local indigenous communities, which resulted in the creation of the Garifuna (Black Caribs) of St. Vincent. In 1676, a report from the governor of St. Vincent estimated that there were approximately 3,000 Black Caribs residing on the island. British encroachment on indigenous/Black Carib lands resulted in two conflicts: the First Carib War (1772–1773) and the Second Carib War (1795–1796). The first Carib War was resolved relatively peacefully. However, in response to the British land grabs, in 1796 the Black Caribs attacked and destroyed a number of the English settlements. This conflict resulted in a series of forced migrations of the Garifuna, first from St. Vincent to Baliceaux (a small island adjacent to St. Vincent), then to the Bay Islands and eventually to Honduras (see Figures 12.1 and 12.2).

Lester Firschein (1961) conducted the first genetic study of the Black Caribs of Belize. He sampled populations from Stann Creek, Hopkins, and Seine Bight and examined blood group polymorphisms and hemoglobin variation. In 1975–1976, this initial genetic research was followed up by studies of Garifuna volunteers in

Michael H. Crawford, Christine Phillips-Krawczak, Kristine G. Beaty, and Noel Boaz, *Migration of Garifuna* In: *Human Migration*. Edited by: Maria de Lourdes Muñoz-Moreno and Michael H. Crawford, Oxford University Press. © Oxford University Press 2021. DOI: 10.1093/oso/9780190945961.003.0013

Figure 12.1 Map of the Caribbean. The map includes the islands of Dominica and St. Vincent, showing the origins of Afro-Caribbean populations, including a homeland in South America, European and African contact, and the relocation in 1797.

Livingston, Guatemala, and in Stann Creek and Punta Gorda in Belize (Crawford 1983; Crawford et al. 1981; Devor et al. 1984). St. Vincent communities (Sandy Bay, Owia, and Fancy) were first sampled in 1979. The genetic structure of St. Vincent and Central American Garifuna was first characterized by Devor et al. in 1984. The first application of molecular genetics to observed variation in mitochondrial DNA (mtDNA) in Black Carib populations of Belize was conducted by Monsalve and Hagelberg (1997). This research was followed by work by Salas et al. (2005) and Herrera-Paz et al. (2010) on the characterization of mtDNA and autosomal short tandem repeat (STR) variation in Garifuna populations of Honduras. In 2006, Phillips-Krawczak sampled populations of Belize (Punta Gorda and Barranco) and St. Vincent (Fancy, Owia, Sandy Bay, and Greiggs). Benn-Torres et al. (2015) examined mtDNA, Y chromosome, and autosomal STRs in a small sample from St. Vincent.

This chapter focuses on the causes of Garifuna migrations from St. Vincent and their subsequent evolutionary, societal, and demographic consequences. In 1797, this admixed population was forcibly relocated from St. Vincent to Baliceaux, to the Bay Islands, and eventually to the coast of Central America. The Garifuna rapidly expanded from two founding villages (Rio Negro and Santa Fe) in Honduras to 54 villages distributed along the coast of Central America (see Figures 12.3 and 12.4). The initial transplanted Garifuna population contained fewer than 2,000 persons, but the population expanded to more than 300,000 persons, currently residing in coastal villages of Belize, Guatemala, Honduras, and Nicaragua, as well as on St. Vincent (Beaty 2017).

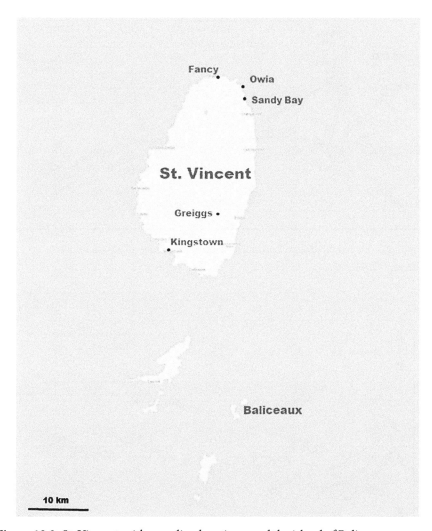

Figure 12.2 St. Vincent, with sampling locations, and the island of Baliceaux.

Causes of Migration

The Garifuna villages of Central America and St. Vincent were established as a result of multiple migrations with numerous causes and consequences. These causes are described below.

Expansion of Native Americans from Mainland South America to the Caribbean

Based on linguistic and archaeological analyses of indigenous populations of Dominica, Taylor (2012) argued that the original inhabitants of the Caribbean Islands were Arawak speakers. He noted that men and women used different

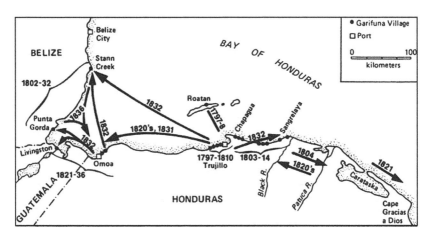

Figure 12.3 Dispersal of the Garifuna in Central America, 1797–1836 (after Davidson 1984).

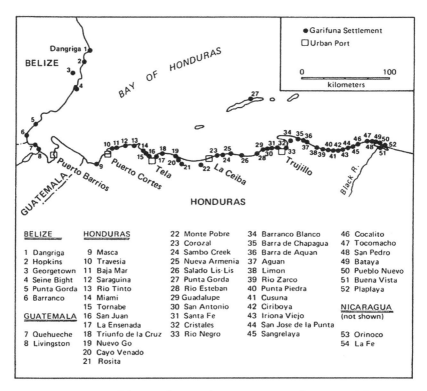

Figure 12.4 Locations of the Garifuna settlements in Central America (after Davidson 1984)

languages (i.e., women spoke Arawak, while men spoke Carib). These linguistic data were originally suggested as evidence of a massive Carib invasion following the Arawak migration. During the invasion, Carib men took Arawak wives. However, Davis and Goodwin (1990) suggested that a male language was utilized for purposes of trade, in interactions with Cariban-speaking people, and that the use of two languages does not necessarily imply invasion by Carib Native Americans. Irrespective of whether two Native American groups interbred or not, the founders of St. Vincent originally migrated from South America.

Forced Migration of African Slaves

In the early 1600s, African slaves were forcibly transported to the Leeward Island slave center of Barbados, and some of these slaves escaped and settled on the adjacent island of St. Vincent. In 1667, an English governor reported that two slave ships from Guinea were wrecked off the coast of St. Vincent, and a number of the slaves survived, escaped, and intermixed with the indigenous population of the island. A government report in 1676 enumerated approximately 3,000 Black Caribs residing on St. Vincent.

Forced Relocation of Garifuna from St. Vincent
to Baliceaux Island

British encroachment on Carib lands resulted in hostilities, followed by two wars between the indigenous residents of St. Vincent and British settlers and military. The First Carib War was settled between the combatants after a brief period of hostilities. However, the Second Carib War of 1795–1796 resulted in the eventual defeat and round-up of the Garifuna by British military. The Garifuna were deported by the British to a small island, Baliceaux, off the southern coast of St. Vincent. In July of 1796, 276 Black Caribs were shipped from St. Vincent to Baliceaux. Within 6 months, the total number of captives grew to 4,633 individuals, including children and women, who were imprisoned on this small island, roughly 0.8 km^2, lacking fresh water and containing little edible vegetation. Black Caribs who successfully avoided capture by the British hid in the mountainous regions of St. Vincent and gave issue to the contemporary Garifuna populations currently residing along the northern coast in three villages—Sandy Bay, Fancy, and Owia (see Figure 12.2). Today, the majority of the Garifuna of St. Vincent reside in these three villages.

Transplantation of Garifuna from Baliceaux to Roatán, Bay Islands

The captive Black Carib population imprisoned on Baliceaux was devastated numerically by a lack of fresh water and resources, followed by a yellow fever or typhus epidemic (Gonzalez, 1984, 1988; Taylor 2012). Given the untenable conditions on the tiny island, the British authorities decided to relocate the prisoner population of Baliceaux. A year later, on February 25, 1797, the HMS *Experiment,* accompanied by naval transport, arrived in Baliceaux and transported the surviving 2,026 Black Caribs to Roatán in the Bay Islands, located off the coast of nduras (Gullick 1984). Figure 12.1 traces the Black Carib movements from South America to St. Vincent and eventually to Central America.

Relocation of the Garifuna from Roatán to Honduras

With a scarcity of resources on the island of Roatán, the Black Caribs convinced the Spanish authorities to transport them to the coast of Honduras and to establish two communities: Rio Negro and Santa Fe. Fewer than 2,500 Black Caribs were transported to Honduras and established these initial settlements. The coast of Honduras already contained some Creole settlements founded earlier by slaves who had been brought to work the mines and fruit plantations. There was some genetic admixture, particularly between the Creoles of Haiti and the Black Caribs during the initial colonization of the Mosquito Coast of Central America. Only 206 Black Caribs decided to remain on Roatán; they are the ancestors of the contemporary populations of the Bay Islands, although more recent migration to Punta Gorda, for jobs on this tourist destination island, may explain some of the diversity seen there today (Taylor 2012).

Population Expansion from Two Villages in Honduras to 54 Villages Geographically Distributed Along the Coast of Central America

Davidson (1984) documented the timeline for a rapid dispersal of the Garifuna along the coast of Central America in five temporal-territorial units (see Figure 12.4):

1. St. Vincent to Roatán Island, Honduras, 1797
2. The Trujillo core, 1797–1810
3. Honduran Mosquitia, 1803–1814
4. Belize, 1802–1832
5. Western Honduras and Guatemala, 1821–1836

These population expansions and migrations were facilitated by favorable environmental conditions along the coast of Central America, a subsistence economy based on a hybrid African/Native American agriculture, and fishing industries. These environmental and cultural conditions resulted in exceptionally high fertility and local admixture, followed by population fission and migration.

Field Investigations

During the summers of 1975, 1976, and 1978, a total of 1,327 blood specimens were collected from Garifuna and Creole volunteers from Belize, Guatemala, and St. Vincent (Crawford et al. 1984). Table 12.1 summarizes the number of participants in the initial anthropological genetic study of the Garifuna of Belize, Guatemala, and St. Vincent. The purpose of the study was to determine the evolutionary effects of forced migration on transplanted Garifuna populations (Crawford 1983) using standard genetic markers (blood groups, proteins, and immunoglobulins). The ethnicity of each participant was based on self-identification (Crawford et al. 1984).

During the 1980s and 1990s, with the methodological developments in DNA extraction, characterization of DNA haplotypes and sequencing permitted much finer-grained analyses of population differentiation and a more informative characterization of evolutionary change (Rubicz et al. 2007). Starting in 2005, a series of follow-up investigations of Garifuna populations were initiated by researchers from the Laboratory of Biological Anthropology (LBA), University of Kansas. Table 12.2 summarizes the samples collected from the Caribs of Dominica and from Garifuna populations of Belize, Honduras, and Roatán, Bay Island.

Table 12.1   This table contains the sample sizes of participants in the initial study using standard genetic markers (blood groups, proteins and immunoglobulins). Ethnicity of each participant was based on self-identification. (Crawford et al, 1984).

Population	N
Mainland Populations (Guatemala and Belize) Garifuna	
Livingston	205
Stann Creek (Dangriga)	354
Punta Gorda	239
Total	798
St. Vincent Garifuna	
Sandy Bay	161
Owia	85
Total	246

Table 12.2 Molecular genetic studies by researchers from the Laboratory of Biological Anthropology (LBA) of Carib and Garifuna Populations of St. Vincent, Belize, Dominica, Honduras, and the Bay Islands

Investigators	Island/Country	Village	Participants (N)
Phillips-Krawzack[1] 2005	St. Vincent	Fancy	24
		Sandy Bay	34
		Owia	22
		Greggs	20
	Belize	Punta Gorda	50
		Barranco	27
Crawford and Boaz 2006	Dominica	Kalinago Reserve	51
Beaty and Herrera-Paz[2] 2015–2016	Honduras	Cristales	44
		Rio Negro	76
		Santa Fe	60
	Roatán, Bay Islands	Punta Gorda	132

[1]Phillips-Krawczak, 2012
[2]Beaty, 2017

Consequences of Migration

How was the genetic diversity of Garifuna populations affected by the multiple migrations? Black Carib populations on St. Vincent and the Caribs of Dominica displayed high genetic diversity and heterozygosity because of the complex patterns of admixture resulting from their unique history. Arawak and Carib populations first settled the Leeward Islands, with possible admixture with Cariban groups, and then African slaves (primarily from the west coast) contributed to the genetic makeup of the Black Carib populations. Based on standard genetic markers (i.e., blood groups and serum and red blood cell proteins), St. Vincent populations had the highest heterozygosity (D = 0.27), with Livingston Garifuna having the lowest D, 0.21. The lower levels of heterozygosity reflect the population fission observed in Central America as the Garifuna rapidly spread from the original two villages in Honduras to the 54 communities distributed throughout the eastern coast of Central America.

Based on mtDNA sequences, Phillips-Krawczak (2012) also detected higher genomic diversity in St. Vincent Black Caribs (H = 0.96) and the Caribs of Dominica (H = 0.99) than in residents of the coastal villages in Belize (H = 0.91). This is indicative of genetic drift interacting with gene flow into the population.

Admixture in Black Carib Populations

The first estimates of admixture, based on 24 classical blood marker polymorphisms, indicated that the coastal Garifuna populations exhibited much greater African ancestry than was estimated for St. Vincent populations. African ancestry in the Garifuna was measured as 70% in Livingston, Guatemala, compared to 46% in populations on St. Vincent (Crawford et al. 1981, 1984). Immunoglobulins (GM and KMs) similarly indicated 71% to 80% African ancestry in Stann Creek and Punta Gorda, Belize, versus 41% to 58% African ancestry in St. Vincent populations. In later studies using smaller samples ($N = 43$), similar results were found based on mtDNA (Torres et al. 2015).

Molecular studies of mtDNA in Dominica began in 2006, with field investigations by Crawford and Boaz in the Kalinago Carib Reserve (Phillips-Krwaczak 2012). This was an attempt to reconstruct the genetic structure of Carib ancestral populations. The Native American contribution to the currently admixed population of the Reserve is 67%, based largely upon the presence of haplogroups C (57%) and A2 (10%). In contrast, Benn-Torres et al. (2007) found that the majority of individuals whom they sampled on Dominica had African mtDNA haplogroups, with L being the most common. They reported that, in Dominica, only 28% of the lineages were Native American, with a small founding population, and high nucleotide diversity indicative of high level of admixture. Benn-Torres et al. (2015) also reported, based on mtDNA, an extremely high African ancestry, 63%, for St. Vincent Black Caribs.

The Central American coastal Garifuna display significantly higher African ancestry than observed for Garifuna of St. Vincent or the Caribs of Kalinago Reserve on Dominica. Garifuna of Honduras have 93% African subclades (L0-L3f1), while Belize Garifuna have 96% African mtDNA.

In addition, the distribution of African subclades varied. Only L2a was found in St. Vincent, Dominica, Punta Gorda, and several coastal communities; L2b and L2e were only found in St. Vincent and Dominica; while L2c was found only in coastal communities and Roatán. A similar pattern was seen with L3 haplotypes, with nearly all L3d and L3e2 haplotypes found in Dominica and St. Vincent, while coastal communities displayed L3f1 and L3e1 haplotypes. Haplotypes belonging to L0 and L1 were not shared between St. Vincent, Dominica, and the Honduran coast, and more similar haplotypes were found in close geographic proximity (Table 12.3).

There was considerable variation in estimates of admixture for the two islands of the Caribbean depending on the geographic locations of sampling sites and which genetic markers were utilized. The LBA sample from Dominica was collected on the east coast of the island at the Kalinago Reserve (Benn-Torres et al. 2015), which exhibits more than twice the Native American ancestry than the sample taken by Benn-Torres et al. (predominantly taken on the southern coast). The frequencies are more similar when comparing Garifuna communities, though a higher frequency was found in the communities sampled by the LBA. Benn-Torres et al. primarily sampled populations from the south and east coast, who displayed fewer

Table 12.3 Comparison of mtDNA haplogroup frequencies from Dominica and St. Vincent. (Torres et al. 2007; Torres et al. 2015; Laboratory of Biological Anthropology (LBA) study, 2019). Native American maternal ancestry included sequences that belonged to haplogroups A, B, and D. African ancestry was indicated by haplogroups L0, L1, L2, and L3

Parental Population	Hg	Dominica		Garifuna of St. Vincent	
		LBA	Benn Torres et al. 2007	LBA	Benn Torres et al. 2015
Native American	A	0.10	0.08	0.10	0.44
	C	0.57	0.04	0.33	0.17
	D		0.16		
African	L0		0.12		
	L1	0.02	0.44	0.03	0.11
	L2	0.14	0.12	0.47	0.06
	L3	0.16	0.04	0.07	
Unclassified		0.02			

Parental Pops.	Dominica		Garifuna - St. Vincent	
	LBA	Torres et al. 2007	LBA	Benn Torres et al. 2015
Native American	60%	28%	43.3%	38.2%
African	40%	72%	56.7%	61.8%

(Torres et al. 2007; Torres et al. 2015; Laboratory of Biological Anthropology (LBA) study, 2019). Native American maternal ancestry included sequences that belonged to haplogroups A, B, and D. African ancestry was indicated by haplogroups L0, L1, L2, and L3

genetic contributions from Native Caribbeans. The use of ancestry informative markers (AIM) instead of mtDNA by Benn-Torres et al. (2013) gave lower Native American estimates for Dominica (16.2%) and St. Vincent (6.5%).

Evolutionary Success of Migrants

Under favorable environmental conditions and little competition for food resources, the Garifuna expanded rapidly from two initial villages in Honduras to 54 communities distributed from Belize to Nicaragua. During the first few generations of Garifuna expansion, an exceptional fertility level was reached—10.9 children per childbearing woman at the completion of her reproductive career (Brennan 1983). Firschein (1961) found that females heterozygous for sickle cell trait (Hb AS) had an advantage of 1.45 to 1 when compared to the fertility of Hb AS homozygotes in an environment with endemic malaria. He posited that the differential fertility of females with Hb AS contributed to the genetic success of the colonizing Garifuna populations. While the high fertility estimate of 10.9 children per childbearing woman is probably skewed by the effects of local admixture fusion, the rapid expansion of the Garifuna signals a favorable environment for settlement.

Selection and Malaria

Malaria was introduced into the Leeward Islands of the Caribbean and Central America by African slaves brought to work the mines and plantations. There is no evolutionary evidence that Native Americans had a history of malarial infection. None of the genetic markers associated with genetic adaptation to malaria (hemoglobinopathies S and C, Duffy null, G-6-PD deficiency) have been detected in pre-Columbian or contemporary populations. On the other hand, Central American populations of African origins or those having experienced gene flow from slave populations have polymorphic genetic markers that are associated with resistance to *Plasmodium falciparum* and *Plasmodium vivax*.

There is considerable evidence that the coast of Central America had a relatively high incidence of malaria. Custodio and Huntsman (1984) observed in 1973–1978 that Honduras had a high incidence of malarial infections: "These people have remained until recently under heavy malarial pressure both from *Plasmodium vivax* and *Plasmodium falciparum*" (Custodio and Huntsman 1984, 335).

The incidence of malarial infections varied from 7.2 to 37.5 positives for *P. vivax* and 0.02 to 0.86 positives for *P. falciparum*.

Firschein (1961) sampled Garifuna from Stann Creek, Hopkins, and Seine Bight, Belize, and found a high incidence of Hb AS: 0.24. However, Indigenous Native American populations had no evolutionary experience with malaria until European/African contact. Because of their evolutionary past, Native Americans did not exhibit any polymorphisms that would provide protection against any malarial species. As a result, once the Central American mosquitoes were infected by malarial organisms, the indigenous populations experienced endemic malaria and dramatic population decline. Some of the coastal Mayan populations relocated to the highlands to avoid malarial infection.

Because of their African ancestry, the Garifuna brought with them genes that gave resistance to malaria. For example, Custodio and Huntsman (1984) found that the incidence of Hb AS varied from 6.1% to 18.9% (Seine Bight) in five populations of Garifuna in Honduras and from 14.6% to 18.9% in the Garifuna of Belize. Crawford et al. (1981, 1984) found the Hb^s allele at a lower frequency of 7% to 8% in coastal populations of Guatemala and Belize. Similarly, Duffy null (Fy allele ranges from 78% to 93% in Garifuna populations), was brought to Central America by the transplanted Garifuna. In contrast, the Duffy null allele occurs at low frequency (21%) in populations on Dominica, indicating low African admixture (Harvey et al. 1969).

Demographic and Social Consequences

The Garifuna population of Central America increased from fewer than 2,000 persons in 1800 to currently more than 300,000. This increase was due to exceptional fertility rates, with estimates as high as 10.9 children per woman by the completion of her reproductive career, and admixture with other local groups. The Garifuna communities also experienced a shift from patrilocal to matrilocal households (Gonzales 1969). This shift in household composition was necessitated by economic factors, namely temporary jobs being available in distant regions, such as lumbering in Belize, which drew many males away from households in other regions of the Caribbean. These economic conditions resulted in households primarily composed of females—i.e., mothers and grandmothers—as well as children (plus visiting males) for prolonged periods of time. This pattern meant resources for the families were provided by male household members located in various geographic regions. As new employment became available, the males of the households relocated for these positions, while new males replaced earlier family residents. One result of this consecutive monogamy was a population with a high frequency of matrilocal households containing numerous half sibs who had the same mothers but different fathers.

Conclusion

The causes and effects of migration are complex and result from the interactions of the genetic structure of the parental populations, the cultures of the ancestral groups, unique historical events, and the actions of the forces of evolution. The Garifuna migrations and relocations were influenced by interethnic marriages of Garifuna and Creole populations and the introduction of European genes into the Garifuna populations. The establishment of this triracially admixed population resulted in a high level of genetic diversity. The presence of hemoglobinopathies at relatively high frequencies gave the migrants an evolutionary advantage over Native American populations, and this contributed to survivorship, high fertility, and a massive geographical expansion of the Garifuna along the coast of Central America.

The result of this confluence of historical events can be viewed as a unique evolutionary success story that is paralleled by the patterns of human migration out of Africa and occupation of most of the world by African migrants.

Acknowledgments

This research was supported in part by the following grants: PHS Career Development Award, K04 DE028-05; NIDR #DE04115-02; Biomedical Sciences

Research Grants #4349-5706, 4309-5706 and 4932-x706-4; and KU Faculty Senate Research Grant #3507-x038.

References

Beaty, K. 2017. "Forced Migration and Population Expansion: The Genetic Story of the Garifuna." PhD diss., University of Kansas.

Benn-Torres, J., R. A. Kittles, and A. C. Stone. 2007. "Mitochondrial and Y chromosome diversity in the English-Speaking Caribbean." *Annals of Human Genetics* 71, no. 6: 782–790.

Benn-Torres, J., A. C. Stone, and R. Kittles. 2013. "An anthropological genetic perspective on Creolizaion in the Anglophone Caribbean." *American Journal of Physical Anthropology* 15: 135–143.

Benn-Torres, J. , M. G. Vilar, G. A. Torres, J. B. Gaieski, R. B. Hernandez, Z. E. Browne, . . . The Genographic Consortium. 2015. "Genetic diversity in the Lesser Antilles and its implications for the settlement of the Caribbean Basin." *PLOS One* 10, no. 10: e0139192.

Brennan, E. 1983. "Factors underlying decreasing fertility among the Garifuna of Honduras." *American Journal of Physical Anthropology* 60: 177.

Crawford, M. H. 1983. "The anthropological genetics of the Black Caribs (Garifuna) of Central America and the Caribbean." *Yearbook of Physical Anthropology* 26: 161–192.

Crawford, M. H. 1984. "Problems and Hypotheses: An Introduction." In *Current Developments in Anthropological Genetics: Volume 3, Black Caribs—A Case Study in Biocultural Adaptation,* edited by M. H. Crawford, 1–9. New York: Plenum Press.

Crawford, M. H., D. D. Dykes, K. Skradsky, and H. Polesky. 1984. "Blood Group, Serum Protein and Red Cell Enzyme Polymorphisms, and Admixture among the Black Caribs and Creoles of Central America and the Caribbean." In *Current Developments in Anthropological Genetics: Volume 3, Black Caribs—A Case Study in Biocultural Adaptation,* edited by M. H. Crawford, 303–333. New York: Plenum Press.

Crawford, M. H., N. L. Gonzalez, M. S. Schanfield, D. D. Dykes, K. Skradski, and H. F. Polesky. 1981. "The Black Caribs (Garifuna) of Livingston, Guatemala: Genetic markers and admixture estimates." *Human Biology* 53, no. 1: 87–103.

Custodio, R., and R. Huntsman. 1984. "Abnormal Hemoglobins among the Black Caribs." In *Current Developments in Anthropological Genetics: Volume 3, Black Caribs—A Case Study in Biocultural Adaptation,* edited by M. H. Crawford, 335–343. New York: Plenum Press.

Custodio, R., R. Huntsman, R. Newton, D. Tills, H. Weymes, A. Warlo, . . . J. Lord. 1984. "Blood Group, Hemoglobin, and Plasma Protein Polymorphisms in Black Carib Populations." In *Current Developments in Anthropological Genetics: Volume 3, Black Caribs—A Case Study in Biocultural Adaptation,* edited by M. H. Crawford, 289–301. New York: Plenum Press.

Davidson, W. V. 1984. "The Garifuna in Central America: Ethnohistorical and Geographical Foundations." In *Current Developments in Anthropological Genetics: Volume 3, Black Caribs—A Case Study in Biocultural Adaptation,* edited by M. H. Crawford, 13–35. New York: Plenum Press.

Davis, D. D., and C. Goodwin. 1990. "Island Carib Origins: Evidence and Nonevidence." *American Antiquity* 55, no. 1: 37–48.

Devor, E. J., M. H. Crawford, and V. Bach-Enciso. 1984. "Genetic Population Structure of the Black Caribs and Creoles." In *Current Developments in Anthropological Genetics: Volume 3, Black Caribs—A Case Study in Biocultural Adaptation,* edited by M. H. Crawford, 365–380. New York: Plenum Press.

Firschein, L. I. 1961. "Population dynamics of the sickle-cell trait in the Black Caribs of British Honduras, Central America." *American Journal of Human Genetics* 13, no. 2: 233–254.

Fitzpatrick, S. W. 2011. "Verification of an Archaic Age occupation on Barbados, Southern Lesser Antilles." *Radiocarbon* 53: 595–604.

Gonzalez, N. L. S. 1969. *Black Carib Household Structure.* Seattle: University of Washington Press.

Gonzalez, N. L. S. 1988. *Sojourners of the Caribbean: Ethnogenesis and Ethnohistory of the Garifuna.* Urbana: University of Illinois Press.

Gullick, C. J. 1984. "The Changing Vincentian Carib Population." In *Current Developments in Anthropological Genetics: Volume 3, Black Caribs—A Case Study in Biocultural Adaptation,* edited by M. H. Crawford, 37–50. New York: Plenum Press.

Harvey, R. G., M. J. Godber, A. C. Kopec, A. E. Mourant, and D. Tills. 1969. "Frequency of genetic traits in the Caribs of Dominica." *Human Biology* 41: 342–364.

Herrera-Paz, E. F., M. Mattamoros, and A. Carracedo. 2010. "The Garifuna (Black Carib) people of the Atlantic Coasts of Honduras: Population dynamics, structure, and phylogenetic relations inferred from genetic data, migration matrices and isonymy," *American Journal of Human Biology* 22: 36–44.

Monsalve, M. V., and E. Hagelberg. 1997. "Mitochondrial DNA polymorphisms in Carib people of Belize." *Proceedings Royal Society London B* 264: 1217–1224.

Phillips-Krawczak, C. 2012. "Origins and Genetic Structure of the Garifuna Population of Central America." PhD diss., University of Kansas.

Rubicz, R., P. Melton, and M. H. Crawford. 2007. "Molecular Markers in Anthropological Genetics." In *Anthropological Genetics: Theory, Methods and Applications,* edited by M. H. Crawford, 141–186. Cambridge: Cambridge University Press.

Salas, A., M. Richards, M-V. Lareu, et al. 2006. "Shipwrecks and founder effects: Divergent demographic histories reflected in Caribbean mtDNA." *American Journal of Physical Anthropology* 128: 855–860.

Schanfield, M. S., R. Brown, and M. H. Crawford 1984. "Immunoglobulin Allotypes in the Black Caribs and Creoles of Belize and St. Vincent." In *Current Developments in Anthropological Genetics: Volume 3, Black Caribs—A Case Study in Biocultural Adaptation,* edited by M. H. Crawford, 345–363. New York: Plenum Press.

Taylor, C. 2012. *The Black Carib Wars: Freedom, Survival and the Making of the Garifuna.* Oxford: Signal/Books.

PART IV
CULTURE AND MIGRATION

13

Out of Africa, Again

Leaving Africa; or Why Do African Youth Migrate?

Abdelmajid Hannoum

Introduction

Migration is part of the human condition. Humans have always migrated, for a variety of reasons. In modern history, European colonialism was by far the largest human migration. More than two thirds of the globe were ruled by Europeans, and huge numbers of Europeans migrated to settle and work in parts of Asia, Africa, Australia, and the Americas. But rarely was this movement of people termed "migration." Then, in the postcolonial era, migration from south to north increased significantly due to the dire need for labor in Western countries, especially after the massive destruction of World War II, and in part due to the dependency of the former colonies on their colonizers. The deep fascination with the dominant that is often felt by the dominated cannot be underestimated. This migration seems to have intensified since the 1990s, with the implementation of neoliberal policies that made moving easier, necessary, and more complicated all at the same time. Neoliberal policies consisted mainly of state policies dictated to the state by the market and more specifically by corporations.

The new migration is not only massive, but also draws youth willing to pay the ultimate cost to make it to Europe. On the shores of the southern Mediterranean, a new tragic figure emerged for the first time, the *harrag*, the North African child, an adventurer who inspires both admiration and fear (Hannoum 2009). This youth is willing to die, and many of his ilk do, but nothing deters him. He continues to gamble on his life: *rabha walla dabha* ("make it or die"). The case of the North African harrag is not unique. A multitude of youth from sub-Saharan Africa head daily to the shores of the Mediterranean, often in groups, or rather in teams, to cross the Mediterranean in hope of finding a new life. It is a dream, a dangerous dream, one worth sacrificing everything for, including life itself. (It should be noted that Africans not only migrate to Europe, but also migrate within Africa itself, although intracontinental migration poses different challenges and obeys different rules that are not within the scope of this chapter.) This chapter deals with migration to Europe mainly because it speaks to larger issues, and also because it has a characteristic that seems to be new: the willingness of today's African youth to face

Abdelmajid Hannoum, *Out of Africa, Again* In: *Human Migration*. Edited by: Maria de Lourdes Muñoz-Moreno and Michael H. Crawford, Oxford University Press. © Oxford University Press 2021. DOI: 10.1093/oso/9780190945961.003.0014 tremendous danger, including a high risk of death, to migrate for the dream of a better life.

Causes of African Migration

I conducted fieldwork with young Africans from 2008 continuously until the summer of 2016. My ethnographic site was the city of Tangier, but I also conducted interviews in several other Moroccan cities, especially Larache, Rabat, and Meknés. Migrants are concentrated in the major Moroccan cities, especially the coastal ones. I interviewed Moroccan youth, including a large number of underage Moroccan children between the ages of 6 and 16. I also conducted interviews with sub-Saharan Africans almost exclusively from West Africa. The questions I posed attempted to elicit information: Why do African youths migrate? Why do they withstand the great danger and high risk associated with the endeavor? What makes a young man risk his life rather than stay safe at home?

Despite the images of death associated with crossing the Mediterranean, during my interviews, death was rarely mentioned as part of the risk calculation. Instead, the young migrants expressed a shared belief that death is a destiny awaiting us all, and its timing is not up to us. Common wisdom has it that the cause of a large part of African youth migration is poverty. While poverty cannot be excluded as a factor, the concept itself should be problematized to gain a more nuanced view of the causes of migration. Second, poverty's cure is not necessarily migration, even for the poor person himself. Undertaking the enterprise of migration is not cheap. The poorest simply cannot afford the associated costs, which usually amount to about $7,000, a small fortune that a poor person cannot save in an African country. A smuggler's fee alone is 1,200 euros, the price of a visa is 120 euros, and the cost of everyday existence for a period of 6 months can exceed $4,000. However, these expenditures can be explained by the fact that the enterprise of migrating is rarely an individual one. Usually, it involves the entire family, who harness their resources to help one member (almost always a young male) to undertake the enterprise (Piot 2010, 85; Vigh 2006, 106). And so African migration is not, it turns out, as much a singular and desperate attempt at getting out of poverty as it is a family enterprise, an attempt to invest collective resources in something that will benefit the entire family. A family member living abroad in a Western country is undoubtedly a source of income, prestige, and pride. Living abroad is a status-changer in and of itself.

However, it should be mentioned that although migration is often a family enterprise, individual considerations are also behind the motivation to move. Europe, especially in the age of globalization, exerts a tyrannical power over the imaginations of youth everywhere. Images of modern cities, with their hyper technology, luxurious goods, and attractive lifestyles, offer impetus to undertake the enterprise. These images are not only digital but also derived from real-life contact with those who "made it" and who often exaggerate a life of success—or may even invent one. This invention is often backed by real material signs—cars, clothes, fancy cell phones, money—and symbolic ones, a plethora of stories about life there, its beauty and bounty, its pleasures and comfort, its women, and, occasionally, in the case of women migrants, its men (although, as has been noted, migration is mostly undertaken by males; Hannoum 2019; Lahlou 2005). In other words, in the minds of the youth, there is an entire world called Europe, made up of stories and images. The abundance of images is not colonial, but is instead the product of globalization, where images are produced and spread almost at the speed of light. There is, then, the urge to leave the ordinary for the extraordinary, the prosaic for the magical, to travel from Earth to Heaven, from Africa to Europe.

There is also the human need to move, the human urge to refuse being stuck. The need to move may be more human than the need to feel rooted. It speaks to the human need for freedom. The folkloric theme of the hero who leaves and comes back prosperous and triumphant is a universal tale that the cinema has explored with wonderful effects. The desire, or rather the urge, to move is also a specifically youthful desire—to experience the world fully, to experience it firsthand, and to live it body and soul. Youth migration is thus a fight against the feeling of being stuck, of being "unfree," especially in the presence of a whole other world out there, where success is easy and where life is larger and better—or so it seems from the other side. This fight is, after all, a fight for life. Often I was told during my interviews with Moroccan harraga, including the very young: "I want to live my life," "I want to see that world." Moroccan youth told me frequently, "This is not a life, it is as if we are dead here" (Hannoum 2009, 2019). The idea of leaving is about freedom— the freedom to see the world, and also the freedom to come back. I have never encountered a young migrant who just wants to go there, to Europe, and not return. They do not want to go there to get lost, they want to go there so that they can come back, with a car, money, and clothes, and be able to help their families have better lives. In other words, they express the need to be free and the need to be rooted. In Morocco, they have families, they have "homes."

Neoliberalism and Migration

The postcolonial condition is similar to the one that prevailed during the colonial period, when Europe used Africa as a "reservoir of men" supplying labor and manpower to Europe. However, since the early 1990s, African migration to Europe has drastically increased.

Commonly called globalization or the age of globalization, this new era is undoubtedly marked by an intense migration from north to south, with the exact pathway often decided by the prior attachment of a nation-state to a colony, and the movement being most often in the direction of a former colony to its former metropole. North Africans are most likely to migrate to Spain or France, while Nigerians are most likely to think of the United Kingdom as an ideal destination. This new pattern of migration can first be explained by the old pattern from which it emerged, the one following African independence in the 1960s and 1970s, when Europe was still in dire need of labor to reconstruct its economy damaged by World War II. However, by the 1990s, migration, instead of being sought after and encouraged by Europe, began to be seen as a burden on its economy and, for political parties and the media, also as a threat to national identity. Later, after 9/11, migration was seen as a danger to national security. In fact, the new pattern of migration is characterized by what is commonly called "illegality"—meaning that the migration of youth is addressed by an array of laws that make it a crime, despite the fact that in the United Nation's Universal Declaration of Human Rights, "Everyone has the right to leave any county, including his own, and to return to his country" (1948). But this right is at odds with the right of states to control who enters and who leaves.

Be that as it may, illegality itself increased the risk and danger for migrants. It forced the would-be migrant to find ways and to deploy strategies to circumvent the law, and in so doing he is labeled illegal. Indeed, the number of deaths reported in the Mediterranean every year is remarkably high, and yet this figure does not account for the deaths in the desert, which are rarely reported by the media despite their frequency in the narratives of migrant interlocutors. In any case, the new pattern of migration is also marked by the ways the European Union responds to it. In addition to the array of laws against it, drastic measures have been firmly put in place to halt migration: drastic laws that criminalize migration and an entire structure of arrestability: "gates, corridors, hallways, zones, stop signs, detention facilities, and so forth" (Van Reekum and Schinkel 2016, 38).

In developing countries, now called the Global South, consisting mainly of countries that were subjected to colonial rule, where the educational system is prey to privatization, and where the public schools increasingly produce unemployed graduates, life elsewhere is shown, especially on social media, to be better and more fulfilling. Stories of success told by returning migrants confirm this. But these are also fed by narratives (often exaggerated or even made up) that contribute to shaping the imagination. Migration has offered new possibilities of life; it has opened new horizons and has transformed lives—for better and for worse. Rarely evoked in the discourse of globalization are the tragic aspects of migration: rampant xenophobia; endemic racism; social exclusion; social deviance; criminality, itself produced by state laws or by social exclusion of, violence against, and occasionally murder of, migrants; and intense symbolic violence by political parties and the media. The topic of migration has become an important political card in the hands of political parties willing to mobilize a population taught that migration constitutes a direct danger. The media and political discourse often hold migration responsible for unemployment and for crimes and insecurity, even terrorism, despite a dearth of evidence.[1]

---

[1] One needs to make the distinction between a migrant and a terrorist disguised as a migrant.

The intensity of migration cannot be explained without the new order of neo-liberalism. This order of globalization is often defined by the speed and amount of human flows, despite the existence also of sequestration, as evident in thousands of African migrants stuck in northern African coastal cities. Thus, sequestration and being stuck are also marks of the global age. At the peak of the discourse on globalization, we also find an unprecedented building of walls and/or fences (and/or the promise to build them) especially, though not exclusively, around the so-called Western world—at the edge of Europe, at the very edge of Africa, and at the intersection between the northern and southern Americas. In North Africa, the sea that separates the continent from Europe was no longer a sufficient barrier; the European Union also built fences in the early 1990s around the Spanish enclaves of Ceuta and Mellila to discourage migration (Anderson 2014; Gold 2000).

Numbers and the Geography of Migration

Migrations are not random or accidental, they "do not just involve any possible combination of countries, they are patterned" (Sassen 1998, 56). Generally speaking, young migrants from western Africa take the road via Mali and Algeria, often to Morocco and also, to a lesser extent, to Tunisia. In Morocco, the journey is taken either from the city of Tangier via boats (private or of smugglers) or from Ceuta and Melilla by climbing a 6-meter fence. West Africans may also choose to take the Tunisian road, especially given the proximity of the Italian islands to the Tunisian shores. This road remains, for practical reasons, the preference of migrants from central Africa and even migrants from eastern Africa. Until 2011, Libya was also a highly attractive transit point for central and eastern Africans. The fall of the Gaddafi regime turned the country into a war zone that remains extremely dangerous. But even this degree of danger does not entirely deter migrants from trying their luck—in fact, it offers something of an "opportunity." Unlike in the time of Gaddafi, who was helping the European Union to halt migration, in the absence of the state, the route became lawless and unguarded, despite, or because of, the danger of war. The absence of the state meant to thousands of African migrants the absence of the police, and thus freedom to move and cross. The dangers that migrants encounter in stateless Libya are the same they encounter in the desert: gangs seeking to rob them, but not police wanting them to go back. Yet, Libya is only one route for migrants, especially for migrants from central and eastern Africa. There are two other main routes. The second route is from West Africa through Morocco to Spain; the third is from central Africa to Algeria—but this one is less frequented and constitutes a road for Algerian youth to migrate to France. Algeria is the main point of departure for Algerian youth themselves, through either regular or irregular means. Libya and Morocco remain the main roads for African migration.

In the map in Figure 13.1, one can see the existence of several routes of African migration to Europe. Again, they may be summarized as: from west and central

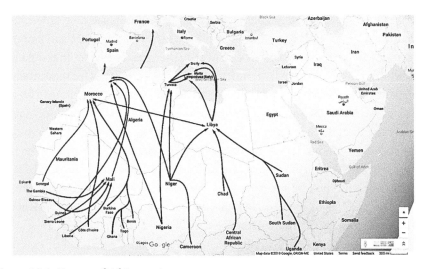

Figure 13.1   Routes of African migrants.

Source: Map courtesy Google Earth. Routes drawn by Randy David based on https://www.unhcr.org/
5aa78775c.pdf

Africa to Morocco and from Morocco to Spain; from central Africa to Tunisia
and from Tunisia to Italy; and from central Africa and eastern Africa to Libya,
and from Libya to Italy. Neither Spain nor Italy is necessarily the final destination
of the migrants. The first route, to Morocco, is the most popular, especially for
West Africans. In Morocco, the road bifurcates: one takes the migrants to Tangier
and from Tangier either to the Spanish cities of Tarifa or Algeciras; the second
takes the migrants to the enclave of Melilla, which, though situated in northern
Africa, is considered a Spanish territory, and thus part of the European Union.
It is difficult to say which one of these two routes is the most common. Melilla
has attracted much media attention because migrants are seen in front of or on
top of a 6-meter fence created by the European Union. The road from Tangier is
more secretive and its activities happen at night. The road from central Africa to
Tunisia is also a favorite for migrants who choose Italy. The third road has also become more common (although also more dangerous) since the fall of the Gaddafi
regime in 2011.

The cartography of migration can be best understood numerically. The numbers are undoubtedly alarming for the European Union. It is hoped the numbers
will decrease with the increase of strict and draconian laws and drastic measures
to curb the flow, or rather to stop it at the gate of Europe, on the other side of the
Mediterranean shores, inside the countries of the Maghreb. But, as the numbers in
Figure 13.2 show, these measures have not stopped the flow of migrants. Despite
the EU's efforts, the number of crossers continued to increase. With the increase of
crossers, there is an increase in the death toll. Thus, at the peak of draconian measures in 2015, reports mentioned 1,015,078 crossers with the death toll just at sea
estimated at 3,771. This number was higher than the death toll of the previous year,

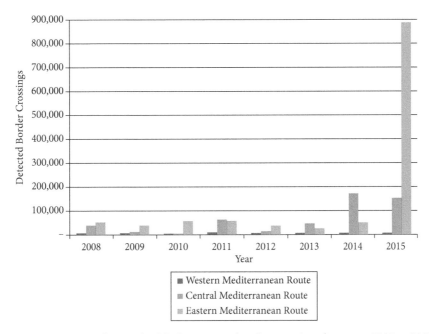

Figure 13.2 Detected irregular Mediterranean border crossings by route, 2008 to 2015.

Source: Originally published in Caitlin Katsiaficas, "Asylum Seeker and Migrant Flows in the Mediterranean Adapt Rapidly to Changing Conditions," Migration Information Source, June 22, 2016. Reprinted with permission.

which was 3,538 deaths out of 216,054 crossers, which made the Mediterranean "the deadliest road"[1] for migrants. The peak in 2015 reflects the intensity of the wars in Syria and Libya, which created a flood of migrants and refugees (even from North Africa), and migrants disguised as (Syrian) refugees in many cases.

Figure 13.2 shows the number of migrants from all routes to Europe. The western Mediterranean route includes not only migrants from central and especially eastern Africa but also migrants from Syria and Turkey. One can see that the number of migrants was ~ 100,000 in 2008 and had greatly increased by 2015. The total number of migrants from all routes remained less than 100,000 until 2013, but it increased in 2014 due to migrants who took the central Mediterranean route, and then in 2015 migration skyrocketed the eastern Mediterranean route, mainly because of the intensity of the war in Syria and the rise of ISIS that forced millions to flee their homes and find refuge in Europe. However, African migrant flows taking the central and western routes also increased, exceeding 100,000 in 2014 and 2015.

Using data drawn from the UNHCR (2014, 2015, 2016, 2017, 2018), Figure 13.3 shows the numbers of migrant arrivals and the numbers of the dead and missing. One can see that, in 2017, 172,301 migrants crossed and 3,139 died. In 2018, there were 138,882 crossers and there were 2,275 deaths. Although 2014 witnessed the crossing of a reported 216,054 persons and a death toll of 3,538, the percentage of deaths was smaller than in 2017 and 2018, when the percentage of deaths increased

YEARS	ARRIVALS	DEAD AND MISSING
2014	216054	3538
2015	1015078	3771
2016	362753	5096
2017	172301	3139
2018	138882	2275

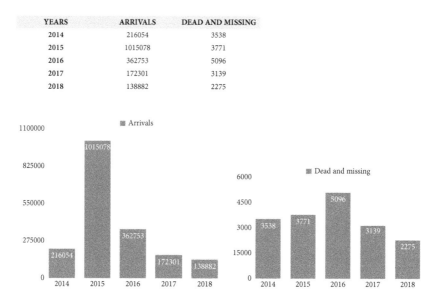

Figure 13.3 Statistics for migrant crossings, arrivals, and dead/missing.

Source: Data drawn from UNHCR, 2014, 2015, 2016, 2017, 2018. Courtesy Nejat Brahmi.

to 1.82% and 1.62%, respectively, compared to 1.64% in 2014. How to explain the increased death rate despite a decrease in the number of crossers? Luck, or rather bad luck, by itself cannot explain the death toll. While the weather and the conditions of the sea are factors, there are also the risks taken by the migrants. The migrants know that there is a high season and low season for crossing. Summer is the high season, when police patrols in the sea are also at their highest, but the view is clear and movements in the sea are frequent. Winter, by contrast, is when police patrols are at the lowest, because patrols are more difficult in the fog. When migrants calculate the risk of crossing, they are more likely to choose winter, when the vigilance of the border police is much reduced, and when the bad weather can be as much a friend as a deadly enemy.

Consequences of Migration

As a postcolonial phenomenon that marks the age of globalization, migration has consequences for Africa as well as for Europe. The new pattern of migration from Africa to Europe has undoubtedly caused important changes both in North African societies, where many migrants get stuck for years, and maybe even forever, and in Europe, where many migrants arrive.

For North African societies, the consequences are both societal and political. Migration, which is seen as a huge challenge for the EU, has become a diplomatic card in the states of the Global South. In many diplomatic meetings, the issue of "cooperation" between the EU and this or that state of North Africa is often at the top of the agenda. Migration has become the issue around which states negotiate and speak in terms of cooperation and challenge.

The cooperation (especially sought from the South by the North) does not happen, and cannot happen, without demands from the South on financial and political registers. In 2010, Gaddafi was able to obtain an apology and a reparation from Italy in exchange for securing the borders between Libya and Italy, then a main migration gate (Reuters, August 30, 2008). In 2019, Morocco was able to secure a promise of 110 million euros from the EU in exchange for "fighting illegal migration". However, these same demands put these countries at odds with the states in which the migrants originate, and especially with the African Union. The three countries of the Maghreb are expected to cooperate in ways that curb, if not halt, the migratory movements. This cooperation is often seen as the racial targeting of young sub-Saharan Africans, whose only crime is their desire to make a new life. Yet, turning a blind eye, especially in Tunisia and to a great extent in Morocco, does not provide a sustainable answer. For instance, to ward off the accusation of racism, the Moroccan state granted amnesty and papers to sub-Saharan African migrants on three occasions. However, these diplomatic gestures are of little importance to the migrants, who often, and even with residency cards, are unable to find jobs in a country whose youth suffer from serious unemployment. The residency cards themselves do not protect the migrant populations against the daily racism, social exclusion, and occasional violence they face in North African societies. Thus, their increased presence has highlighted the issues of race and color in North African societies (Hannoum 2019).

For African countries, the migration of their youth, or rather the despair that causes the youth to migrate, underscores the inability of the state to control its population and its borders. This inability is itself the result of neoliberal policies that weakened states and strengthened markets. This weakness causes the state to sacrifice part of its youth, the one that is most energetic, entrepreneurial, and adventurous. Paradoxically, these countries could also benefit from incentives from the EU to curb migration, because they also benefit from the remittances of migrants who "make it" to Europe.

For Europe, the new patterns of migration have prompted the EU to take drastic measures that sometimes ignore its fundamental values of hospitality and humanitarianism. In addition to the long fence, or wall, as the migrants call it, the EU resorted to laws to criminalize migration, and put in place an entire structure of arrestability at odds with the history of Europe as a refuge for migrants and refugees. In sum, migration made Europe a fortress. Worse, the frontiers of the EU are no longer within Europe itself, but are extended beyond it, especially to North Africa, where an entire infrastructure of "measures of security" is in place. This has not been successful: despite the effort, migrants and refugees alike still make it to Europe and to the sites of camps—in Germany, France, Italy, and other European countries (see, for the example of Italy, Smith and Tondo 2018). One may also wonder whether the walls—the physical ones as well as the metaphorical ones, the old ones and the promised ones—are becoming the "nomos of the modern," to use Agamben's words (1996, 166). Beyond the walls, there are men and women who live permanently in states of exception, in war zones, in zones ravaged by poverty, and in dictatorial polities, "outside of the boundaries of humanity" (Agamben 1996). Looking at the causes of migration requires an examination of all the factors that made young people engage in this dangerous endeavor. Examining its consequences requires us also to take into account the historical evolution of colonial regimes to postcolonial, neoliberal ones that have shaped our current cultural and political order. The patterns of migration have historical roots that must be explored in order to reach a better understanding of how old patterns gave birth to new ones.

Conclusion

People's movement from Africa to Europe and from Europe to Africa has a long history and shows different patterns. In modern times, Africa witnessed an intense movement of Europeans to Africa, with some countries (such as Algeria and South Africa) populated by European migrants. The logic of this movement was the capitalist logic of colonization, searching for new land, new resources, and new human labor. But prior to colonization, Africa witnessed a forced migration of its population to Europe and to the New World in the slave trade (Grenouilleau 2004). During colonial times, it also witnessed quasi-forced migration to Europe, especially during World War I and World War II to supply soldiers and laborers to serve in the wars and to help with reconstruction (Hannoum 2013; Sassen 1999). In the post-independence era, that is, by the 1950s and 1960s, the migration from Africa to Europe continued for a variety of reasons, including the need for labor to reconstruct Europe after World War II and the dependency of African nations on Europe that resulted from the long European colonization. By the early 1990s, the new pattern of migration emerged that this chapter discusses.

This new pattern is marked by the fact that the migrant population consists, first, of a young population. It is also marked by what is usually called illegality, in the sense that the migration transgresses new rules about migration and mobility imposed by the market. It is also marked by an exceptionally tragic aspect, because it results in thousands of deaths every year. These factors are not specific to Africa and can be observed in different parts of the world where new borders have emerged in the last three decades or so. Scholars often speak about globalization as a condition where people and ideas move with ease and unprecedented speed. Migration, documented and undocumented, surely confirms that these movements are massive and swift, but it is important to note that the pace differs from population to population. African migratory movement is slow and fraught with danger, like its sister at the United States–Mexico border. Clearly, not everybody moves easily—and definitively not the migrants from the Global South, and this is not only at the Mediterranean border or the Mexican one (Albahari 2015; Anderson 2014; DeGenova 2005; Hannoum 2019). Globalization has produced the global migrant, who may be different from border to border, but is in many places the same: a young person in search of a new life, since the one left behind is either not livable or not worth living. And in many places this youth will be considered illegal, due to the same laws that regulate the movement of capital, and goods, and certain—but not all—people.

References

Albahari, Maurizio. 2015. *Crimes of Peace: Mediterranean Migrations at the World's Deadliest Border*. Philadelphia: University of Pennsylvania Press.

Andersson, Ruben. 2014. *Illegality, Inc*. Oakland: University of California Press.

Degenova, Nicholas. 2005. *Working the Boundaries: Race, Space, and "Illegality" in Mexican Chicago*. Durham: Duke University Press.

Gold, Peter. 2000. *Europe or Africa? A Contemporary Study of the Spanish North African Enclaves of Ceuta and Melilla*. Liverpool: Liverpool University Press.

Grenouilleau, Olivier. 2004. *Les traites négrières: Essai d'histoire globale [The slave trade: An essay on global history]*. Paris: Gallimard.

Hannoum, Abdelmajid. 2009. "The *harraga* of Tangiers." *Encounters* 1, no. 1: 231–246.

Hannoum, Abdelmajid. 2013. *DDR Before DDR: North African Veterans from Colonial Mobilization to National Integration*. Washington, DC: The World Bank.

Hannoum, Abdelmajid. 2019. *Living Tangier: Migration, Race, and Illegality in a Moroccan City*. Philadelphia: University of Pennsylvania Press.

Lahlou, Mehdi. 2005. "Les migrations irrégulières entre le Maghreb et l'Union européenne: évolutions récentes" (CARIM Research Report 2005/03). Florence: Institut Universitaire Europeen. http://cadmus.eui.eu/bitstream/handle/1814/6278/?sequence=1

Piot, Charles. 2010. *Nostalgia for the Future: West Africa After the Cold War*. Chicago: University of Chicago Press.

Sassen, Saskia. 1998. *Globalization and Its Discontents*. New York: New Press.

Sassen, Saskia. 1999. *Guests and Aliens*. New York: New Press.

Smith, Sean, and Lorenzo Tondo. 2018. "Shattered Dreams: Life in Italy's Migrant Camp—A Photo Essay." *Guardian*, October 10, 2018.

Vigh, Henrik. 2006. *Navigating Terrains of War: Youth and Soldiering in Guinea-Bissau*. Oxford: Berghahn Books.

14

A Sociogenetic Approach to Migration and Urbanization in Peruvian Amazonia

Implications for Population Architecture

Randy E. David and Bartholomew Dean

Introduction: Yurimaguas and the Huallaga River Valley

Both sociocultural and molecular biological research on the peopling of the New World has underscored the continents' complex patterns of human mobility, particularly in Beringia, the Great Plains, Mesoamerica, and the central Andes (Achilli et al. 2013; Crawford 1998; González-Martín et al. 2015; Harris et al. 2018; O'Rourke, Hayes, and Carlyle 2000). In contrast, significantly less attention has been accorded to assessing population movements in other key areas of the Americas, particularly Upper Amazonia (Alexiades 2009; Arias et al. 2018; de Jong, Lye, and Abe 2006; Díaz-Matallana et al. 2016; Gómez-Carballa 2018; Rothhammer et al. 2017). Limited scholarship has been dedicated to exploring the sociocultural, ecological, and molecular evolutionary consequences of migration and urbanization in this region. This absence of research is surprising given that urbanization has been a defining aspect of both colonial and postcolonial America. Sociogenetics is the field of research that examines how social processes affect genetic expression in an integrated fashion. While clearly consequential in terms of natural resource management, urban planning, public health initiatives, economic development, and social inclusion, the sociogenetic dynamics of contemporary Amazonian population mobility and urbanization have remained overlooked.

Given its rich archaeological and historical record, Peruvian Amazonia offers an exceptional opportunity for the study of the *longue durée* of human migration, as well as useful comparative information for understanding what V. G. Childe (1950) termed the "urban revolution"—arguably the most consequential social disruption known to humanity (Lefebvre 2014). Although Peru is most often emplaced in popular Western imaginaries as bucolic, more than three quarters (76%) of Peru's population reside in urban areas (INEI 2007). This includes historically important urban centers like Yurimaguas, the focus of this study. Although Amazonia comprises nearly 60% of Peru's national territory, most scholarship on internal migration has

Randy E. David and Bartholomew Dean, *A Sociogenetic Approach to Migration and Urbanization in Peruvian Amazonia* In: *Human Migration*. Edited by: Maria de Lourdes Muñoz-Moreno and Michael H. Crawford, Oxford University Press. © Oxford University Press 2021. DOI: 10.1093/oso/9780190945961.003.0015 emphasized population movements in the Andean highlands (*sierra*) and along the Pacific coast (*costa*).

Notwithstanding its status as terra incognita, the *selva baja* (80–400 m above sea level), or the lowland forested region of Peru's Huallaga Valley, is particularly important for studies on human migration and neotropical urbanization. Unlike other regions of Spanish colonial dominion, the area under consideration drew relatively scant attention—even by national elites, foreign explorers, and would-be settlers, all eager to establish footholds in the broader Huallaga Valley region. The *selva baja* is significant precisely because it flanks the transition zone between the Andean Cordillera and the Amazon Basin, a region critically important for understanding South America's history of peopling.

A geographically consequential crossroads of migration between the tropical Andean intermontane valleys and plateaus of the *selva alta* (401–1,000 m above sea level) and the *selva baja*, the Huallaga Valley is characterized by its degree of ethnic diversity among both indigenous and immigrant populations. Study of the Huallaga Valley draws attention to the complexity and heterogeneity of contemporary Amazonian society, not to mention the importance of drawing from an interdisciplinary repertoire of research methods and theories. Scholars have described Yurimaguas as a patchwork of ethnic diversity comprised of regional Amerindian and mestizo communities as well as foreign migrants. Indigenous peoples of the Huallaga Valley include Quechua-speaking populations (Kichwa Lamista, Kichwa del Pastaza); Jivaroan speakers (Achuar [Shiwiar, Jivaro], Aguaruna [Awajún], Kandozi, and Wampis); Tupi-Guaraní speakers (Kukama-Kukamira [Cocama-Cocamilla]); Cahuapanan speakers (Chayahuita [Shawi] and Jebero); Arawakan speakers (Chamicuro); and speakers of the linguistic isolate, Urarina (Kachá; Justice, Dean, and Crawford 2012). Following the Rubber Boom of the late 19th to early 20th centuries, Ashkenazi Jews, Portuguese, North Americans, Spaniards, Italians, British, Cantonese, Japanese, and others emigrated to the region (Cook 2016; Dean 2009; Harrison 1995; Takenaka 2004).

Oscillatory or cyclical migration is prevalent in Peruvian Amazonia. The population is constantly on the move—be it by canoe, horseback, motorcycle, bus, or automobile. The region's agrarian economy, extractive modes of production, and exploitative labor conditions have led to frequent regional population turnover. In the Huallaga Valley, individuals and families often leave their home communities for several weeks or months at a time to engage in various economic activities in other areas, especially in the regions of Loreto (*selva baja*) and San Martín (*selva alta*).

Causes of Migration

Environment and Political Economy

Migration enables people to effectively respond to dynamic ecosystems and complex agricultural cycles. Synchronized with seasonal fluctuations in levels of rivers and the migratory patterns of fauna, there is a constant flow of people and forest resources, which in turn sustains communal lifeways and individual livelihoods in the Huallaga Valley.

Recognized primarily for its commercial rather than political significance, contemporary Yurimaguas is comprised of a combination of indigenous peoples and the descendants of migrant populations that emigrated to the region during international economic boom phases related to rubber (~ 1880–1915), barbasco (~ 1940s) and petroleum (~ 1970–2000), as well as the demand for hides, cascarilla, chicle, cotton, and palm oil. Numerous illicit economies developed, attracting communities to participate in hardwood logging (Bennett et al. 2018; Dean 2002) and coca leaf cultivation (Dean 2013a, 2013b, 2015a). Many of these opportunities were the result of the emergence and consolidation of market economies in the region coupled with the Peruvian state's long-standing "civilizing projects" aimed at incorporating Amazonia and its autochthonous peoples into the "modern" nation-state (Dean 2019). The constant movement of people is vital to agrarian and extractive economies, especially given that scarcity of labor in the Huallaga Valley is a primary productive constraint.

Infrastructure

In the lowland transition region of the Huallaga Valley, human migration is fashioned by the interface of rivers and roadways. As river levels rise in the *selva baja* and riverine villages are inundated, people move seasonally to the *vareza* (upland areas), where they use roadways to access markets and urban-based social services (in Yurimaguas, Tarapoto, and Moyobamba). Linked to major Andean and coastal cities by road and to the rest of the Amazon Basin via the Huallaga and Paranapura Rivers, Yurimaguas has become a thriving population center in the *selva baja*. Over the past 50 years, migrants from rural Peru have relocated to Yurimaguas due to the development of novel transport modes (motorboat, bus, *motokar*, etc.) and infrastructure in the Lower Huallaga Valley.

Migration to Yurimaguas continues to burgeon as this once backwater expands its commercial influence in the region, with trade links to the largest Peruvian cities via *Carretera IIRSA Norte*, a modern transnational thoroughfare, and Moisés Benzaquén Rengifo airfield. The recent construction of a 110-m, 40-ton capacity suspension bridge now permits the passage of transoceanic ships to Puerto Yurimaguas—Nueva Reforma, a new international river port (Desarrollo Peruano 2014). Such infrastructural improvements have made Yurimaguas a major regional transshipment point for goods crossing the South American continent. New peri-urban communities (e.g., Pampa Hermosa, Nueva Alianza, and Pongo de Caynarachi) have developed along recently constructed roadways. ("Peri-urbanization" refers to those processes of expansive growth that produce hybrid landscapes with fragmented urban and rural features.) Infrastructural undertakings also lead to the influx of laborers, skilled and unskilled, who cause an increase in the need for municipal services and lead to greater economic activity.

Lifeways

Between 1961 and 1971, the urban population of Alto Amazonas Province (Loreto Region) grew from 8,057 to 17,624, with many abandoning their riverine and interfluvial communities to live in the city (INEI 1981). In 1981, the population of Yurimaguas District (Alto Amazonas Province) was 36,417, 63% of whom (~ 23,000) lived in an urban environment (Rhoades and Bidegaray 1987). By 2007, the population had nearly doubled, to 63,345, 77% of whom lived in an urban environment (INEI 2007). At this time, 26.5% of the Yurimaguas District population had arrived in only the past 2 years (INEI 2007). The Instituto Nacional de Estadística e Informática (INEI, National Institute of Statistics and Informatics) reported that "small urban centers" (e.g., Yurimaguas) went from comprising 19.8% of the total Peruvian populace in 1993 to 22.2% in 2007.

In addition to an increase in economic opportunities and greater transportation infrastructure, many peoples from surrounding rural regions emigrated to the more densely populated city of Yurimaguas due to family ties (chain migration) and greater access to social services, such as education and healthcare (Dean 2015b). In his influential essay "Urbanism as a Way of Life" (1938), sociologist Louis Wirth emphasized that urban life is characterized by detached and instrumental interactions that liberate people from the strictures of membership in communal groups, such as extended families. Albeit not without pitfalls (i.e., alienation, random violence, marginalization, toxicity), urban lifeways provide novel, exciting forms of self-actualization, access to more extensive social networks, opportunities for socioeconomic mobility, and greater freedom.

Violence and Social Upheaval

Many of the recent migrants to Yurimaguas have been displaced from their rural homes by neoliberal economic policy and the ravages of an internal armed conflict that took the lives of nearly 70,000 (1980 to present; Starn and La Serna 2019). Families that previously lived on agricultural plots (*chacras*) without formal land tenure were displaced by agroindustrial development and expanding narcoeconomies (Dean 2002, 2013a, 2013b, 2015a). To divide feudal estates, agrarian reforms implemented by the military-led government of President Juan Velasco (1968–1975) caused a rapid increase in rural-to-urban migration. The formation of a national indigenous rights movement led to social mobilization and the recognition of claims of indigenous peoples to collective territory (Dean and Levi 2003). This, in turn, made the city of Yurimaguas a center of political action for indigenous peoples residing in the region.

Data and Sample Collection

Interview-based data and 182 buccal DNA swabs were collected from individuals in two communities of Yurimaguas District, Alto Amazonas, Loreto, Peru. One collection site was Hospital Santa Gema de Yurimaguas, located in the central business district of Yurimaguas. The second sampling site was Hospital Santa Gema de Yurimaguas—Munichis, located in a small peri-urban community within Yurimaguas District but outside the borders of the city of Yurimaguas proper. Government hospital clinics were utilized as a means of data collection due to their familiarity, geographic centrality, and relatively inclusive philosophy toward all strata of the community. Moreover, such clinics provided greater hygiene and privacy than the potential alternatives.

Demographic data and migration histories were collected via standardized one-on-one, in-person interviews. "Migration" was defined as either permanent relocation or relocation for a period of ≥ 1 year for any reason other than an immediate existential threat. Interviews took approximately 30 minutes to complete, with responses recorded via predesigned survey sheets (replete with both structured response formats [producing categorical, ordinal, and interval data] and unstructured response formats [producing fully descriptive qualitative or numeric data]). Each study participant had buccal cells collected from their inner cheek using Isohelix™ (Harrietsham, U.K.) SK–1 swab kits. According to kit directions, swabs were transferred to pre-sterilized tubes and had an Isohelix™ "dri-capsule" added for long-term preservation. Only individuals over 18 years of age were asked to participate. Of the 182 sampled participants, 79 (43.3%) identified as men, 97 (53.3%) identified as women, and 6 (3.3%) chose to skip this question. The mean age of the sample population was 41.5 years (42.8 years for men and 40.7 years for women). Respondents included 99.4% Hispanophones, with other languages spoken including Chayahuita (3.43%), Kichwa (*Lamista* and *del Pastaza* variants; 2.86%), Kukama-Kukamira (1.14%), Jebero (0.57%), Italian (0.57%), and Portuguese (0.57%). Due to an interest in migratory processes, individuals were not kept from participating based on ancestral claims.

Genomic DNA was extracted and isolated, producing yields ranging from 4.20 ng/μL to 182.00 ng/μL. Forward and reverse primers were designed using Primer3 Input 0.4.0 software (Koressaar and Remm 2007) for the amplification of the rCRS mitochondrial control region, carried out in the Laboratories of Biological Anthropology, University of Kansas. PCR products were tested for successful amplification via electrophoresis, purified, and Sanger sequenced (Sanger et al. 1977) by Genewiz (South Plainfield, NJ). Sequences were edited using Gene Codes Corporation's (Ann Arbor, MI) Sequencher® 5.4.6 software. Mitochondrial haplogroup assignation was conducted using HaploGrep v.2.1.1 software (Kloss-Brandstätter et al. 2011; Weissensteiner et al. 2016) in conjunction with Phylotree Build 17 (van Oven 2015; van Oven and Kayser 2009) and confirmed using the EMMA algorithm (Röck et al. 2013). Bioinformatic calculations were performed using Arlequin version 3.5 (Excoffier and Lischer 2010).

Consequences

Corresponding to the shift from agrarian to industrial lifeways and the rural-to-urban flow of migration, there was a general movement of people from the *selva alta* (including the border between San Martín and Loreto Regions) to the *selva baja* city of Yurimaguas (Figure 14.1). This invariably is a consequence of growth in regional economic activities. Moreover, the rate of migration has been intensified by decreasing access to arable lands in the Andean highlands, resulting in systemic land grabbing and urban "invasions" in the Huallaga Valley in and around Yurimaguas. The majority of migrants arriving in Yurimaguas generally come from Amazonian (high- and low-altitude) regions, including San Martín, Loreto, Amazonas, and Ucayali, although noteworthy numbers of migrants also come from the Andean region of Cajamarca and the coastal region of Lima (including Lima Province). Note the migratory flow of peoples from coastal, Andean, and intermontane Peru to the lowland forested Huallaga Valley, a pattern of great precolonial importance for human movements, trade patterns, and communication networks. This historically important migratory route was most likely followed due to the pathway's lack of active volcanos, lower altitude Andean peaks, and the ameliorated angle of ascent of the Nazca tectonic plate, southward from the Panama isthmus (Justice et al. 2012).

In the study sample, the cities that account for the most migrants to Yurimaguas are Lagunas, Loreto (10 migrants, 89 km northeastward), Lamas, San Martín (5 migrants, 152 km west-southwestward), Santa Cruz de Succhabamba, Cajamarca (5 migrants, 741 km west-southwestward), Pucallpa, Ucayali (4 migrants, 820 km south-southeastward) and Lima (3 migrants, 1,098 km southwestward). Distances were calculated according to roadway travel, except for the city of Lagunas, which was calculated by direct distance due to a lack of roadways linking the city with Yurimaguas. Migrants traveled a mean of 194 km to Yurimaguas, and on average lived in two other locations before residing in Yurimaguas. In Loreto Region, where few roadways exist, many migrants come from riverine communities connected to Yurimaguas via rivers, such as the Huallaga, Paranapura, Marañon, and Amazon. Among study participants, there was a similar proportion of men and women who engaged in migration, 41.8% for men and 43.6% for women. Likewise, those who speak an autochthonous language (45.5%) migrated at a similar rate as those who do not speak an autochthonous language (42.6%). Migration commonly entails family units moving together. This is evident from the example of chain migration of the five kin who relocated from Santa Cruz de Succhabamba to Yurimaguas.

The study sample shows that migrants relocating *to* Yurimaguas came from more communities, and in lesser numbers per community, than those relocating *from* Yurimaguas. This trend supports a rural-to-urban pattern of migration, whereby individuals relocate in small numbers from pastoral communities to regional cities, such as Yurimaguas, and eventually aggregate in urban metropolises. Lima, the most populous city in Peru (pop. 10,554,000; United Nations 2018) had three migrants relocate *to* Yurimaguas, but it absorbed 24 migrants *from* Yurimaguas.

Figure 14.1 Choropleth of locales emigrated from (to Yurimaguas).

The community that provided the most migrants to Yurimaguas (10), Lagunas (pop. 12,827; INEI 2005), absorbed only two residents *from* Yurimaguas. Study results indicate that individuals from the peri-urban community of Munichis were more likely to travel further (856 km) to reach their migratory destination than those from Yurimaguas proper (585 km, $P = 0.047^*$). No study participants came to Yurimaguas from abroad, although two individuals from Yurimaguas relocated, one to Quito, Ecuador, and the other to Milan, Italy. The aforementioned migratory and urbanization patterns discernable from Figure 14.1 have genetic consequences in terms of human population architecture. Figures 14.2 and 14.3 and Tables 14.1 and 14.2 illustrate a number of these critical consequences.

A notable consequence of migration in the Peruvian Amazon is an increase in genetic diversity in the area's populations. Mitochondrial DNA (mtDNA), an

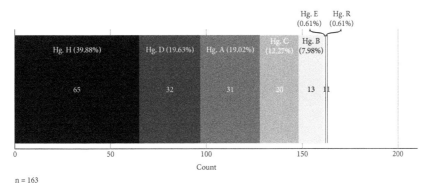

Figure 14.2 Mitochondrial haplogroup (Hg.) distribution.

ancestry-informing marker generally inherited maternally and not subject to re-combination, was assessed in study participants. Biodemographic algorithms were used to divide individuals into haplogroups (and subclades) according to mitochon-drial control region polymorphisms and their worldwide geographic associations.

Figure 14.2 depicts the haplogroup distribution of the Yurimaguas popula-tion. Haplogroup H comprises the plurality (39.88%), followed by D (19.63%), A (19.02%), C (12.27%), B (7.98%), and, last, E (0.61%) and R (0.61%). Of the 65 individuals assigned to haplogroup H, one possessed subclade H8a, seven pos-sessed subclade H1 (H1bs and H1cd) and the remaining 57 possessed subclade H2 (H2a2a1). Subclade H8a is associated with the Druzes of present-day Israel/Jordan/Lebanon/Syria (Shlush et al. 2008). H1 subclades arose in the Franco-Cantabrian region of southern France/northern Spain and are found in greater frequency among Basques than other regional populations (Behar et al. 2012a). Subclade H2a has been found at significant frequencies (i.e., ~ 1%–10%) among a number of mostly European populations, including Swedes, "Caucasian Americans," Basques, Austrians, Finns, Estonians, Eastern Slavs, and Turks (Álvarez-Iglesias et al. 2009; Behar et al. 2012b; Brotherton et al. 2014; Fendt et al. 2011; Greenspan 2005a, 2005b; Loogväli et al. 2004).

Subclades D1 and D4h3 are the only haplogroup D subclades indigenous to South America (Bodner et al. 2012; van Oven 2015). All haplogroup D participants in the study sample possessed subclade D1, except for two, who possessed subclade D4 (one, indigenous subclade D4h3, and the other could not be differentiated be-yond D4). All haplogroup A individuals possessed subclade A2, which is found al-most exclusively among indigenous American populations. Subclade A2 has been reported in isolated pockets of the Chukotka Autonomous Okrug and Kamchatka Krai of eastern Russia and has been interpreted as evidence of early migrations from northeastern Asia across Beringia (Hoffecker, Elias, and O'Rourke 2010; Volodko et al. 2008). Consequently, all A2 subclades were confirmed for indigenous American ancestry at the haplotype level.

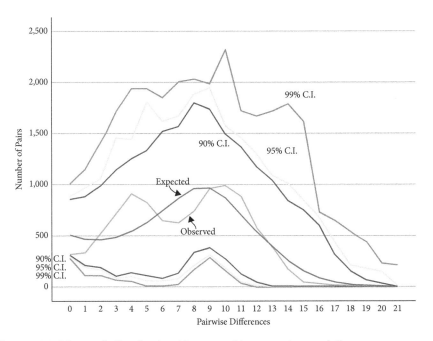

Figure 14.3 Mismatch distribution (demographic expansion model).

Among the 20 study participants assigned to haplogroup C, 10 possessed indigenous subclade C1b, three possessed indigenous subclade C1c, five possessed subclade C1 (could not be differentiated further), one possessed non-indigenous subclade C4b, and one possessed subclade C4 (could not be differentiated further; Fagundes et al. 2008; Hooshiar Kashani et al. 2012). All except one of the 13 study participants from the Yurimaguas sample assigned to haplogroup B possessed subclade B2, found exclusively in indigenous Americans (Just et al. 2008; Starikovskaya et al. 2005; Tabbada et al. 2010; Tamm et al. 2007). One participant was assigned to subclade B4a1c3, which is found at its greatest frequency in Japan (Bilal et al. 2008; Takenaka 2004). One participant was assigned to haplogroup E (subclade E1a1a), which is most closely associated with maritime Southeast Asia (Soares et al. 2008; Tabbada et al. 2010). Last, one participant was assigned to haplogroup R (subclade R30a1b), found at its greatest frequency in southern India (Palanichamy et al. 2004; Rani et al. 2010).

Ninety-five of the 163 individuals who underwent haplogroup assignation were of indigenous American maternal ancestry. The presence of all mitochondrial subclades native to South America (A2, B2, C1 [and in rare cases C4c], and D1 [and in rare cases D4h3]) speaks to precolonial and intracontinental migration. The presence and relatively high frequency of the H haplogroup is evidence of waves of European colonial and chain migration to the Huallaga Valley since the mid-16th century. Early conquistadors were commonly of Iberian, including Basque, origins (Dean and David 2017). Later, further European migration occurred, including the mass movement of Tyrolese (who possess a high frequency of mitochondrial haplogroup H) to the nearby Pozuzo Valley following the War of the Pacific (1879–1884; Barbieri et al. 2014). The discovery of Japanese and Southeast Asian ancestry is likely a product of subsequent migrations to Peru for socioeconomic opportunities by the Chinese (beginning in 1849) and Japanese (beginning in 1899; Lausent-Herrera 2011; Takenaka 2004).

Table 14.1 shows a comparison of various genetic diversity measures between the study sample and a number of populations that share a geographic (proximity favored), environmental (lowland Amazonia favored), and/or population size (urban favored over indigenous community) profile with Yurimaguas. Generally, genetic diversity has been found to increase with urbanization and decrease while moving eastward across the South American continent (Fuselli et al. 2003). Percentage of unique haplotypes is second greatest in the Yurimaguas sample, registering at 0.786 (78.6%). The only listed sample with a greater percentage of unique haplotypes (0.885 [88.5%]) contains only peri-urban *barriada* residents of Yurimaguas and is therefore a subpopulation (Justice et al. 2012). Thidiversity (*H*), 0.994 ± 0.002, followed closely by the Yurimaguas *barriada* subpopulation, at 0.989 (Justice et al. 2012). According to the mean number of pairwise differences metric (*Π*)), the Yurimaguas sample is the fifth most diverse (7.239 ± 3.411) of the 15 populations included in Table 14.1, after the Huancas (10.245 ± 4.872), Chachapoya (9.824 ± 4.513), Cajamarca population (8.832 ± 4.176), and Jivaro (8.014 ± 3.792). Like percentage of unique haplotypes, nucleotide diversity (π), 0.020 ± 0.011, is second greatest in the Yurimaguas sample, after only the Yurimaguas *barriada* subpopulation, at 0.022.

The diversity indices that Justice et al. (2012) reported within the Yurimaguas *barriada* subpopulation appear to be the most similar to those of our study sample listed in Table 14.1, lending credence to the findings of this study. Yurimaguas has an extensive pre- and postcolonial history of gene flow and is currently a significant cosmopolitan center in the region. Nearly half (6 of 14) of the comparative populations are isolated indigenous groups. Others are relatively exclusive cities (e.g., Lima, Peru; La Paz, Bolivia; Santiago, Chile; Iquique, Chile) that are often highly segregated by socioeconomic factors, potentially skewing the sample. Nevertheless, genetic diversity metrics corroborate the extensive genetic variation found in Yurimaguas, a product of the movement of peoples and gene flow.

Table 14.2 and Figure 14.3 assess population expansion in the Yurimaguas sample. Table 14.2 provides Tajima's *D* and Fu's F_S neutrality test calculations. Both values are negative, indicative of a population that is numerically expanding (or undergoing purifying selection). This is explained by a relatively high proportion of low-frequency polymorphisms in the population (Fu 1997; Tajima 1989). The significance of Tajima's *D* and Fu's F_S statistics were calculated via coalescent simulations under the infinite-site model (Hudson 1991; Nordborg 2003). Note that Fu's F_S is highly significant (*P* = 0.000***), whereas Tajima's *D* is not (*P* = 0.080). Chakraborty's test of amalgamation produced a nonsignificant result (*P* = 0.093), indicating that given sample size, number of potential populations encompassed,

Table 14.1 Comparative mitochondrial DNA diversity indices

Population	Living Environment (Pop.)	n	K	% Unique	H	Π	π	Source
Cajamarca, Peru	Urban (< 500K)	34	22	0.647	0.972 (±0.014)	8.832 (±4.176)	0.014 (±0.007)	Guevara et al. 2016
Chachapoya (intermontane Peru)	Indigenous	245	99	0.404	0.967 (±0.005)	9.824 (±4.513)	0.016 (±0.008)	Guevara et al. 2016
Cochabamba, Bolivia	Urban (> 500K)	103	71	0.689	0.981 (±0.007)	6.677 —	0.019 (±0.001)	Taboada-Echalar et al. 2013
Huancas (intermontane Peru)	Indigenous	21	12	0.571	0.929 (±0.033)	10.245 (±4.872)	0.016 (±0.009)	Guevara et al. 2016
Iquique, Chile	Urban (< 500K)	189	90	0.476	0.975 (±0.005)	6.903 (±3.262)	0.017 (±0.009)	Gomez-Carballa et al. 2016
Jivaro (lowland Peru)	Indigenous	46	19	0.413	0.934 (±0.018)	8.014 (±3.792)	0.013 (±0.007)	Guevara et al. 2016
La Paz, Bolivia	Urban (> 500K)	253	121	0.478	0.960 (±0.007)	5.106 —	0.019 (±0.001)	Taboada-Echalar et al. 2013
Lima + 33 urban locales, Peru	Urban (> 500K)	119	70	0.588	0.971 (±0.007)	6.299 (±3.009)	0.016 (±0.008)	Messina et al. 2018
Santiago, Chile	Urban (> 500K)	167	85	0.509	0.969 (±0.007)	6.831 (±3.233)	0.017 (±0.009)	Gomez-Carballa et al. 2016
Tupi-Guaraní/Jê Speakers (lowland Brazil)	Indigenous	237	61	0.257	0.959 (±0.005)	5.094 (±2.480)	0.013 (±0.007)	Ramallo et al. 2013; Messina et al. 2018
Uros (Andean Peru)	Indigenous	99	64	0.646	0.986 (±0.004)	6.136 (±2.943)	0.015 (±0.008)	Messina et al. 2018; Sandoval et al. 2013
Yanesha (high-altitude intermontane Peru)	Indigenous	111	42	0.378	0.9509 —	—	0.014 —	Barbieri et al. 2014; Di Corcia et al. 2017
Yungay, Peru	Village	36	20	0.556	0.954 —	—	0.018 —	Lewis et al. 2007
Yurimaguas (*barriada* only), Peru	Urban (< 500K)	52	46	0.885	0.989 —	—	0.022 —	Justice et al. 2012
Yurimaguas, Peru	Urban (< 500K)	140	110	0.786	0.994 (±0.002)	7.239 (±3.411)	0.020 (±0.011)	Present Study

Table 14.2 Tajima's D, Fu's F_S, and Chakraborty's amalgamation neutrality tests

Tajima's D	Fu's F_S	Chakraborty's Test of Amalgamation
−1.311 ($P = 0.080$)	−24.757 ($P = 0.000$***)	$P = 0.093$

and the genetic divergence of potential constituent populations, an amalgamation effect could not be established with an $\alpha = 0.05$ (Chakraborty, Smouse, and Neel 1988).

Mismatch distributions, or comparative distributions of pairwise differences, are based upon a generalized least-squares approach (Schneider and Excoffier 1999). A population with a stationary population size will display a ragged mismatch distribution, resulting in a relatively high sum of square deviations (*SSD*) and Harpending's raggedness index (*rg*), while a numerically expanding population results in a smooth, unimodal mismatch distribution (Harpending 1994). Figure 14.3 visually portrays a bimodal distribution of observed pairwise differences, typically indicating a numerically stationary population, particularly when this distribution varies in its modality from the expected distribution (in this case unimodal versus bimodal). The *SSD* P-value is, however, nonsignificant, indicating that the "observed" distribution pattern is not significantly different from the "expected" distribution pattern (Excoffier 2004; Kusza et al. 2018; Ray, Currat, and Excoffier 2003; Slatkin and Hudson 1991). This denotes that the hypothesis of demographic population expansion cannot be rejected. The P-value of *rg* represents confidence in the shape (raggedness) of the "observed" distribution. Statistical nonsignificance indicates a population that is not numerically stationary (Rogers and Harpending 1992; Slatkin and Hudson 1991).

Conclusion

The "urban revolution" has had a significant effect on Amazonian populations. As demonstrated in this chapter, adaptation to diverse ecosystems, political-economic constraints and opportunities, novel infrastructure, changing lifeways, and violence and social upheaval have all shaped the migratory histories of the peoples of Yurimaguas and the Huallaga Valley. This is evident from the biodemographic and genetic data assessed in this chapter. Study of the complex causes and consequences of migration and urbanization reveals the implications of population architecture in contemporary Peruvian Amazonia.

Acknowledgments

The authors acknowledge the generous support of the following sources: Fulbright Scholars Award—Peru (2011–2012); Summer Research Award, International Programs, University of Kansas (2015); General Research Fund Award #2301063, University of Kansas (2015–2017); Graduate Studies Summer Fellowship, University of Kansas (2017); and Institute for Policy and Social Research, Center for Migration Research, Seed Grant Program, University of Kansas (2017).

References

Achilli, A., U. A. Perego, H. Lancioni, et al. 2013. "Reconciling migration models to the Americas with the variation of North American Native mitogenomes." *PNAS* 110, no. 35: 14308–14313.

Alexiades, M., ed. 2009. *Mobility and Migration in Indigenous Amazonia: Contemporary Ethnoecological Perspectives.* New York: Berghahn Books.

Álvarez-Iglesias, V., A. Mosquera-Miguel, M. Cerezo, et al. 2009. "New population and phylogenetic features of the internal variation within mitochondrial DNA macro-haplogroup R0." *PLOS One* 4, no. 4: e5112.

Arias, L., C. Barbieri, G. Barreto, et al. 2018. "High-resolution mitochondrial DNA analysis sheds light on human diversity, cultural interactions, and population mobility in northwestern Amazonia." *American Journal of Physical Anthropology* 165, no. 2: 238–255.

Barbieri, C., P. Heggarty, D. Y. Yao, et al. 2014. "Between Andes and Amazon: The genetic profile of the Arawak-Speaking Yanesha." *American Journal of Physical Anthropology* 155, no. 4: 600–609.

Behar, D. M., C. Harmant, J. Manry, et al. 2012a. "The Basque paradigm: Genetic evidence of a maternal continuity in the Franco-Cantabrian region since pre-Neolithic times." *American Journal of Human Genetics* 90, no. 3: 486–493.

Behar, D. M., M. van Oven, S. Rosset, et al. 2012b. "A 'Copernican' reassessment of the human mitochondrial DNA tree from its root." *American Journal of Human Genetics* 90, no. 4: 675–684.

Bennett, A., A. Ravikumar, and P. Cronkleton. 2018. "The effects of rural development policy on land rights distribution and land use scenarios: The case of oil palm in the Peruvian Amazon." *Land Use Policy* 70: 84–93.

Bilal, E., R. Rabadan, G. Alexe, et al. 2008. "Mitochondrial DNA haplogroup D4a is a marker for extreme longevity in Japan." *PLOS One* 3, no. 6: e2421.

Bodner, M., U. A. Perego, G. Huber, et al. 2012. "Rapid coastal spread of First Americans: Novel insights from South America's Southern Cone mitochondrial genomes." *Genome Research* 22, no. 5: 811–820.

Brotherton, P., W. Haak, J. Templeton, et al. 2014. "Neolithic mitochondrial haplogroup H genomes and the genetic origins of Europeans." *Nature Communications* 4, no. 1764.

Chakraborty, R., P. E. Smouse, and J. V. Neel. 1988. "Population amalgamation and genetic variation: Observations on artificially agglomerated tribal populations of Central and South America." *American Journal of Human Genetics* 43, no. 5: 709–725.

Childe, V. G. "The urban revolution." 1950. *The Town Planning Review* 21, no. 1: 3–17.

Cook, K. P. 2016. *Forbidden Passages: Muslims and Moriscos in Colonial Spanish America.* Philadelphia: University of Pennsylvania Press.

Crawford, M. 1998. *The Origins of Native Americans: Evidence from Anthropological Genetics.* Cambridge: Cambridge University Press.

David, R. E., and B. Dean. 2019. "*A sociogenetic approach to migration and urbanization in Peruvian Amazonia.*" In *Proceedings of the Center for Migration Research*, Invited Speakers. Lawrence: University of Kansas.

de Jong, W., T-P. Lye, and K-I. Abe. 2006. *The Social Ecology of Tropical Forests: Migration, Populations and Frontiers.* Kyoto: Kyoto University Press.

Dean, B. 2002. "State Power and Indigenous Peoples in Peruvian Amazonia: A Lost Decade, 1990–2000." In *The Politics of Ethnicity: Indigenous Peoples in Latin American States*, edited by D. Maybury-Lewis, 199–238. Cambridge: Harvard University Press.

Dean, B. 2009. *Urarina Society, Cosmology and History in Peruvian Amazonia.* Gainesville: University Press of Florida.

Dean, B. 2013a. "The transgressive allure of white gold in Peruvian Amazonia: Towards a genealogy of coca capitalisms and social dread." *ID: International Dialogue, A Multidisciplinary Journal of World Affairs* 3: 74–91.

Dean, B. 2013b. "The Huallaga Valley & trauma in the endless war on drugs." *Panoramas: Foro: Commentario Latino Americano.* www.ucis.pit.edu/panoramas

Dean, B. 2015a. "Narratives, resilience and violence in Peruvian Amazonia: The Huallaga Valley, 1980–2015." *The Open Anthropology Journal* 1, no. 1: 1–2.

Dean, B. 2015b. "Indigenous Identity and Education in Peruvian Amazonia." In *Indigenous Education: Language, Culture, and Identity*, edited by W. J. Jacob, S. Y. Cheng, and M. K. Porter, 429–446. Dordrecht: Springer.

Dean, B. 2019. *The State and the Aguaruna: Frontier Expansion in the Upper Amazon, 1541–1990.* San Diego: Cognella Academic Publishing.

Dean, B., and R. E. David. 2017. "Yurimaguas and the Lower Huallaga River Valley: A Biocultural Approach to Disruptive Patterns of Migration and Urbanization in Peruvian Amazonia—An Ideal Location for a Migration-Based Approach to Obesity." In *Proceedings of the Second International Human Migration Conference*, Mexico City, Mexico.

Dean, B., and Levi, J., eds. 2003. *At the Risk of Being Heard*: Identity, Indigenous Rights and Postcolonial States. Ann Arbor: University of Michigan Press.

Dean, B., S. Silverstein, M. Reamer, et al. 2011. "The new urban jungle." *Cultural Survival Quarterly* 35, no. 2: 34–45.

Desarrollo Peruano. 2014. "Infraestructura Peruana: Puente Paranapura" *Desarrollo Peruano*, April 20, 2014. http://infraestructuraperuana.blogspot.com/2014/04/puente-paranapura.html

Díaz-Matallana, M., A. Gómez, I. Briceño, et al. 2016. "Genetic analysis of Paleo-Colombians from Nemocón, Cundinamarca provides insights on the early peopling of northwestern South America." *Revista de la Academia Colombiana de Ciencias Exactas, Físicas y Naturales* 40, no. 156: 461–483.

Di Corcia, T., C. Sanchez Mellado, T. J. Davila Francia, et al. 2017. "East of the Andes: The genetic profile of the Peruvian Amazon populations." *American Journal of Physical Anthropology* 163, no. 2: 328–338.

Excoffier, L. 2004. "Patterns of DNA sequence diversity and genetic structure after a range expansion: Lessons from the infinite-island model." *Molecular Ecology* 13, no. 4: 853–864.

Excoffier, L., and H. E. L. Lischer. 2010. "Arlequin Suite ver 3.5: A new series of programs to perform population genetics analysis under Linux and Windows." *Molecular Ecology Resources* 10, no. 3: 564–567.

Fagundes, N. J., R. Kanitz, R. Eckert, et al. 2008. "Mitochondrial population genomics supports a single pre-Clovis origin with a coastal route for the peopling of the Americas." *American Journal of Human Genetics* 82, no. 3: 583–592.

Fendt, L., H. Niederstätter, G. Huber, et al. 2011. "Accumulation of mutations over the entire mitochondrial genome of breast cancer cells obtained by tissue microdissection." *Breast Cancer Research and Treatment* 128, no. 2: 327–336.

Fu, Y-K. 1997. "Statistical tests of neutrality of mutations against population growth, hitchhiking and background selection." *Genetics* 147, no. 2: 915–925.

Fuselli, S., E. Tarazona-Santos, I. Dupanloup, et al. 2003. "Mitochondrial DNA diversity in South America and the genetic history of Andean highlanders." *Molecular Biology and Evolution* 20, no. 10: 1682–1691.

Gómez-Carballa, A., J. Pardo-Seco, S. Brandini, et al. 2018. "The peopling of South America and the trans-Andean gene flow of the first settlers." *Genome Research* 29, no. 1: 767–779.

González-Martín, A., A. Gorostiza, L. Regalado-Liu, et al. 2015. "Demographic history of indigenous populations in Mesoamerica based on mtDNA sequence data." *PLOS One* 10, no. 8: e0131791.

Greenspan, B. 2005a. "Genbank Accession EU157923 (September 18, 2007)." In D. A. Benson, I. Karsch-Mizrachi, D. J. Lipman, J. Ostell, and D. Wheeler, eds. *Nucleic Acids Research* 33 (Database Issue).

Greenspan, B. 2005b. "Genbank Accession EU795361 (June 6, 2007)." In D. A. Benson, I. Karsch-Mizrachi, D. J. Lipman, J. Ostell, and D. Wheeler, eds. *Nucleic Acids Research* 33 (Database Issue).

Guevara, E. K., J. U. Palo, S. Guillén, et al. 2016. "MtDNA and Y-chromosomal diversity in the Chachapoya, a population from the northeast Peruvian Andes-Amazon divide." *American Journal of Human Biology* 28, no. 6: 857–867.

Harpending, H. C. 1994. "Signature of ancient population growth in a low-resolution mitochondrial DNA mismatch distribution." *Human Biology* 66, no. 4: 591–600.

Harris, D. N., W. Song, A. C. Shetty, et al. 2018. "Evolutionary genomic dynamics of Peruvians before, during, and after the Inca Empire." *PNAS* 115, no. 28: 6526–6535.

Harrison, R. 1995. "The language and rhetoric of conversation in the Viceroyalty of Peru." *Poetics Today* 16, no. 1: 1–27.

Hoffecker, J. F., S. A. Elias, and D. H. O'Rourke. 2014. "Out of Beringia?" *Science* 343, no. 6174: 979–980.

Hooshiar Kashani, B., U. A. Perego, A. Olivieri, et al. 2012. "Mitochondrial haplogroup C4c: A rare lineage entering America through the ice-free corridor?" *American Journal of Physical Anthropology* 147, no. 1: 35–39.

Hudson, R. R. 1991. "Gene Genealogies and the Coalescent Process." In *Oxford Surveys in Evolutionary Biology,* Vol. 8, edited by D. J. Futuyma and J. D. Antonovics. New York: Oxford University Press.

INEI (Instituto Nacional de Estadística e Informática). 1981. *Censos Nacionales 1981 — VIII de Población y III de vivienda.* Lima: Instituto Nacional de Estadística e Informática,

INEI (Instituto Nacional de Estadística e Informática). 2005. *Censos Nacionales 2005 — X de Población y V de Vivienda.* Lima: Instituto Nacional de Estadística e Informática,

INEI (Instituto Nacional de Estadística e Informática). 2007. *Perfiles de Sociodemográficos, Resultados Censos de los Nacionales.* Lima: Instituto Nacional de Estadística e Informática,

Just, R. S., T. M. Diegoli, J. L. Saunier, et al. 2008. "Complete mitochondrial genome sequences for 265 African American and U.S. 'Hispanic' individuals." *Forensic Science International: Genetics* 2, no. 3: e45–e48.

Justice, A., B. Dean, and M. H. Crawford. 2012. "Molecular Consequences of Migration and Urbanization in Peruvian Amazonia." In *Causes and Consequences of Human Migration,* edited by M. H. Crawford and B. C. Campbell, 449–472. Cambridge: Cambridge University Press.

Kloss-Brandstätter, A., D. Pacher, S. Schönherr, et al. 2011. "HaploGrep: A fast and reliable algorithm for automatic classification of mitochondrial DNA haplogroups." *Human Mutation* 32, no. 1: 25–32.

Koressaar, T., and M. Remm. 2007. "Enhancements and modifications of primer design program Primer3." *Bioinformatics* 23, no. 10: 1289–1291.

Kusza, S., F. Suchentrunk, H. Pucher, et al. 2018. "High levels of mitochondrial genetic diversity in Asian elephants (*Elephas maximus*) from Myanmar." *Hystrix, the Italian Journal of Mammalogy* 29, no. 1: 152–154.

Lausent-Herrera, I. 2011. "The Chinatown in Peru and the changing Peruvian Chinese communities." *Journal of Chinese Overseas* 7, no. 1: 69–113.

Lefebvre, H. 2014. *The Urban Revolution.* Minneapolis: University of Minnesota Press.

Loogväli, E. L., U. Roostalu, B. A. Malyarchuk, et al. 2004. "Disuniting uniformity: A pied cladistic canvas of mtDNA haplogroup H in Eurasia." *Molecular Biology and Evolution* 21, no. 11: 2012–2021.

Messina, F., T. Di Corcia, M. Ragazzo, et al. 2018. "Signs of continental ancestry in urban populations of Peru through autosomal STR loci and mitochondrial DNA typing." *PLOS One* 13, no. 7: e0200796.

Nei, M. 1987. *Molecular Evolutionary Genetics.* New York: Columbia University Press.

Nei, M., and W. H. Li. 1979. "Mathematical model for studying genetic variation in terms of restriction endonucleases." *PNAS* 76, no. 10: 5269–5273.

Nordborg, M. 2003. "Coalescent Theory." In *Handbook of Statistical Genetics,* 2nd ed., edited by D. Balding, M. Bishop, and C. Canninga. New York: John Wiley & Sons.

O'Rourke, D. H., M. G. Hayes, and S. W. Carlyle. 2000. "Spatial and temporal stability of mtDNA haplogroup frequency in Native North America." *Human Biology* 72, no. 1: 15–34.

Palanichamy, M. G., C. Sun, S. Agrawal, et al. 2004. "Phylogeny of mitochondrial DNA macrohaplogroup N in India, based on complete sequencing: Implications for the peopling of South Asia." *American Journal of Human Genetics* 75, no. 6: 966–978.

Ramallo, V., R. Bisso-Machado, C. Bravi, et al. 2013. "Demographic expansions in South America: Enlightening a complex scenario with genetic and linguistic data." *American Journal of Physical Anthropology* 150: 453–463.

Rani, D. S., P. S. Dhandapany, P. Nallari, et al. 2010. "Mitochondrial DNA haplogroup 'R' is associated with Noonan syndrome of South India." *Mitochondrion* 10, no. 2: 166–173.

Ray, N., M. Currat, and L. Excoffier. 2003. "Intra-deme molecular diversity in spatially expanding populations." *Molecular Biology and Evolution* 20, no. 1: 76–86.

Rhoades, R. E., and P. Bidegaray. 1987. *The Farmers of Yurimaguas: Land Use and Cropping Strategies in the Peruvian Jungle.* Lima: International Potato Center.

Röck, A. W., A. Dür, M. van Oven, et al. 2013. "Concept for estimating mitochondrial DNA haplogroups using a maximum likelihood approach (EMMA)." *Forensic Science International: Genetics* 7, no. 6: 601–609.

Rogers, A. R., and H. Harpending. 1992. Population growth makes waves in the distribution of pairwise genetic differences." *Molecular Biology and Evolution* 9, no. 3: 552–569.

Rothhammer, F., L. Fehren-Schmitz, G. Puddu, et al. 2017. "Mitochondrial DNA haplogroup variation of contemporary mixed South Americans reveals prehistoric displacements linked to archaeologically-derived culture history." *American Journal of Human Biology* 29, no. 6: e23029.

Sandoval, J. R., A. Salazar-Granara, O. Acosta, et al. 2013. "Tracing the genomic ancestry of Peruvians reveals a major legacy of pre-Columbian ancestors." *Journal of Human Genetics* 58, no. 9: 627–634.

Sanger, F., S. Micklen, and A. R. Coulson. 1977. "DNA sequencing and chain-terminating inhibitors." *PNAS* 74: 5463–5467.

Schneider, S., and L. Excoffier. 1999. "Estimation of demographic parameters from the distribution of pairwise differences when the mutation rates vary among sites: Application to human mitochondrial DNA." *Genetics* 152, no. 3: 1079–1089.

Shlush, L. I., D. M. Behar, G. Yudkovsky, et al. 2008. "The Druze: A population genetic refugium of the Near East." *PLOS One* 3, no. 5: e2105.

Slatkin, M., and R. R. Hudson. 1991. "Pairwise comparisons of mitochondrial DNA sequences in stable and exponentially growing populations." *Genetics* 129, no. 2: 555–562.

Soares, P., J. A. Trejaut, J. H. Loo, et al. 2008. "Climate change and postglacial human dispersals in Southeast Asia." *Molecular Biology and Evolution* 25, no. 6: 1209–1218.

Starikovskaya, E. B., R. I. Sukernik, O. A. Derbeneva, et al. 2005. "Mitochondrial DNA diversity in indigenous populations of the southern extent of Siberia, and the origins of Native American haplogroups." *Annals of Human Genetics* 69, no. 1: 67–89.

Starn, O., and M. La Serna. 2019. *The Shining Path: Love, Madness, and Revolution in the Andes.* New York: W. W. Norton.

Tabbada, K. A., J. Trejaut, J. H. Loo, et al. 2010. "Philippine mitochondrial DNA diversity: A populated viaduct between Taiwan and Indonesia?" *Molecular Biology and Evolution* 27, no. 1: 21–31.

Taboada-Echalar, P., V. Álvarez-Iglesias, T. Heinz, et al. 2013. "The genetic legacy of the pre-colonial period in contemporary Bolivians." *PLOS One* 8, no. 3: e58980.

Tajima, F. 1989. "Statistical method for testing the neutral mutation hypothesis by DNA polymorphism." *Genetics* 123, no. 3: 585–595.

Takenaka, A. 2004. "The Japanese in Peru: History of immigration, settlement, and racialization." *Latin American Perspectives* 31, no. 3: 77–98.

Tamm, E., T. Kivisild, M. Reidla, et al. 2007. "Beringian standstill and spread of Native American founders." *PLOS One* 2, no. 9: e829.

United Nations, Population Division of the Department of Economic and Social Affairs. 2018. *2018 Revision of World Urbanization Prospects.* https:\\population.un.org\wup\

van Oven, M. 2015. "PhyloTree Build 17: Growing the human mitochondrial DNA tree." *Forensic Science International*, Genetics Supplement Series 5: 392–394.

van Oven, M., and M. Kayser. 2009. "Updated comprehensive phylogenetic tree of global human mitochondrial DNA variation." *Human Mutation* 30, no. 2: 386–394.

Volodko, N. V., E. B. Starikovskaya, I. O. Mazunin, et al. 2008. "Mitochondrial genome diversity in Arctic Siberians, with particular reference to the evolutionary history of Beringia and Pleistocenic peopling of the Americas." *American Journal of Human Genetics* 82, no. 5: 1084–1100.

Weissensteiner, H., D. Pacher, A. Kloss-Brandstätter, et al. 2016. "HaploGrep 2: Mitochondrial haplogroup classification in the era of high-throughput sequencing." *Nucleic Acids Research* 44, no. W1: W58–W63.

Wirth, L. 1938. "Urbanism as a way of life." *American Journal of Sociology* 44, no. 1: 1–24.

15

Causes of Migration to and from the Ch'orti' Maya Area of Guatemala, Honduras, and El Salvador

Brent E. Metz

When one sees the Ch'orti' area on a map, it is difficult not to imagine a territory occupied by a transhistorical "people." When one investigates the migrations to and from the region over the past two millennia, this image dissolves. Archeological, linguistic, epigraphic, historical, ethnographic, and genetic research over the past three decades provides enough pieces of the historical puzzle to refute the notion of ethnic and demographic stability, while also documenting some remarkable continuities. The maelstrom of population movements to, from, and within the Ch'orti' region has patterns. Climate change, population increase, invasion, and political oppression have been key drivers of migration. The consequences are more much complex and difficult to identify, and in the space afforded here, causes are emphasized.

Nowadays, the point of uncovering the pieces is not simply to know, but to provide history to people from who it was taken. For semiliterate people like the Ch'orti' who have no oral history beyond 100 years ago, the return of historical puzzle pieces is positively transformative. So, the goals of the investigations described here are both academic and to enhance the means by which self-identified Ch'orti's and others in the Ch'orti' region can more accurately understand themselves.

Pre-Columbian Ethnic Articulation, Disarticulation, and Genesis

The Ch'orti' area coincides with the southeastern Ch'olan Maya region during the Classic Period (200–900 AD). At the time of the Spanish invasion in the 1510–1520s, two closely related eastern Ch'olan languages—Apay (Ch'orti') and Ch'olti'—were spoken from northwestern El Salvador to southern Mexico, with a possible third being Torquegua, spoken on the Caribbean coast of Guatemala and Honduras (Feldman 1998). The languages, including their western Ch'olan cousins Ch'ol and Chontal, were so similar that some Spaniards considered them dialects (Thompson 1970, 89–102).

Brent E. Metz, *Causes of Migration to and from the Ch'orti' Maya Area of Guatemala, Honduras, and El Salvador* In: *Human Migration*. Edited by: Maria de Lourdes Muñoz-Moreno and Michael H. Crawford, Oxford University Press. © Oxford University Press 2021. DOI: 10.1093/oso/9780190945961.003.0016

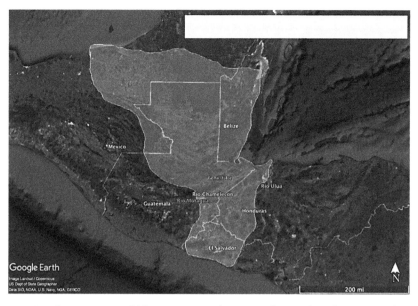

Figure 15.1 Approximate Ch'olan Maya-speaking area during the Classic Period (200–900 AD).

Beginning in 100 AD, the lowland Maya area was colonized by Ch'olans (Figure 15.1), who dispersed after the drying of Lake Miraflores into Kaminaljuyú in highland central Guatemala (Sharer 2009). The Copán dynasty, which dominated the southeast, was officially founded in 426 AD and wrote its last hieroglyph in 822 AD when the site was burned (Fash 2001). Why the dynasty ended and the population dispersed is debated, and archeobotanical evidence no longer supports the once-popular theory that the explanation was the overuse of natural resources and deforestation (McNeil 2009). Historical and archeological records do show increasingly violent warfare throughout southern Mesoamerica, but their causes are debated. One possibility is a combination of climate change and political-religious delegitimization, which would mirror the disintegration of the Guatemalan Ch'orti' political-religious systems in the 20th century (Fought 1969; Metz 2006). Thereafter, populations and power dispersed (McNeil, Burney, and Burney 2009), and Nahua-speaking Pipils from central Mexico invaded the Pacific coast and pushed north into the Ch'orti' area (Fowler 2002, 134, 147). Why the Pipils migrated is uncertain, although it may be no accident that they colonized the most productive cocoa-producing areas in Mesoamerica, since cocaa was a crop associated with prestige and trade (De Solano 1974, in Dary 2003, 38–39).

When the Spanish and their Nahuatl and Maya allies arrived in the early 1500s, Ch'orti' was known as *apay* or *apayac*[1] and a related language was variably known

---

[1] *Apay* was spoken in the Guatemalan departments of Chiquimula, Zacapa, El Progreso, and Jalapa, the Honduran departments of Santa Rosa de Copán, Ocotepeque, Santa Bárbara, and Cortés, and the Salvadoran departments of northern Santa Ana and western Chalatenango (García de Palacio 1985/1576 11–13, fn. 3,

as *alaguilac*. The -*ac* suffix means "tongue" in Ch'orti', and *apay* in Ch'olti' meant "your friend" or "your relative" (Robertson et al. 2010, 30), just as *ap(a)ya'r* does in Ch'orti' today. Given (a) *alaguilac*'s suffix, (b) *alaguilac*'s geographic distribution in and around *apay* populations (the Department of Zacapa and Lakes Guija and Metapán in El Salvador), (c) *apay*'s linguistic simplification as lingua franca (Quizar 2018), and (d) the use of core Nahua loan words in *apay*/Ch'orti', it suggests that both *alaguilac* and *apayac* were Ch'orti'-Nahua hybrids. Moreover, both *alaguilac* and *apay* speakers united behind Copán Kalel[2] in the Spanish invasion (Fuentes y Guzmán (1933/1699 119–122, 169–182, 204–209), and *apay*-, Nahua-, and *alaguilac*-speaking polities shared the same religious center, Mitlán, where they worshiped the Mexican Quetzalcoatl and the Maya Itzqueye, or mother/ moon-goddess (García de Palacio 1985/1576, 36–38). The natives called the region Chiquimula, a likely Nahua-*apay* hybrid, while the Pipils called it Payaquí (García de Palacio 1985/1576, 51–52). Many Lenca terms in Ch'orti' and Lenca place names in the Ch'orti' area can be found as well, probably due to proto-Apays absorbing Lencas into their polities in the Classic Period (see McNeil 2009; Sharer 2009; Manahan and Canuto 2009; Maca 2009).[3] Johnson et al. (2019), in fact, make a compelling historical argument that the Department of Ocotepeque was occupied by mixed Lenca, Pipil, and Ch'orti' groups, with the latter two concentrated mostly in the Ocotepeque valley.

The Spanish-Indigenous Conquest

The Spanish and their Nahuatl and Maya allies rounded up the defeated Apays (Figure 15.2), branded them with hot irons (RAHM CM A/105 4849 f.22 1527 and Sherman 1979, 384 n.1 in Newson 1986, 109–110), and exported many to the Caribbean to mine gold, to Panama to serve as porters, and to Peru to be miners (CDIU 11, 396–397; Pedraza 1544; AGI AG 164; Pedraza 1.1.1547 in Newson 1986, 109–110). The slave trade in Honduras, which initially had easily exploitable gold deposits, displaced 500,000 "Indians" (Newson 1986, 4, 89; Sherman 1979). Ch'orti' war captives were also forced to mine a variety of minerals and gems in the region (Browning 1975, 146–147; Chinchilla Aguilar 1994, 446; Dary 2003, 48; Juarros

46–48, 51–52; Gómez Zúñiga 2013, 69; Herranz 1994, 87; Lardé y Larín 1979). *Chorti*, meaning "the language of the corn farmers or country folk" (Girard 1962, 1; Morán 1935/1695; Robertson et al. 2010, 9; Wisdom 1940, 6), did not appear in colonial records until the second half of the 1700s (Cortés y Larraz 1958/1773; Juarros in Gates 1920, 607, in Robertson et al. 2010, 30).

[2] His name was itself a Nahua/Ch'ol hybrid, as *copantl* is bridge in Nahua and *karer* means drunkenness in Ch'orti' (*kalel* in Ch'olti').

[3] Ochoa Lugo et al.'s (2017) review of lowland and highland Maya ancient and contemporary mitochondrial DNA shows a predominance of A2, C1, B2, and D1 haplotypes. The contemporary Ch'orti' (Justice, Johnson, and Crawford 2011) are anomalous for having neither B2 or D1 haplotypes, and ancient Copán, for whom only nine remains have been tested, were anomalous for having neither A2 nor B2. This also distinguishes them from contemporary non-Mesoamerican Central American groups like the Rama, who exhibit mostly B2, some A2, and no C1 or D1 (Baldi-Salas, 2013).

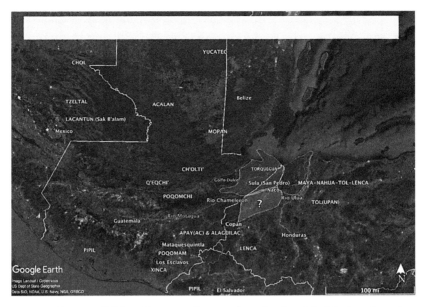

Figure 15.2 Eastern Ch'olan Maya-speaking areas at the Spanish-Mexican invasion, early 1500s.

1936, 34–38; Metz 2006, 52; Museo Nacional "David J. Guzmán" 1973, 147, 151–152; Ramírez Vargas 1994, 611). Some Apays managed to escape to the Ch'olti's in the northern jungles (Gavarrette 1980, 41–42; Ximénez in Milla 1879 II, 246), with whom they shared not only linguistic similarities but some cultural ones as well. Those populations eventually disappeared in the early 1700s due to a combination of epidemics, Spanish forced exile, and predations by British-allied, gun-toting Miskitu slave-raiders from Nicaragua (Feldman 2000).

Besides warfare and slave-trading, in the early decades of colonialism, epidemics, disruptions to agricultural systems, plagues of crop-eating cattle, and famine devastated the Apays and their neighbors. In western and central Honduras, the indigenous population of 851,260 in the 1520s fell to 32,000 by 1550 (Newson 1986, 91, 126, 292, 330; Newson and MacLeod 1990, 50–51 in Euraque 2004, 138–139), and the once heavily populated Tegucigalpa and Choluteca mining regions had to import indigenous workers, including Guatemalan Apays (Newson 1986, 111–113, 133, 151, 158). The Guatemalan Apay population fell from 100,000 to 10,000 (Pérez 1997, 106), and in El Salvador, the communities of Angué, Ostuá, Zacualpa, Guijar, and Santiago de Metapán disappeared due to brutal mining conditions and malaria, yellow fever, smallpox, measles, and tuberculosis epidemics (Comité Progresista Metapaneco 1996; CTMPIES 2003, 19–20; Lardé y Larín 1955, 19, 69). In Chalatenango, the Apays were reduced to four towns: Tejutla (60 homes), Citalá (100 homes), Chiconhueso, and Tepeagua (Lardé y Larín 1979, 138; 2000, 172–174).

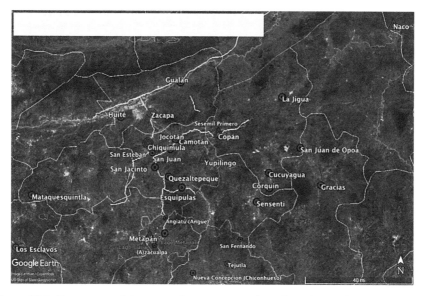

Figure 15.3 Documented Apay-speaking towns in the colonial period, 1524–1821.

The Colonial Period

After being concentrated in settlements, Apays were forced to pay tribute, do compulsory labor (*repartimiento*), financially support the Catholic Church, weave and haul cotton cloth, and buy Spanish goods and sell their own at artificially high and low prices, respectively. Some towns, such as Chiquimula, Zacapa, Esquipulas, and Camotán, were founded on precolonial settlements, and the latter may have been founded by Apays displaced from Tejutla by Pipils during the Postclassic Period (Figure 15.3). Other towns were new, such as Jocotán and San Juan, and the settlers seem to have been forcibly moved there from Copán (Dary, Elías, and Reyna 1998, 37; Feldman 1985, 59; Galindo 1834 in Flores 2004, 8, and Morley 1920, 603; Girard 1949, 54; cf. Torres 1994, 19).

The natives could mitigate their oppression somewhat by controlling their own communal lands and electing their own "governors" and town councils (*cabildos*), whom the Spanish made responsible for tribute, forced labor turns, and local law and order (Newson 1986, 225–226, 228). The governors, however, often took bribes from Spaniards for illegal labor distribution and absconded with tributes (Fry 1988, 23–28; Newson 1986, 165, 199). The Indians were also allowed to form religious brotherhoods (*cofradías*) for the worship of particular saints, which they converted to mutual aid societies replete with tribute-free earnings from communal lands and livestock that they used for emergency expenses like legal fees and burial costs, not to mention worshiping deities in ways that made sense to them. Governors, town councils, and priests incessantly attempted to steal the brotherhoods' funds (Brewer 2009, 142; Dary et al. 1998, 47–49; Documentos 1978; Fry 1988, 31–32, 111, 127,

130; MacLeod 1973, 343–345; Newson 1986, 212, 215–216, 237; van Oss 1986, 45, 109–115).

Neither limited sovereignty nor ethnic loyalty was enough to prevent many Apays from fleeing to mountain forests. Others slipped away to expanding haciendas, where their tributes and forced labor were often covered by the owners. In the late 1500s, Spanish elites began investing in haciendas, particularly in the Motagua valley, the main trade artery with Spain. There, pasturage could be found for cattle, burros, and oxen and land for commodities like sugarcane, tobacco, indigo, cocoa, and cotton. Poor Spaniards and Africans (manumitted or the offspring of slaves) followed the money to set up homesteads or to manage the haciendas. As a result, many Apays and Alaguilacs in the Motagua valley became absorbed as peons or emigrated eastward to join other Apays (Brewer 2002, 58, 83, 159–160; Dary 1990, 1995, 2003, 56; Feldman 1983, 157; 1985; 1998, 38–40; 2009; Fry 1988, 32–33, 38–39, 100; Lauria-Santiago 1999, 31; Lutz n.d. in Smith 2004, 602; Muñoz 1994, 562, in Dary 2003, 57; Newson 1986, 223–229; Ramírez Vargas 1994, 612, 614; Richards and Richards 1994, 355; Taracena Arriola 2004, 94; Terga 1980, 26, 42, 43, 61–64, 75–76, 90–92; Torres Moss 1994, 3–5). The Spaniards also pushed from the east, as western Honduras was amenable for cattle and tobacco production (Documentos 1978, 2; Fuentes y Guzmán 1699/1933, 210; Girard 1949, 54; Lara Pinto 2002; Martínez 1980 in Mortensen 2005, 112–113). Consequently, the Jocotán Parish, which may have been absorbing the displaced, grew from 1,775 Indians in 1680 to 18,140 by 1750 (Metz 2006, 45–46).

The last 70 years of the colonial period (1750–1821) were especially harsh. Greater exploitation, particularly in indigo and *nascacolote* dye production, epidemics, and natural disasters devastated the population again (AGCA 1696–1697; Brewer 2002, 68; Browning 1975, 124-142; Dary 2003, 46; Dary et al. 1998, 67–69; Documentos 1978; Feldman 1982, 146–168; 1983, 157–158, 160; Lauria-Santiago 1999, 23, 32; MacLeod 2008; Newson 1986, 144–147, 190–191, 193, 198; Rubio Sánchez 1976, 17, 23). Forced labor in indigo was so horrendous that many committed suicide and infanticide to avoid it (Cabezas Carcache 1994b, 433–435).

Because indigo and tobacco required little land and capital, a wave of poor Ladinos[4] invaded the peripheries of Indian and hacienda lands in the 1700s. Ladino and Indian smallholders, in fact, produced one half to two thirds of all indigo, and the industry also stimulated local cattle production, as leather was needed for packaging and meat was needed to feed the workers. Still, they earned very little of the profit, because merchants in central Guatemala monopolized lending, distribution, and labor law enforcement (Cabezas Carcache 1994a, 394; 1994b, 433–435; Lauria-Santiago 1999, 18–21, 29, 32; Lutz 1988, 24–26, 40–41; 1999, 130; Pinto Soria 1986, 16–17, 84, 92; Proyecto Ch'orti' 2004, ix; Torres 1994, 27, 64–70). The 2-decade-long

---

[4] Initially, *ladino* referred to non-Europeans, particularly Indian nobles, who converted to Spanish Catholic ways, Indian wives of Spaniards, tribute-exempt Indian craftsmen, and successful Indian entrepreneurs (Newson 1986, 196; Smith 2004, 602; Taracena Arriola 2004, 91).

construction of the Esquipulas basilica for the famous miracle-granting Black Christ image also attracted Ladino craftsmen (Horst 1998a, 137). Some Ch'orti's abandoned their community obligations and ethnicity to intermarry with Ladinos and to settle among them, especially in El Salvador (Browning 1975, 77–100, 124–142; Cortés y Larraz 1958/1773 in Montes 1977, 181, 183–185; Fry 1988, 52, 59, 93).

The Post-Independence Period: Warfare, Population Growth, Liberalism, and Privatization

After Independence, Liberals, who sought a modern, class-based (not ethnic-based) capitalist nation, nonetheless were strapped for funds due to the decline of the indigo market and wars with Conservatives, and preyed on "Indians" as second-class citizens, especially in the Ch'orti' area (Ingersoll 1972, 19). In 1823, the government of the Central American federation expropriated the *cofradía* community's coffers, supposedly to pay for public schooling, but much was diverted to state infrastructure projects (Ingersoll 1972, 54–55). In 1825, the government instituted a hated tax on artisanal production and Indian communal lands and replaced town councils with politically appointed officials, some of whom were Ladinos (Proyecto Ch'orti' 2004, xxiii–xxv). In 1832, the dreaded head tax was re-instituted and sparked Ch'orti' revolts in El Salvador (Browning 1975, 241; Ingersoll 1972, 53; Lauria-Santiago 1999, 106). Indigenous "vacant" lands were also put up for sale, which included *cofradía* property and forests used for hunting, gathering, construction materials, and medicine. Well-positioned Ladinos and Ch'orti's expropriated much of this land, eventually leading to riots in 1837 (Proyecto Ch'orti' 2004, xxiv–xxv; Ingersoll 1972, 64). One of the most hated policies of all was the granting of forests to the British timber company Meany Bennet, which simply clear-cut them. Several Guatemalan Ch'orti' communities rebelled, and the government arrested them and added insult to injury by forcing them to work for the company (Fry 1988, 206; Ingersoll 1972, 60–64). The Liberals also tried to prohibit Maya dress, to make school attendance mandatory, and to force all rural men to work without pay 3 days per month in repairing roads (Handy 1984, 44, 48–50; Jonas 1974, 122–126; Weaver 1999; Woodward 1985, 96–104). The state monopolized tobacco production in 1836, and only a couple of towns—none Ch'orti'—were given licenses (Pinto Soria 1986, 211–214). When the state enforced a quarantine during a cholera epidemic in 1837, peasants throughout eastern Guatemala united behind an illiterate Ladino, Rafael Carrera. Carrera outfoxed the Liberal forces in protracted warfare and ushered in 33 years of paternalistic Conservative rule (Ingersoll 1972; Woodward 1985, 98–112). Guatemalan and Honduran Ch'orti' populations then began a long period of expansion that continues to this day (Davidson 2002).

The death knell for Ch'orti' ethnicity in many towns came when Liberals regained power in the last decades of the 1800s, decreeing that communal lands be privatized (cf. Lauria-Santiago 1999, 61). Communal land was the focal point of Ch'orti' identities, and without it only weakening memories, traditions, and defense against discrimination united them. Modern weaponry and military training allowed Ladinos and Creoles to try such privatization and the re-implementation of forced labor (Cambranes 1985, 149–150; Handy 1984, 60–83; Jonas 1974, 131–148; McCreery 1988; Metz 2006, 55–56; Woodward 1985, 149–201). Increasing populations and flight from forced labor and land loss pushed Ch'orti's north along the Guatemalan-Honduran border, and in the 20th century they reached all the way to eastern Honduras and the northern Guatemala. Still, the core Ch'orti' population in the Department of Chiquimula grew from 36,937 in 1800 to 84,942 by 1900 (Bonilla Alarcón 2003, 9, 12–13; Dary et al. 1998, 51, 62–63, 108; Pérez 1997, 107–108; Proyecto Ch'orti' 2004, xxx–xxxv). The language, however, was spoken only in a few isolated towns (Dary 1995, 2, in Molina Loza 2007, 484; MacLeod 1973, 386; Newson 1986, 224). When Honduras opened communal and "vacant" lands to privatization in the 1870s, Creoles and Ladinos gobbled it up for tobacco, cattle, and coffee production. Ocotepeque Indians lost 28 km^2 and control over appointing their mayor and schoolteacher. In Copán, when a border dispute with Guatemala was settled in the early 1930s, land barons like the Guerras, Cuevas, and Huesos seized title to nearly all the valley land, including that already occupied by Ch'orti's (Herranz 1998, 53–54; Flores 2002, 19; see Loker 2005).

By the 1800s in El Salvador, Ladinos used their numerical superiority and political connections to expropriate and let their cattle graze on indigenous lands (Browning 1975, 268–269; Lauria-Santiago 1999, 50, 56, 62–63, 75). The Ch'orti's became more Ladino and some began to produce for the market instead of subsistence (Lauria-Santiago 1999, 45, 57–58). By the end of century, just a few small Ch'orti' communal settlements remained around Metapán, Tejutla, and Citalá (Browning 1975, 314, 316, 324–326; Conozcamos Tejutla n.d.; Lauria-Santiago 1999, 49, 86) and would soon disappear.

The Past Century: Global Political Economics, Failed States, and Desperation

Over the past century, migratory and ethnic trajectories have headed in different directions in the respective countries. Guatemalan and Honduran Ch'orti's have continued to migrate east and north into the last forests, driven by population increase, lack of land titles, and the quest for fertile virgin forest lands. Also contributing greatly to migration, especially in Guatemala, have been warfare and military oppression fanned by global politics. Thousands of Guatemalan Ch'orti's suffered forced labor and political oppression in the 1930s to 1940s under the U.S.-backed Ubico military state, and violence continued between peasants and large landholders during the decade of democratic opening (1944–1954), which came to a crescendo with a land reform in which unused plantation lands were forcibly bought by the state and distributed to land-poor peasants. After the

CIA-orchestrated the overthrow of democracy in 1954, Ch'orti's who had taken part in the land reform were massacred. Selective violence against peasant leaders continued through the 1950s and became entangled with the guerrilla war from 1960 to 1996, during which thousands were massacred and selectively tortured and assassinated. Many took refuge in Honduras and El Salvador (Metz 2009). The war between El Salvador and Honduras in 1969 also displaced countless peasants, especially in Ocotepeque, after Salvadoran troops pillaged Honduran households. The Salvadoran civil war (1979–1992), which hit Chalatenango especially hard, drove youth from nearly every community to the United States, where some formed gangs to defend themselves from other gangs. When the United States deported them, they brought gang culture with them, which now drives victims to migrate to the United States. Elite mafias, who violently market narcotics, timber, children, organs, and more, also motivate people to flee north (Metz, Mariano, and López García 2010). In Guatemala, young Ch'orti's began trickling northward around 2000, but local officials estimate that 4,000 to 5,000 Ch'orti's in Jocotán alone migrated to the United States between 2014 and 2018. Adding to the sense of desperation has been undemocratic and corrupt Honduran and Guatemalan governments supported by the United States.

Economics also spurs emigration. After the CAFTA-DR came into effect in 2005 (the negotiations for which excluded peasants), U.S. corn and soybean exports to Central America and the Dominican Republic doubled in 8 years, and peasant prices for selling crops were undercut. Central American and Dominican agroindustry exports to the United States have increased even more, demonstrating for whom the "free" trade pact was meant (https://citizen.typepad.com/eyesontrade/2014/08/central-america-crisis-belies-caftas-empty-promises.html, April 27, 2019). Wage labor, mostly in coffee, is paid so poorly—$5 to $8/day— that workers can't afford to buy food, whereas they can early 10 to 20 times more in the United States. Adding to peasants' woes is climate change, in which droughts and monsoons in El Niño and La Niña years are more extreme than ever (Pons et al. 2018). In the summer of 2018, rain did not fall for 42 consecutive days in the growing season, causing the loss of 80% of the corn crop. Other factors driving migrants to the United States are desires for decent education and healthcare, women's and homosexuals' escape from domestic abuse and gender discrimination, social capital earned by young men, and unification of families.

Conclusion

Some migratory patterns are evident. Immigrants, like Ch'orti's, Pipils, Spaniards, and Ladinos, have been attracted to the region for its minerals, agricultural, and transportation potential. The reasons for leaving are more varied and include flight from oppression and economic exploitation, lack of opportunities, epidemics, climate change, criminality, and climbing populations. Today, the elephant in the room is the viability of peasant agriculture. Most Ch'orti's consider semi-subsistence agriculture to be righteous and satisfying, but the above forces have chipped away at their ability to provide for themselves and to acquire new commodities and services they, especially the youth, now deem highly desirable. Peasant agriculture is largely manual, which motivates families to reproduce more hands, but peasants are generally losing land, not gaining it, while what remains is already severely degraded.

The demographic dynamism in the Ch'orti' area is not unusual in Mesoamerica or many other places in the world. The more we investigate what we think are discrete populations, the more such discreteness crumbles. This raises the question, who are Ch'orti' Mayas? So many people have come, gone, and blended in the region that labeling it with any ethno-national moniker seems essentialist. Notwithstanding, the Ch'orti' presence has been the most pervasive over the past 1800 years, and we have seen Ch'orti's articulate a common identity at key historic moments—during the Classic Period, the Spanish invasion, colonial rebellions, the post-Independence wars, and recently in the Maya Movement. We could learn much more about migrations to and from the region with more ethnographic, historical, and especially genetic research. Justice (2011) and colleagues (Justice et al. 2015) have already demonstrated reduced genetic diversity due to population implosions in Jocotán-Camotán, as well strong Euro-Mediterranean and African admixtures (26%) in Ch'orti' male lines, but more testing throughout the region could tell us the rates of admixture and thus the history of Ch'orti's with Lencas, Pipils, Tolupáns, Europeans, and Africans.

References

AGCA (Archivo General de Centroamérica). 1696–97. Chiquimula de la Sierra.

This also distinguishes them from contemporary non-Mesoamerican Central American groups like the Rama, who exhibit mostly B2, some A2, and no C1 or D1 varieties (Baldí 2013).

Baldi-Salas, N. F. 2013. "Genetic Structure and Biodemography of the Rama Amerindians from the Southern Caribbean Coast of Nicaragua." Doctoral dissertation, Department of Anthropology, University of Kansas. Defended April 5.

Bonilla Alarcón, Carlos Renaldo. 2003. "Evaluación Económica de la Certeza Jurídica de la Tierra en el Municipio de Jocotán, Chiquimula [Economic Evaluation of the Legal Certainty of the Land in the Municipality of Jocotán, Chiquimula]." *Magister en Gerencia de la Agricultura Sostenible y los Recursos Naturales*, Ciencias Agrícolas y Ambientalistas, Universidad Rafael Landivar.

Brewer, Stewart W. 2002. "Spanish Imperialism and the Ch'orti' Maya, (1524–1700): Institutional effects of the Spanish Colonial System on Spanish and Indian Communities in the Corregimiento of Chiquimula de la Sierra, Guatemala." PhD diss., Department of History, State University of New York at Albany.

Brewer, Stewart W. 2009. "The Ch'orti' Maya of Eastern Guatemala under Imperial Spain." In *The Ch'orti' Maya Area, Past and Present*, edited by Cameron L. McNeil, Brent E. Metz, and Kerry M. Hull, 137–147. Gainesville: University Press of Florida.

Browning, David. 1975. *El Salvador: La tierra y el hombre [El Salvador: The Earth and the Man]*. San Salvador: Ministerio de Educación.

Cabezas Carcache, Horacio. 1994a. "Regimen Regulador del Trabajo Indigena [Regulatory Regime of Indigenous Labor]." In *Dominación española: Desde la conquista hasta 1700*, edited by Ernesto Chinchilla Aguilar, 387–397. Guatemala City: Asociación de Amigos del País [Association of Friends of the Country].

Cabezas Carcache, Horacio. 1994b. "Agricultura." In *Dominación española: Desde la conquista hasta 1700*, edited by Ernesto Chinchilla Aguilar, 421–442. Guatemala City: Asociación de Amigos del País.

Cambranes, J. C. 1985. *Coffee and the Peasants in Guatemala*. Stockholm: Institute of Latin American Studies.

Chinchilla Aguila, Ernesto. 1994. "Lavaderos de oro y la minería [Gold washhouses and mining]." In *Dominación española: Desde la conquista hasta 1700*, edited by Ernesto Chinchilla Aguilar, 443–450. Guatemala City: Asociación de Amigos del País.

Comité Progresista Metapaneco. 1996. *Metapán: Su Historia, Sus Hombres [Metapán: Its History, Its Men]*. Metapán, El Salvador: Casa de Cultura de Metapán, Asociación Internacional de Apoyo al Parlamento Centroamericano y Plan Trifinio, Consejo Municipal, Cemento de el Salvador.

Cortés y Larraz, Pedro. 1958[1773]. *Descripción geografico-moral de la diocesis de Goathemala [Geographical-moral description of the diocese of Goathemala]*. Vol. XX, Tomo I & II. Guatemala City: Sociedad de Geografía e Historia Guatemala.

CTMPIES (Comité Técnico Multisectoral para los Pueblos Indígenas de El Salvador). 2003. *Perfil de los Pueblos Indígenas de El Salvador [Profile of the Indigenous Peoples of El Salvador]*. San Salvador: CONCULTURA, Ministerio de Educación.

Dary, Claudia. 1990. "Los asentamientos humanos de origen español en el oriente de Guatemala y el origen de Ia literatura popular. [Human settlements of Spanish origin in eastern Guatemala and the origin of popular literature]" *Encuentros*, 28: 9-17 .

Dary, Claudia. 1995. "Chortís, negros y ladinos de San Miguel Gualán, Zacapa [Chortís, blacks and ladinos from San Miguel Gualán, Zacapa]." *La Tradición Popular, Centro de Estudios Folklóricos, Universidad de San Carlos de Guatemala* (103).

Dary, Claudia. 2003. *Los Comuneros Orientales: Identidades étnicas y tierras comunales en Jalapa [The oriental commoners: Ethnic identities and communal lands in Jalapa], Guatemala*. Guatemala City: IDEI, Universidad de San Carlos.

Dary, Claudia, and Silvel Elías y Violeta Reyna. 1998. *Estrategias de Sobrevivencia Campesina en Ecosistemas Frágiles [Peasant Survival Strategies in Fragile Ecosystems.]*. Guatemala: FLACSO.

Davidson, William. 2002. *Honduras: Estructura territorial y estadística según el censo 1895 [Territorial and statistical structure according to the census1895]*. Tegucigalpa: Academia Hondureña de Geografía e Historia.

Documentos Históricos de la Parroquia de Santiago Jocotán, Departamento de Chiquimula. 1978. Jocotán, Guatemala.

Euraque, Dario. 2004. *Conversaciones históricas con el mestizaje y su identidad nacional en Honduras[Historical conversations with miscegenation and their national identity in Honduras]*. San Pedro Sula: Centro Editorial.

Fash, William L. 2001. *Scribes, Warriors and Kings*. London: Thames and Hudson.

Feldman, Lawrence H. 1982. *Colonial Manuscripts of Chiquimula, El Progreso, and Zacapa Departments, Guatemala*. Columbia: University of Missouri Museum of Anthropology.

Feldman, Lawrence H. 1983. "Un reconocimiento de los recursos de Centroamérica en manuscritos Chortí [A recognition of the resources of Central America in Chortí manuscripts]." In *Introducción a la Arqueología de Copán, Honduras, Tomo I*, edited by Claude Baudez, 143–194. Tegucigalpa: Ministerio de Cultura y Turismo.

Feldman, Lawrence H. 1985. *A Tumpline Economy*. Culver City: Labyrinthos Press.

Feldman, Lawrence H. 1998. *Motagua Colonial*. Raleigh: Boson Books.

Feldman, Lawrence H. 2000. *Lost Shores, Forgotten Peoples: Spanish Explorations of the South East Mayan Lowlands, Latin America in Translation/en Traducción/em Tradução*. Durham: Duke University Press.

Flores, Lázaro H. 2004. *Memorial Chortí*. Comayalgueya, Honduras: Multigráficos Flores.

Fought, John G. 1969. "Chortí (Mayan) ceremonial organization." *American Anthropologist* 71: 472–476.

Fowler, William R., Jr. 2002. *El Salvador: Antiguas Civilizaciones [El Salvador: Ancient Civilizations]*. San Salvador: Banco Agrícola Comercial.

Fry, Michael Forrest. 1988. "Agrarian Society in La Montaña, 1700–1840." PhD diss., Tulane University.

Fuentes y Guzmán, Francisco Antonio de. 1933[1699]. *Recordación florida: Discurso historial y demostración natural, material, militar y política del Reino de Guatemala [flowery remembranzas: Historical speech and natural, material, military and political demonstration of the Kingdom of Guatemala]*, Vol. 2. Guatemala City: Sociedad de Geografía e Historia Guatemala.

Garcia de Palacio, Diego. 1985[1576]. *Letter to the King of Spain, Translated and with Notes by Ephraim G. Squier [1859], w/ Additional Notes by Alexander von Frantzius and Frank E. Comparato*. Culver City: Labyrinthos.

Gavarrete Escobar, Juan. 1980. *Anales para la historia de Guatemala [Annals for the history of Guatemala] (1497–1811)*. Guatemala: Editorial Jose De Pineda Ibarra.

Girard, Rafael. 1949. *Los chortis ante el problema maya: Historia de las culturas indígenas de América, desde su origen hasta hoy [The Chortis in the face of the Mayan problem: History of the indigenous cultures of America, from their origin to today]*. 5 vols. Mexico City: Antigua Librería Robredo.

Girard, Rafael. 1962. *Los Mayas Eternos [The eternal Mayas]*. Mexico City: Libro Mex Editores.

Gómez Zúñiga, Pastor Rodolfo. 2013. "Los indígenas de la Gobernación de Higueras-Honduras: Una visión interdisciplinar sobre la frontera entre Mesoamérica y el Área Intermedia en el siglo XVI [The indigenous people of the Government of Higueras-Honduras: An interdisciplinary vision of the border between Mesoamerica and the Intermediate Area in the XVI century]." PhD diss., Universidad Nacional de Educación a Distancia.

Handy, James. 1984. *Gift of the Devil: A History of Guatemala*. Boston: South End Press.

Herranz, Atanasio. 1994. "Los mayas-chortíes de Honduras [The maya-chorties from Honduras]." *Mayab* 9: 87–92.

Herranz, Atanasio. 1998. "Estado, iglesia y marginalidad lenca [State, church, and lenca marginality]." In *Rompiendo el espejo: visiones sobre los pueblos indígenas y negros en Honduras*, edited by Marvin Barahona and Ramón D. Rivas, 43–58. Tegucigalpa: Servicio Holandés de Cooperación al Desarrollo (SNV) and Editorial Guaymuras.

Horst, Oscar H. 1998. "Building blocks of a legendary belief: The Black Christ of Esquipulas, 1595–1995." *The Pennsylvania Geographer* 36, no.1: 135–147.

Ingersoll, Hazel Marylyn Bennet. 1972. "The War of the Mountain: A Study of Reactionary Peasant Insurgency in Guatemala, 1837–1873." Ph.D. diss., George Washington University.

Johnson, Erlend M., Pastor Gómez Zúñiga, and Mary Kate Kelly. 2019. "Ch'orti', Lenca, and Pipil: An onomastic approach to redefining the sixteenth-century southeastern Maya frontier." *Ethnohistory* 66, no. 2: 301–328.

Jonas, Susanne. 1974. "Guatemala: The Land of Eternal Struggle." In *Latin America: The Struggle with Dependency and Beyond*, edited by Ronald H. Chilcote and Ronald H. Edelstein, 93–219. New York: John Wiley and Sons.

Juarros, Domingo. 1936 [1808–1818]. *Compendio de la historia de la ciudad de Guatemala [Compendium of the history of Guatemala City]*. Guatemala: Tipografia Nacional.

Justice, Anne, Stephen Johnson, and M. H. Crawford. 2011. "Y-chromosome variation of a Ch'orti' Maya population in eastern Guatemala." *American Journal of Physical Anthropology* S52: 181.

Justice, Anne Elizabeth. 2011. "Genetic Structure of the Maya in Guatemala: Perspectives on the Population History of the Maya using mtDNA and Y-chromosome Markers." PhD diss., University of Kansas.

Lara Pinto, Gloria. 2002. *Perfil de los pueblos indígenas y negros de Honduras [Profile of the indigenous and black towns from Honduras]*. Tegucigalpa: RUTA/Banco Mundial.

Lardé y Larín, Jorge. 1955. *Monografías históricas del Departamento de Santa Ana [Historical Monographs of the Department of Santa Ana]*. San Salvador: Ministerio del Interior.

Lardé y Larín, Jorge. 1979. "Los chontales de El Salvador precolombino [The Chontales from el Salvador pre-columbian]." *Revista Cultura* 64, no. 1: 125–139.

Lardé y Larín, Jorge. 2000. *El Salvador: Descrubrimiento, conquista, y colonialización [El Salvador: Discovery, conquest, and colonialization]*. San Salvador: CONCULTURA.

Lauria-Santiago, Aldo A. 1999. *An Agrarian Republic: Commercial Agriculture and the Politics of Peasant Communities in El Salvador, 1823–1914*. Pittsburgh: University of Pittsburgh Press.

Lehmann, Walter. 1920. *Zentral-Amerika*. Vol. II. Berlin: Dietrich Reimer.

Loker, William M. 2005. "The rise and fall of flue-cured tobacco in the Copán Valley and its environmental and social consequences." *Human Ecology* 33, no. 3: 299–327.

Lutz, Christopher H. 1988. "Guatemala's Non-Spanish and Non-Indian Population: Its Spread and Demographic Evolution, 1700–1821." Paper presented at the Guatemalan History and Development Conference, University of Guelph, Ontario, Canada.

Lutz, Christopher H. 1999. "Evolución demográfica de la población no indígena [Demographic evolution of the non-indigenous population]." In *Siglo XVIII hasta la independencia*, edited by Ernesto Chinchilla Aguilar. Guatemala: Asociación de Amigos del País.

Maca, Allan L. 2009. "Ethnographic Analogy and the Archeological Construction of Maya Identity at Copán, Honduras." In *The Ch'orti' Maya Area, Past and Present*, edited by Brent E. Metz, Cameron L. McNeil, and Kerry M. Hull, 90–107. Gainesville: University Press of Florida.

MacLeod, Murdo J. 1973. *Spanish Central America: A Socioeconomic History,1520–1720*. Berkeley: University of California Press.

MacLeod, Murdo J. 2008. "Spain and America: The Atlantic Trade, 1492–1720." In *Europe and America*, 341–388. Cambridge: Cambridge University Press.

Manahan, T. Kam, and Marcello Canuto. 2009. "Bracketing the Copan Dynasty: Late Preclassic and Early Postclassic settlements at Copan, Honduras." *Latin American Antiquity* 20, no. 4: 553–580.

Martínez Girón, Eric Jorge. 1980. "El Valle de Copán en la época colonial [The Copán Valley in colonial times]." *Yaxkin* 3, no. 4: 215–236.

McCreery, David. 1988. "Land, Labor and Violence in Highland Guatemala: San Juan Ixchoy (Huehuetenango), 1893–1945." Guatemala History and Development, University of Guelph, Ontario, Canada.

McNeil, Cameron L. 2009. "The Environmental Record of Human Populations and Migrations in the Copán Valley, Honduras." In *The Ch'orti' Maya Area, Past and Present*, edited by Brent E. Metz, Cameron L. McNeil, and Kerry M. Hull, 47–60. Gainesville: University Press of Florida.

McNeil, Cameron L., David A. Burney, and Lida Pigott Burney. 2010. "Evidence disputing deforestation as the cause for the collapse of the Ancient Maya polity of Copan, Honduras." *PNAS* 107(3):1017-22 .

Metz, Brent E. 2006. *Ch'orti' Maya Survival in Eastern Guatemala: Indigeneity in Transition*. Albuquerque: University of New Mexico Press.

Metz, Brent E. 2009. "Las 'ruinas' olvidadas en el área ch'orti': Apuntes para una historia de la violencia en el oriente de Guatemala [The forgotten 'ruins' in the Ch'orti area ': Notes for a history of violence in eastern Guatemala]." In *Guatemala: Violencias desbordadas*, edited by Julián López García, Santiago Bastos, and Manuela Camus, 65–92. Córdoba, Spain: FLACSO (Facultad Latinoamericano de Ciencas Sociales) and Universidad de Córdoba.

Metz, Brent E., Lorenzo Mariano, and Julián López García. 2010. "The violence after La Violencia in the Ch'orti' region of eastern Guatemala." *Journal of Latin American and Caribbean Anthropology* 15, no. 1: 16–41.

Milla, D. José. 1879. *Historia de la América Central [History of Central America]*. Vol. I. Guatemala City: El Progreso.

Molina Loza, Jorge Estuardo. 2007. "'¡En Estanzuela no hay indios!' Identidad ladina en un pueblo del oriente de Guatemala [There are no Indians in Estanzuela! '"Ladino identity in a town in eastern Guatemala]." In *Mayanización y vida cotidiana: La ideología multicultural en la sociedad guatemalteca*, edited by Santiago Bastos and Aura Cumes, 469–493. Guatemala City: FLACSO, CIRMA, and Cholsamaj.

Morán, Francisco. 1695[1935]. *Arte en lengua choltí, que quiere decir lengua de milperos [Art in the Choltí language, which means the language of milperos]*, edited by Gates W. New Orleans: Tulane University. Photographic copy.

Morley, Sylvanus G. 1920. *The Inscriptions at Copan*, Publication 219. Washington, DC: Carnegie Institution of Washington.

Mortensen, Lena. 2005. "Constructing Heritage at Copan, Honduras: An Ethnography of the Archaeology Industry." PhD dissertation Indiana University.

Museo Nacional "David J. Guzmán." 1973. *Exploración etnográfica en el Departamento de Santa Ana [Ethnographic exploration in the Department of Santa Ana]*. San Salvador: Ministerio de Educación, Dirección de Cultura.

Newson, Linda. 1986. *The Cost of Conquest: Indian Decline in Honduras under Spanish Colonial Rule*. Boulder: Westview.

OchoaLugo, M I, G.Pérez-Ramírez, J. Cervini-Silva, M. Moreno-Galeana, E. Ramos, A Romano-Pacheco and M. de L. Muñ. 2017. "*Study of Maternal Lineage and Mitochondrial Genetic Diversity of the Prehispanic Mayan Populations of the Palenque Archaeological Sites in Chiapas and El Rey in Quintana Roo.*" Second International Human Migration Conference: What Can Genomic and Culture Diversity Tell Us about the Migration? CINVESTAV-IPN, Mexico City, October 18.

Pérez, Hector. 1997. "Estimates of the Indigenous Population of Central America (16th to 20th Centuries)." In *Demographic Diversity and Change in the Central American Isthmus*, edited by Anne R. Pebley and Luis Rosero-Bixby, 97–115. Santa Monica: Rand.

Pinto Soria, J. C. 1986. *Centroamérica, de la colonia al Estado nacional (1800–1840) [Central America, from the colony to the national state (1800--1840).]*. Guatemala: Editorial Universitaria, Universidad de San Carlos.

Pons, D., E. Castellanos, D. Conde, J. Brincker, D. Incer, and A. López. 2018. "Escenarios de aridez para Guatemala para los años 2030, 2050 y 2070 utilizando modelos de cambio climático [Aridity scenarios for Guatemala for the years 2030, 2050 and 2070 using climate change models]." *Yu'am: Revista Mesoamericana de Biodiversidad y Cambio Climático* 2, no. 4: 4–16.

Proyecto Ch'orti'. 2004. *Esta tierra es nuestra: Compendio de fuentes históricos sobre denuncias, medidas y remedidas, composiciones, titulaciones, usurpaciones, desmembraciones, litigios, transacciones y remates de tierra (Años 1610–1946), Departamento de Chiquimula [This land is ours: Compendium of historical sources on complaints, measures and remedies, compositions, titles, usurpations, dismemberments, litigation, transactions and land auctions (Years 1610--1946), Department of Chiquimula]*. Vol. IV. Guatemala: Horizont 3000 and Proyecto Ch'orti'.

Public Citizen. 2014. "Central America crisis belies CAFTA's empty promises." Eyes on Trade: Public Citizen on Globalization and Trade. This website is shared by Public Citizen Inc. and Public Citizen Foundation.

Quizar, Robin. 2018. "Ch'olti' and Ch'orti': Separate Language Variations during the Classic Maya." Form and Analysis in Mayan Linguistics, Antigua, Guatemala.

Ramírez Vargas, Margarita. 1994. "El Corregimiento de Chiquimula de la Sierra [The Corregimiento de Chiquimula from the Sierra]." In *Dominación española: Desde la conquista hasta 1700*, edited by Ernesto Chinchilla Aguilar, 611–620. Guatemala City: Asociación de Amigos del País.

Richards, Michael, and Julia Richards. 1994. "Lenguas indigenas y procesos linguísticos [Indigenous languages and linguistic processes]." In *Dominación española: Desde la conquista hasta 1700*, edited by Ernesto Chinchilla Aguilar, 345–360. Guatemala City: Asociación de Amigos del País.

Robertson, John S., Danny Law, and Robbie A. Haertel. 2010. *Colonial Ch'olti': The Seventeenth-century Morán Manuscript*. Norman: University of Oklahoma Press.

Rubio Sánchez, Manuel. 1976. *Historia del añil o xiquilite en Centro América. 2 tomos [History of the indigo or xiquilite in Central America. 2 volumes]*. San Salvador: Ministerio de Educación, Dirección de Publicaciones.

Sharer, Robert J. 2009. "The Ch'orti' Past: An Archeological Perspective." In *The Ch'orti' Maya Area, Past and Present*, edited by Brent E. Metz, Cameron L. McNeil and Kerry M. Hull, 124–133. Gainesville: University Press of Florida.

Sherman, William L. 1979. *Forced Native Labor in Sixteenth-Century Central America*. Lincoln: University of Nebraska Press.

Smith, Carol A. 2004. "Las contradicciones del mestizaje en Centroamerica [The contradictions of miscegenation in Central America]." In *Memorias del mestizaje: Cultura política en Centroamérica de 1920 al presente*, edited by Darío Euraque, Jeffrey Gould, and Charles Hale, 579–617. Antigua: CIRMA.

Taracena Arriola, Arturo. 2004. "Guatemala: el debate historiografico en torno al mestizaje, 1970-2000 [Guatemala: the historiographical debate around miscegenation, 1970-2000]." In *Memorias del mestizaje: Cultura política en Centroamérica de 1920 al presente*, edited by Jeffrey Gould Darío Euraque, and Charles Hale, 77–110. Antigua: CIRMA.

Tejutla, Conozcamos. n.d. Conozcamos Tejutla.

Terga, Ricardo. 1980. "El valle bañado por el río de plata [The valley bathed by the silver river]." *Guatemala Indígena* 15, no. 1-2: 1–100.

Thompson, J. Eric S. 1970. *Maya History and Religion*. Norman: University of Oklahoma Press.

Torres Moss, José Clodoveo. 1994. *Apuntes para la historia de Jocotán [Notes for the history of Jocotán]*. Ed Guatemala UMG (Universidad Mariano Gálvez de Guatemala). Guatemala.

van Oss, Adrian C. 1986. *Catholic Colonialism: A Parish History of Guatemala, 1524–1821*. New York: Cambridge University Press.

Weaver, Frederick Stirton. 1999. "Reform and (counter)revolution in post-independence Guatemala: Liberalism, conservativism, and postmodern controversies." *Latin American Perspectives* 26, no. 2: 129–158.

Wisdom, Charles. 1940. *The Chorti Indians of Guatamala*. Chicago: The University of Chicago Press.

Woodward, Ralph Lee. 1985. *Central America: A Nation Divided*. New York: Oxford University Press.

Ximénez, Fray Francisco. 1929–1931[1722]. *História de la Provincia de San Vicente de Chiapa y Guatemala, Orden de Predicatores [History of the Province of San Vicente de Chiapa and Guatemala, Order of Preachers]*. Guatemala City: Sociedad de Geografía y História de Guatemala.

16

Evidence of Human Migration

Xibalbá in the Puyil Cave, Puxcatán, Tabasco

Enrique Alcalá-Castañeda

Introduction

The practice of ritual ceremonies inside caves occurred among the oldest civilizations. From domestic use of the rocky coats and shallow caverns, one can see the cultural manifestations of these practices in wall paintings or in the abandonment of domestic utensils. Even the deposition of ritual objects as offerings to satisfy the social and economic needs of an inclusive group accompanies human remains in certain cases, indicating funerary practices.

The Puyil cave in Tabasco has a rich pre-Hispanic context and contains burial evidence of more than 40 mortuary remains, distributed in small cavities, sills, and spaces of the floor surface that make up the chambers and galleries formed by carbonate rocks, such as limestone, from leaks and underground rivers. Furthermore, this evidence indicates that the cave was used as a ceremonial and funerary location.

The ornamental elements of the ritual objects that accompany the remains indicate burial practices of pre-Columbian tradition. These findings were obtained in the first phase of a research project carried out in November 2007. Exploration and the expansion of our knowledge about the context of the cave were conducted through the registration and gathering of the uncovered materials. Significant numbers of remains were recorded from the cave, such as samples of carbon and ceramic surface materials, as well as cultural traces made of different materials that were associated with bone deposits.

The locations of the funeral deposits suggest that there was a hierarchy associated with the deposition of the osteological remains. This hierarchy suggests an assimilation of mythological traditions of Xibalbá from the Archaic and Maya populations of the Late Classic period. In addition, modern ceremonial tradition is currently practiced in Puyil.

These events align with Maya beliefs, which consider caves and mountains to be sacred places associated with their ancestors and gods, inhabitants of Xibalbá, and the origin of life.

Evidence from the Puyil cave suggests the association of the contents of the cave with lineages born for the political control of different cities. The cave may also be

Enrique Alcalá-Castañeda, *Evidence of Human Migration* In: *Human Migration*. Edited by: Maria de Lourdes Muñoz-Moreno and Michael H. Crawford, Oxford University Press. © Oxford University Press 2021. DOI: 10.1093/oso/9780190945961.003.0017 associated with Olmec, Maya, or Zoque ancestors, which suggests that the traditions of the pre-Hispanic population predate the Preclassic Period. Nevertheless, the lineage of the individuals buried in the Puyil cave is still unknown.

Background

In 2004, cavers specialists carried out the initial exploration of the Puyil cave and reported to the municipal president of Tacotalpa, Tabasco (Martos-López 2014, 405).

Later, in March 2005, archaeologist Luis Alberto Martos-López (2014) conducted an inspection of the site at the request of the INAH Regional Center, Tabasco, initiating the archaeological project "Cueva de San Felipe," which aimed to elucidate the cultural relationship of the more than 40 burial deposits inside the last three chambers of the cave.

In November 2007, the first stage of the project, conducted by Martos-López, was carried out by archaeologists L. I. Rodríguez, O. Olivo, C. Topete, and E. Alcalá-Castañeda, and the physical anthropologist Eusebio D. Susano-Gómez.

The cave's context, including the material information, was registered. Skeletal remains, as well as samples of coal, ceramic surface materials, and material remains, were collected for laboratory study. There were 251 associated items, including beads, axes, and earrings of jadeite or green stone (28 items), beads and shell earrings (184 items), flint knives (4 items), prismatic knives (6 items), pyrite mirrors (2 items), pieces made in animal bone (3 items), ceramic objects (2 items), and beads and lithic earrings (22 items).

A description of the materials associated with funerary practices was compiled by E. Alcalá–Castañeda (2009), who aimed to establish a methodological systematic analysis to facilitate the interpretation of the archaeological context in the cave, the symbolism of the cave (Topete 2014), and some of the physical characteristics, including sex, age, genetic relationship, and dating, of some of the individuals studied that are important to determine the ethnic origin of individuals (Susano Gómez 2007; Navarro-Romero et al. 2020; Chapter 6 and possible migration routes.

Location, Description of the Cave Context, and Technical Procedures

The Puyil cave is in a mountain named San Felipe, at an elevation of the Sierra Nava Mountains . Puyil is in the far north but within the political division of the state of Tabasco (Figure 16.1). Cave access lies on the northern slope of the elevation, which belongs to the boundaries of the village of Puxcatán in the municipality of Tacotalpa, south of the same state.

The cave has an approximate depth of 175 m (Figure 16.2A), to the point where the bony deposits of the last chamber are found (Chamber 6). The descent from

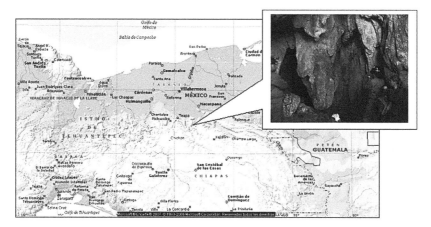

Figure 16.1 Map with the location of the Puyil cave, state of Tabasco, Mexico.

Source: Physical maps, Encarta.

Figure 16.2 A: Drawing of a map of the cave (drawing made by Enrique Alcalá). Chambers 4, 5, and 6 contained the cultural material and bone remains. B: Access to the first chamber of the Puyil cave is made by descending with the help of a rope. C: Calcareous structure in a "cob" shape in the second chamber, along with the remains of a candle from contemporary rituals.

the entrance to the last point is 58 m on average. From the foot of the mountain to the cave access is 28 m; getting to the cave means climbing a rocky escarpment, made basically of limestone and with abundant vegetation, with an inclination of approximately 55° of the north slope where the ascent begins to reach the entrance of the cave.

At the foot of the mountain, on the same north face, a spring of sulfurous water is located. Sulfur is abundant from September to January; the remainder of the year the water is clearer. Some cultural pieces of evidence probably exist underwater, since offering precious objects to the god of water formed part of ancient pre-Hispanic traditions. Similar archaeological contexts exist in many Mexican pre-Hispanic sites.

The development of the cave extends to the southwest-northeast and to the north, through narrow corridors that connect the six chambers that make up the whole cavern. The last chambers are located very close to each other, probably below the origin of the spring mentioned above.

The geomorphological characteristics inside the cave are based on calcareous structures, such as stalactites and stalagmites, some in the process of formation, with abundant sediments on the floor in high-humidity conditions; these circumstances prevent carrying out a systematic search and survey by reticle. Therefore, photography was our alternative resource for the graphical explanation of the context. The photographic material was used to generate some drawings to distinguish important details that are not always clear to the naked eye, which may confuse the material remains with bones or stony surfaces due to the color of the cavern sediment.

According to the field record generated in the first season and the lab study of 29 individual skeletons, the funerary remains correspond to 10 male, 9 female, and 10 undetermined individuals (Gómez 2007, 43). However, will be necessary to analyze each sample genetically to determine the sex of each osteological remain. Several remains were in an advanced state of decomposition due to the degree of humidity inside the cavern.

Most of the individual osteological remains were located in the three deepest chambers of the cave (chambers 4, 5, and 6) and generally were arranged with ornamental elements that probably served as social distinctions or were unique to the funerary rituals practiced in this cavern; these elements included pectorals, earmuffs, necklaces, bracelets, and applications, as well as other ceremonial objects, including knives, leaflets, prismatic knives, axes, turtle shell plates, vases, mirrors, and cinnabar (Alcalá Castañeda 2009). The elaborate material varied between shell, jadeite or green stone, pyrite, obsidian, flint, turtle bone, deer bone, and ceramics. These elements may contribute to the identification of the origin of these individuals and, therefore, to establish if they emigrated from different regions of the American Continent. The access to the cave is through a small hollow that allows entrance to the first chamber. The largest dimension is 60 m (north–south), the width is approximately 30 m, and the height from floor to ceiling is 15 m. The entrance is ornamented with large natural formations of stalactites and stalagmites. Getting into this space requires the aid of ropes, which descend approximately 8 m before encountering a long and slippery ramp due to humidity (Figure 16.2B). In the first and second chambers were found contemporary ceremonial objects, bottles of some beverage, candles, and pre-Columbian ceramic remains from the Late Classic (600–900 AD) and Postclassic (900–1500 AD) periods.

In the second chamber, the most significant manifestation of contemporary activities was a floor to ceiling structure, approximately 7 m high, that resembles a corncob (Figure 16.2C).

The skeletal remains, such as incomplete and disassociated limbs, are present in the third chamber. These remains were most likely taken from a deeper area and deposited in this area, since the most complete primary and secondary ossuaries are concentrated in the fourth chamber.

Chamber 4 is divided into three staggered sections: the first section contains vestiges of highly deteriorated bones in different floor spaces along the corridor. The first and second sections contain bones in small cavities formed by the collapse of large calcareous structures, and only some skeletons have maintained an anatomical position (Figure 16.3 A and B, and Figure 16.4A). Because some remains displayed joint detachment, they were most likely primary deposits, which may be from high-ranking individuals, since they contain ornamental elements, beads of green stone, earmuffs, and applications of shell, as well as flint knives, axes of green stone, and river shell These items are associated with funerary rites (Suarez-Diez, 2002).

Following the steep corridor in this chamber, one descends to the third section by a very slippery 2-m ramp. Five skulls were located on a rocky surface that was 3.5 m long and 2 m wide; they may belong to the skeletons without skulls located in Chamber 6. Samples of these bony remains were taken for genetic analysis in order to confirm this hypothesis (Chapter 6).

The first skull, located at the bottom of the downhill ramp, had a tabular oblique deformation and was associated with a piece of a base vessel ceramic found alongside

Figure 16.3   A: The distribution of a skeleton inside a sloped and narrow hollow in Chamber 4 (Point 46). B: Detail of the bones of Point 46.

Figure 16.4   A: The skull of the individual of point 46 from the Archaic Period. B: Skulls on the surface of Chamber 4 from the Archaic, Pre-Classic, and Classic Periods (Navarro-Romero et al., Chapter 6).

it; there was a recess encompassing osteologic remains in a high state of degradation due to the intense humidity, a green stone axe, and other elements. This recess also contained a significant number of small discoidal applications of shell. At the bottom of this hollow, there was an additional hollow containing a skull, which was also eroded and petrified by the calcareous sediment. The four skulls located on the great rock to the left extreme of this section contained cranial modifications, two had tabular oblique deformation, and another had a tabular erect type. The last one was only the facial skull of a child (Figure 16.4B). Different types of cranial deformations were practiced by different pre-Hispanic cultures, which means that the cranial deformation will contribute to the identification of the cultural affiliation to which they belong and therefore their probable geographical origin.

There was also an important concentration of osteological elements with abundant clasts, which were scattered at the foot of the mentioned rock. These elements were associated with a circular pectoral of shell in the form of a six-peak star, a shell earmuff, and some beads. One human skull was already in the process of petrification. Next to this skull, at the height of the shelf mentioned above, were the remains of another skeleton, which had maintained an anatomical position in a reduced space difficult to access.

Returning a few meters back, there was a hollow on the floor 1 m in diameter and 3.5 m in depth (Figure 16.5A). Once at the bottom, turning west, there is a small side hollow at the floor level, 50 cm wide by 30 cm high to access a 1.3-m channel connecting to a slightly curved tunnel of 5 m long by 1.20 m high and 0.80 m wide that connect with chamber 5. Once in the chamber, there is a ramp, which is 5 m long and 2.5 m wide and continues to the south. The chamber is reduced in width to approximately 1.5 m; likewise, the roof drops from 3.5 m to 40 cm high, where Chamber 6 is found (Figure 16.5B).

Chambers 5 and 6 contain a considerable amount of wet and glutinous sediment on the floor, which makes it difficult to identify the deposited bones in detail. Chamber 5 contains individuals 10, 11, and 12, facing southwest. A green stone axe, remnants of cinnabar, and under this, a flint knife and a green stone pectoral were

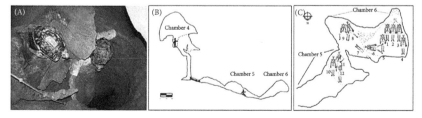

Figure 16.5 A: Depth shot of access to the tunnel to get to Chambers 5 and 6. B: Cross-section of the cave structure from Chambers 4 to 6. Chambers 4, 5, and 6 contained cultural material and bone remains. C: Distribution of the funeral skeletons in Chambers 5 and 6. Drawings were made by Enrique Alcalá. Archaeological Project, cave of San Felipe (2007).

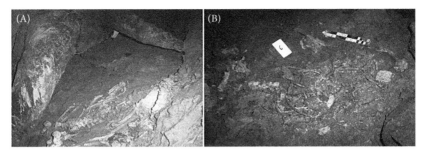

Figure 16.6 A: Skeletons 8 and 9 above an offering of a mirror and a green stone pendant. B: Individual 6, Chamber 6.

found, similar to those that were found with the skeletal remains of individual 12 as part of a funerary offering similar to that of the individual located in Chamber 6 (Figure 16.5C).

Furthermore, the absence of personal ornaments in individuals 8 and 9 was of particularly interest. However, there was an offering (offering 2) on a small ledge on top of the slope and above these deposits, a disc of pyrite with a lithic base and a pectoral of green stone, which was similar to the pectoral found in Chamber 5 (Figure 16.5C and Figure 16.6).

Individuals 6 and 7 were in staggered form, and the bones of individual 7 seemed to have been moved, because the lower extremities were in the foreground some-what far from the top (Figure 16.5C and Figure 16.6). Next to this individual, a green stone-like breastplate was found, as with the previous remains. Skeleton 6 was in an anatomical position, with the same orientation as 7, but the skull was facing the southeastern section (Figure 16.6B).

In the most western section of the chamber, there were five individuals covered with abundant glutinous sediment, one next to the other in anatomical position. The floor of this section had a certain inclination, almost parallel to the roof, which descended abruptly until reaching the floor toward the northern end (Figure 16.7 A and B).

Skeletons 1, 2, and 4 were primary deposits and were better preserved. Two more skeletons, 3 and 5, were not analyzed because they were too degraded. These skeletons did not have skulls. It is likely that the skulls in Chamber 4 belonged to these skeletons.

Materials recovered from Chambers 5 and 6 were enough for initial archaeolog-ical analysis. Individual 1 wore a string of already petrified shell beads; individual 2 had a shell pectoral with a green stone inlay very similar in size and style to the shell pectoral found in chamber 5 in the form of a star. Associated with this pectoral were a green stone bead and two shell beads. The third individual had two rectangular shell applications, an earmuff, two beads of the same material, and a flint knife. Most of these elements were in the abdominal area of each individual, suggesting a ritual or special meaning. In the upper area of individual 5, there was a mirror of

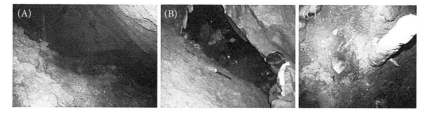

Figure 16.7 Cave of Puyil, A: East–west view of the last section of Chamber 6 with offering 1 and bones 1–5. B: Conditions for exploring Chamber 6's space. C: Offering 1, Chamber 6.

pyrite with a lytic base, similar to the mirror of offering 2. This kind of object was usually worn by privileged warriors or dignitaries (Vela 2016, 30). Ornaments of the fourth individual were not registered because it was on a surface that was difficult to access.

At the end of this section, on a staggered slope and above skeletons 1, 2, and 3 was offering 1, located on a surface of 50 cm by 55 cm, which presides over the context of a greater concentration of individuals in a reserved place in the chamber. The gifts included in this offering represent some special concept that gives meaning to the deposition of these individuals, and perhaps to the whole cave. A flint knife, four prismatic knives, four turtle plates, a grooved bone awl, two river bivalve shells, five axes of green stone, a miniature jug (semiglobular with neck and ears), a shell handle, and some unknown seed remnants were around a phallus-shaped stalagmite 57 cm high by 8 cm in diameter (Figure 16.7A and C).

Discussion

Inside the Puyil cave, there are a series of skeletons from different moments in history that represent different funerary events or different migratory cultural groups. One possible origin of a group is the archaeological site "El Tortuguer", which is located as far as 25 km from the cave (Martos-López 2014). However, according to bone biochemical analysis, the origin of the individuals could have been "La Venta, Malapasito", Comalcalco, or Palenque, since the osteological remains were dated from the Archaic, Preclassic, and Classic Periods (Navarro-Romero et al., 2020; chapter 6)). These findings suggest a probably migration rout from Comalcalco or Palenque to the Puyil cave. Genetic analysis of these osteological remains will contribute to elucidate clearly the migration routes. Since the time span is very long, , it is difficult to confirm which cultural group occupied the cave during the different time periods.

Moreover, the skull deformation in the ancient remains suggests beliefs in the Xibalbá ideology. There are examples of migratory groups that settled in places previously occupied by other populations that adopt the same traditions and ideology to be accepted . Similar processes could account for the use of the cave as a sacred place, such as the cenotes in the Maya area, that were considered the entrance to Xibalbá (de Anda Alanís, 2007; Martos-López, 2007).

Through interpretations that arise from ancient mythology, it has been seen that in pre-Hispanic cities, mainly from the Classic Period (300–900 AD), the ceremonial buildings represented the "sacred mountain," which legitimatized the lineage of the ruler of these cities. In many cases, burial chambers were prepared and built inside the temples for the dignitaries who after death would become the principal means of liaison with the ancestors and with the deities (Carrasco 2004,). According to pre-Hispanic beliefs in the Mesoamerican worldview, caves evoked the vegetable riches of the underworld, water streams, and the powers of growth and reproduction where life itself originates (Manzanilla 1994, 60).

The deposition in the deepest chambers of the Puyil cave suggest a reproduction of the Xibalbá, like that interpreted by Chan Bahlum at Palenque in the temple of the Cross, where the composition and design form a triune. Similarly, in each templeof the Palenque complex, there are three doors that cross the front wall of an interior divided into an antechamber and three subsequent shrines (Schele and Freidel 1990, 296). Similarly, the Puyil cave is divided into three galleries containing three doors to reach the last chamber, where there are the most relevant funeral deposits. These similarities suggest the migration of the people who performed the ceremonies in the Puyil cave to Palenque.

In addition, the triune design also refers to the Olmec culture that preceded the Maya culture. Caves were considered powerful and magical places by the Olmec (Taube, 2007). Mountains that included fountains of water or caves were particularly special, as they had elements of the three levels of the world (De La Fuente 1993; Soustelle 1979). Examples of this cultural tradition are the personage represented inside a cave characterized by the jaws of a jaguar, which represented a dignitary, or Altar 4 of the Preclassic city of La Venta, where the main sculpture of a man is seated at the entrance of a cave depicted through the stylized jaws of a jaguar. What supports the pre-Hispanic population movements from Veracruz to Chiapas and Tabasco. As well as the migrations of the Olmecs to the Mayan area and probably to the region of the Zoques

According to the elements of ornament with the ritual objects that accompany the individuals at Puyil, we know that this is a funerary practice of pre-Hispanic tradition. It is also known that the southern region of the state of Tabasco and northern Chiapas (where the Puyil cave is located) was occupied by Zoque groups, and until the 19th century, they cohabited with Chontales (Ochoa 1997, 27).

In the Puyil cave, different analyses were carried out with the aim of understanding the true sequence of the cultural manifestation in order to develop a formal archaeologic explanation of the cultural groups that used the cave as well as the migration of the pre-Hispanic populations. One advance of paramount importance in the research of Puyil is the molecular analysis practiced by specialists from the Research and Advanced Studies Center of the National Polytechnic Institute

(CINVESTAV-IPN). Muñoz and Navarro-Romero analyzed 10 bone samples from some of the individuals located inside the cave. This analysis will yield information concerning the genetic origin and the possible migratory relationships according to the distribution of mitochondrial DNA (mtDNA) haplogroups and haplotypes within the American continent and Asia.

Therefore, among the more than 40 osteological remaining of the cave, there are individuals who are related to Mexican and Guatemalan groups, as well as the migration of other groups who moved from north to south and vice versa, passing through Chile, Peru, and Cuba (Navarro-Romero 2017; Chapter 6 of this book). In addition, carbon-14 dating performed on six samples demonstrated that the Puyil cave had been used since the Archaic, the Preclassic, and the Late Classic Periods, which demonstrates a cave occupation of 7,000 years, by groups of different migratory origin (Navarro-Romero et al. 2021)..

Conclusions

The context of the Puyil cave, which includes the cultural events, location of the offerings, the cranial deformations, and the epoch of each burial, revealed a probable hierarchy in the deposition of bone remains, as well as burial depositions of more than one ethnic group over a long period of time that were from different origins. All this may suggest different population migrations. In addition, it suggests an assimilation of mythological traditions from the Archaic and between Maya populations of Late Classic and modern ones, practiced until current times at Puyil. Likewise, based on context in the chambers, it also can be proposed that the first two chambers were used first. These events are in agreement with the Maya beliefs that consider caves and mountains sacred places associated with their ancestors and gods, inhabitants of Xibalba, and the origin of life. Evidence in the Puyil cave suggests the association of the inside cave context with linages born for the political control of different cities, and probably they may be also associated with the Olmec, Maya, or Zoque ancestors, which indicates that pre-Hispanic population traditions arose before the Preclassic Period (please see chapter 6 of this book). However, the identification of the lineages of most of the skeletal remains of the Puyil cave is pending, which will confirm with greater precision the migratory routes of the occupants of this cave..

References

Alcalá-Castañeda, E.. 2009. "Catálogo de Materiales de la Cueva de Puylil [Puyil Cave Materials Catalog]." *Informe de análisis de materiales de la cueva de Puyil.* Dirección de Estudios Arqueológicos del INAH, Mexico City.

de Anda-Alanís, G.. 2007. "Los huesos del Cenote Sagrado de Chichen Itza, Yucatán [The bones of the Sacred Cenote of Chichen Itza, Yucatán]." *Revista Arqueología Mexicana* XIV, no. 83: 54–57, Mexico City

Carrasco, R.. 2004. "Ritos funerarios en Calakmul prácticas rituales de los mayas del Clásico [Funerary rites in Calakmul ritual practices of the Classic Maya]." *Culto funerario en la sociedad maya, Memoria de la Cuarta Mesa Redonda de Palenque*, Coordinador Rafael Cobos, Instituto Nacional de Antropología e Historia, Mexico City.

De La Fuente,.. 1993. "El orden y la naturaleza en el arte olmeca [Order and nature in Olmec art]." In *La Antigua América: el arte de los parajes sagrados*, edited by R. Townsand. Chicago: The Art Institute of Chicago, Grupo Azabache.

Manzanilla, L. 1994. "Las Cuevas en el Mundo Mesoamericano [The caves of the Mesoamerican World]." Ciencias, núm. 36, Universidad Nacional Autónoma de México, Facultad de Ciencias, pp. 59–66, Mexico City.

Martos-López, L. A. 2007. "Los cenotes en la actualidad. Entre la veneración y la explotación [The cenotes today. Between veneration and exploitation]," Revista Arqueología Mexicana, Vol. XIV, Numero 83, México,

Martos-López, L. A. 2014 "La cueva de San Felipe o Puyil, Puxcatan, Tacotalpa, Tabasco" [The cave of San Felipe or Puyil, Puxcatan, Tacotalpa, Tabasco], *en Tabasco: Una visión antropológica e histórica*, Coordinadores Miguel A. Rubio et. Al, CONACULTA, Instituto Estatal de Cultura de Tabasco, Universidad Nacional Autónoma de México, 2014.

Navarro-Romero MT, Muñoz ML, Alcala-Castañeda E, Terreros-Espinosa E, Domínguez-de-la-Cruz E, García-Hernández N, Moreno-Galeana M.Á. (2020) A novel method of male sex identification of human ancient skeletal remains. Chromosome Research Dec;28(3-4):277-291. doi: 10.1007/s10577-020-09634-1. Epub 2020 Jul 3. PMID: 32621020..

Navarro-Romero, T. M. 2017. "Secuenciación masiva de muestras Prehispánicas de Puxcatan, Tlacotalpa, Tabasco, México y su significado [Massive sequencing of Prehispanic samples from Puxcatan, Tlacotalpa, Tabasco, México and its meaning]." Conferencia del 28 de febrero en VI seminario de Antropología Molecular, Retos, Logros y Alcances, DAF-MNA, México.

Ochoa, L.. 1997. "Los señoríos prehispánicos en los límites de la imaginación: la concepción Geopolítica de Tabasco al momento del contacto. [Tabasco: apuntes de frontera, The pre-Hispanic lordship in the limits of the imagination: the Geopolitical conception of Tabasco at the time of contact.Tabasco: border notes]" *Tabasco: apuntes de frontera*, Mario Humberto Ruz (coord.), México, Coordinación Nacional de Descentralización del Consejo para la Cultura y las Artes, Programa de Desarrollo Cultural del Usumacinta, Vol. II, pp. 15–40.

Sánchez, L. G.. 2011. "La representación de cuevas y montañas entre los mayas del Período Clásico. Una reflexión desde la historia y la antropología de la imagen [The representation of caves and mountains among the Maya of the Classic Period. A reflection from the history and anthropology of the image]." *VII Jornadas Santiago Wallace de Investigación en Antropología Social*. Sección de Antropología Social. Instituto de Ciencias Antropológicas. Facultad de Filosofía y Letras, UBA, Buenos Aires. 2013.

Schele, L., and David F. 1990. *Una selva de reyes. La asombrosa historia de los antiguos mayas [A jungle of kings. The amazing story of the ancient Maya]*. Fondo de Cultura Económica, Mexico.

Soustelle, J. 1979. *Los olmecas [The Olmecs]*, Fondo de Cultura Económica, México.

Suarez-Diez, L. 2002. *Tipología de los objetos de prehispánicos de concha [Typology of pre-Hispanic shell objects]*, 2nd ed. Conaculta-INAH.

Susano-Gómez, E. D. 2007. *Informe de laboratorio del material óseo de la cueva de San Felipe (Pullil), Puxcatan, en el estado de Tabasco [Laboratory report of the bone material from the cave of San Felipe (Pullil), Puxcatan, in the state of Tabasco]*. Dirección de Estudios Arqueológicos del INAH, Mexico City.

Taube, K.. 2007. "La jadeíta y la cosmovisión de los olmecas [Jadeite and the worldview of the Olmecs]." *Arqueología Mexicana* XV, no. 87: 44, Mexico City.

Topete-Bustamante, C. R. 2014. "Cueva de Puyil: una interpretación de arqueología simbólica [Puyil cave: an interpretation of symbolic archeology]." PhD diss., Escuela Nacional de Antropología e Historia, INAH.

Vela, E.. 2016. La joyería en el México Antiguo [Jewelry in Ancient Mexico]. *Arqueología Mexicana* no. 63: 30.

17

Migration of the Zoques to the Mountain Region of Tabasco, Mexico

Linguistic and Archaeological Perspectives

Eladio Terreros-Espinosa

Introduction

From pre-Hispanic times to the first half of the 20th century, the mountain region of Tabasco was part of a trade network between towns in Chiapas and the coastal plain of the Gulf of Mexico. Much like other regions of Mesoamerica, this exchange of goods and culture was integral to the Zoque economy of the region (Figure 17.1).

Cultural evidence recovered from a rock shelter located in the Ejido Lázaro Cárdenas municipality of Tacotalpa in Tabasco contained one sample of coal that was dated by carbon-14. This coal, combined with vestiges, suggested that the human presence in the Sierra Tabasqueña began at the end of the third millennium BC. The first cultural evidence was scarce and was increased during the second millennium up until the Protoclassic Period (100 BC–200 AD). During the Early Classic period (250–600 AD), there was little pre-Hispanic evidence. In the Late-Terminal-Classic period (600–900 AD), the presence of fine paste pottery indicated pre-Hispanic evidence, which persisted until the Late Postclassic period (1200–1521 AD).

Rands (1967) proposed that the study area is culturally located in the subregion of the Tabasco lowland (1967, 115). The flora and fauna, such as the karst landscape (mogotes), gave the region an impressive splendor that was culturally sustainable.

Geographical Location of the Zoque

Specifying the territory inhabited by the Zoque has been a problem for those who settled on the Gulf Slope, since the Zoque were displaced from their original sites from the mid Preclassic period (500 BC) to the second half of the 20th century the exact territory is problematic. In relation to the above, Donald Bryant and John Clark suggest that, during the late Preclassic, Mayan groups from the lowlands

Eladio Terreros-Espinosa, *Migration of the Zoques to the Mountain Region of Tabasco, Mexico* In: *Human Migration*. Edited by: Maria de Lourdes Muñoz-Moreno and Michael H. Crawford, Oxford University Press. © Oxford University Press 2021. DOI: 10.1093/oso/9780190945961.003.0018

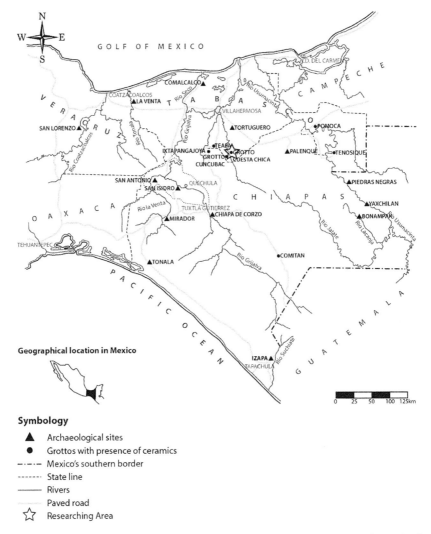

Figure 17.1 Map of the mountain region of Tabasco showing the archaeological zoke area and grottos with the presence of ceramic.

spread to Chiapas, causing the collapse of the Zoque capitals located in the upper basin of the Grijalva River (Bryant et al. 1983, 223–239).

In reference to territorial demarcations, Ochoa (1997) posed:

> If, like the roads of the region, the political contours had been drawn on a skin or a canvas, is possible that climate, time and carelessness prevented their conservation. Therefore, trying to delineate a line of rigid frontiers for the pre-Hispanic Tabasco, similar to that which has been done to delimit the different political territories of the north of the Yucatan Peninsula, can be, as there, rather a result of imagination than adequate treatment of a problem not very different, except for its geography. Additionally, the changes of location of some towns, the abandonment

of others, or the conquest of some more, obviously impacted on the "mobility of the borders" and, therefore, on its "territorial extension." All this leaves strong doubts about the way in which the boundaries of many of the alleged provinces have been determined. (Ochoa 1997, 27).

Similarly, Clark, Hansen, and Pérez (2005) considered that: "In the past, the territories of states and empires were modified according to the political fortune of their rulers. In realistic terms, this means that we cannot be sure of the precise territorial extent of ethnic groups, although this applies to the limits of the territories of the central part" (Clark et al. 2005, 442–444).

Ruz (2000) wrote about the Zoques of the Sierra Tabasqueña: "Zoques are about to disappear as a cultural group. Except for the Chontales from Tabasco, which would seem destined to be diluted in the miscegenation" (Ruz 2000, 38). Therefore, drawing territorial demarcations for the Zoques that settled in Tabasco is problematic. Nevertheless, it is useful to take into consideration what García de León (1971) proposed about the area occupied by the Mixe-Zoque family: "Its scope includes southern Veracruz, western Tabasco and the isthmian part of Oaxaca and Chiapas, a region that corresponds to archaeologically known Olmec area" (García de León 1971, 209). Additionally, Villa Rojas proposed that: "[t]he Zoques province is located in the extreme northwest of the Chiapas state and surrounding areas of the Oaxaca and Tabasco states" (Villa Rojas 1990, 15). Likewise, Lowe (1983) noted that the geographical area inhabited by the Mixe-Zoque culture in the pre-Hispanic period during the different periods of its development ranged from the Grijalva River, down the Angostura canyon to the Pacific Ocean, and then crossed the Isthmus to the west and north of the Gulf of Mexico (Lowe 1983, 127).

Pye and Clark (2006) stated in their work that "The Olmecs are Mixe-Zoques (Lowe, 1983) to the formative archeology"; for additional emphasis, Dr. Lowe stated that "The speakers of Mixe-Zoque-Popoluca spread from the Pacific Coast to the mountains of Los Tuxtlas in Veracruz, including the plains of Tabasco" (Pye et al. 2006).

Lorenzo Ochoa (2011) also stated in his article "News of its millennial history: The pre-Hispanic towns of Tabasco" that "Depression of the Grijalva passed those who reached various points from northwest of Tabasco and even the Usumacinta. There is no certainty about the language spoken by the newcomers. I suggest that they were Mixe-Zoque groups, as proposed by Lowe, Lee and Clark, among other authors" (Ochoa 2011, 207–222).

Otto Schumann (1985) elaborated: "Geographically, the Mixe-Zoques spread through Chiapas, as well as Tabasco; large portions of Veracruz and Oaxaca were occupied by themselves. As far as Tabasco is concerned, the Zoque towns extended from Macuspana to the border with Veracruz, while in the west they reached the present border of the Chiapas state" (Schumann 1985, 114–115).

Norman Thomas (1993) proposed the following about the Zoques in Tabasco: "The Zoques limit on the northwest consists of a line drawn from northwest to southeast through the rivers Teapa, Tacotalpa, and Puxcatán, passing the Zoque towns of Ixtapangajoya, Tapijulapa, and Puxcatán. The main line of Zoque towns that make up the boundary of this border are, from north to southwest: Puxcatán, Oxolotán, and Amatán" (Thomas 1993, 65).

Lorenzo Ochoa (1997) pointed out that, upon the arrival of the Spaniards and in addition to the Chontales in Tabasco, "the Zoque group was still important, not for its number but for its antiquity in the lowlands of Tabasco, where they seem to have arrived more or less three thousand years ago" (Ochoa 1997, 25–27).

In another study, Ochoa and Jaime (2000) commented: "Around the years 1350–1300 BC, from the Pacific coasts of Guatemala and Chiapas, groups identified as Mixe-Zoque, with a background culture in which their ceramics and figurines stand out, arrived in the territory of what is now south of Veracruz and northwest of Tabasco, merging with the local inhabitants" (Ochoa and Jaime 2000, 22–23).

Further, Wichmann, Beliae, and Davletshin (2008) proposed that during 1800 and 1600 BC, the Proto-Mixe-Zoque differed linguistically and extended to the geographical area of the Locona ceramic phase (1450–1350 BC); similarly, they covered a large part of Veracruz and most of the Tabasco territory. In addition, these authors raise the presence of the proto-Zoque group in La Venta, Tabasco (Wichmann et al. 2008, 683).

Based on the comparative method of historical linguistics, Brown, Wichmann, and Beck (2014) revealed that the Chitimacha language of southern Louisiana and the languages of the Totozoqueana family of Mesoamerica are genealogically related (Brown et al. 2014, 425).

According to the cited references, the area inhabited by the Zoques in Tabasco is inscribed in the territory proposed for the Mixe-Zoque. In addition, Wichmann and colleagues described the presence of proto-Zoque groups on the Gulf that would have originated approximately 1600 BC (Wichmann et al. 2008, 667–684).

Finally, Brown, Wichmann, and Beck (2014) proposed the following, based on linguistic data:

"The Automated Similarity Judgment Program (ASJP) provides a method for estimating the latest date at which a proto-language was spoken (Holman et al. 2011). The ASJP date for PCh-Tz is 5582 years before present (BP). ASJP dates are based on automated calculations of lexical similarity for groups of related languages calibrated with historical, epigraphic, and archaeological divergence dates known from the literature for 52 language groups. The discrepancies between estimated and calibration dates are found to be on average 29% as large as the estimated dates themselves, a figure that does not differ significantly among the world's language families. Within this margin of error, the divergence of Ch and Tz from one another would have occurred any time between 7201 and 3963 BP. Chitimacha may have migrated to the LMV at some point within this chronological range, but it could also have happened later if the breakup of the language ancestral to Ch and Tz took place within Mesoamerica". (Brown et al. 2014, 34)

Archaeological Research

Archaeological research on the Zoques from Tabasco is limited. In fact, there is no additional archaeological information except for a brief reference to the visit made by Matthew W. Stirling in 1944 to the populations of Oxolotán, Tapijulapa; a short report on the same towns that Ernesto Vargas and Lorenzo Ochoa described in 1982; the work regarding the convent of Oxolotán by Laura Ledesma Gallegos (1992); and the excavations at the Zoque site of Malpasito that were carried out by Francisco Cuevas Reyes (1991), which in part have addressed the study of the Zoque area of the Gulf.

Due to limited knowledge, we have been exploring the mountain region of Tabasco. A little more than two decades ago, we conducted archaeological research in the municipalities of Tacotalpa and Teapa in Tabasco, where the Zoque area of Tabasco is located. To date, we have conducted research on approximately 600 km^2, covering the abovementioned municipalities; in this area, we located more than 50 archaeological settlements.

As a result of this research, surface surveys within the Project of Archaeological Recognition of Zoque settlements in the municipalities of Tacotalpa and Teapa in Tabasco and Chapultenango in Chiapas (2013–2017), and the analysis of the pre-Hispanic pottery recovered in the mentioned region, we found the following chronology: Early Preclassic (1200–900 BC) and Late Preclassic (400–200 BC). For the Early Classic period (250–600 AD), there are only a few cultural manifestations. Starting in the Late Classic–Terminal (600–900 AD), there is the presence of fine paste pottery, which was maintained until the Late Postclassic (1200–1521 AD).

The dating of collagen in a human tooth with an age of 3033 ± 30 BP and an associated sample of coal with an age of 2135 BC indicate that some flint artifacts were made by the first inhabitants who migrated to the mountain region of Tabasco.

According to the manufacturing characteristics of the ceramics, the conclusion is that most of the ceramics are of local origin, while approximately 1,000 fragments show affinity with some ceramic types of the Olmec area (San Lorenzo Phases 1150–900 BC and Nacaste 900–700 BC; Coe and Diehl 1980, 159–200). The pottery from the Cacahuanó A and B phases from San Isidro in Chiapas (Lowe 1998, 21–47) were pots of Olmecoid designs from the Pomoca site from Balancan Municipality in Tabasco (Ochoa et al. 1991, 8–9).

The more representative pottery recovered from the Early Preclassic period was identified as Calzadas Carved, Limón Incised, Tatagapa Red, Camaño Coarse, Teófilo Punctate, pots with punched decoration, Tapalapa Unslipped, Juventud Red, Sierra Red, and Pobacama Sandy. Cultural manifestations were scarce in the Early Classic period. Fine paste from the Late Classic–Terminal Period (600–900 AD) was also recovered in Altar, Balancán, Tres Naciones, and Achote ceramic groups, as well as a significant number of fragments of incense and incense holders located mainly in caves. The iconography was similar to that exhibited in the Regional Museum of Anthropology, "Carlos Pellicer Cámara" from Villahermosa in Tabasco. Fine paste pottery was widely distributed in the Maya region during the Late Classic–Terminal period and also was reproduced along the Usumacinta River, in northeastern Guatemala, and in the central plateau of Chiapas, among other locations.

The Matillas type, registered in the Early Late Postclassic period (900–1521 AD), is characterized by its fine paste of uniform orange color without temper and dusty to the touch. This type has been reported in several sites in Tabasco, Campeche, Yucatan, and Quintana Roo from Mexico. According to Ball (1978), this pottery continued to be manufactured until the early colonial period.

Due to the abundance of mountains and caverns in the Sierra Tabasqueña, it is feasible to assume the absence of large Zoque buildings in this region. The construction of temples was most likely replaced by use of caverns. Research has proposed that for Mesoamerican societies, pyramidal structures represent mountains and caves; thus, in the case of the Zoques of the Sierra Tabasqueña, it was not necessary to build pyramids. In the geographical environment made up of mountains and caves, there were symbolic temple constructions that were of much more value than if they had been built by hand. The above was documented in codices, sculptures, stelae, and petroglyphs, among other works of Mesoamerican art.

An important finding is the image of an anthropomorphic face of Olmec appearance. This image is on a rocky wall church at the hill in the municipality of Tacotalpa, Tabasco. The face was derived from the morphology of a small speleothem (a structure formed in a cave by the deposition of minerals from water) and delineated with red pigment (iron oxide), which gave volume to the face (Figure 17.2).

The abundant presence of snails (*Pachychylus rovirosai*), freshwater clams (*Proptera alata* and *Rangia cuneate*), and river crayfish (*Potamocarcinus* spp. and *Maxillipes* spp.) suggests that the rock shelter was occupied during periods of extreme rainfall. Furthermore, the skeletal remains of mammals, such as deer and wild boar, indicate that the inhabitants of the rock shelter hunted these animals and captured armadillos and turtles. The lithic tools indicate specific hunting activities. The bone punches could have been used in leatherworking activities.

Although prehistoric evidence does not exist for the wild resources that are still consumed today (i.e., sea bass, tenhuayaca, bobo, sardine, pihua, pejelagarto, and mojarras) or for the plants of the region, including chapaya (*Astrocarium mexicanum*), guaya (*Chamaedorea* spp.), chapaya, cacate, castañas, chayas, hierba mora, and fungi, it is reasonable to assume that these resources were also part of the diet of the first settlers.

Conclusion

The archaeological evidence recovered in the mountain region of Tabasco indicates that human occupation began at the end of the third millennium BC. According to linguistic data, it is possible that the first inhabitants arrived approximately

Figure 17.2 Image of an anthropomorphic face of Olmec appearance depicted on a rocky wall found at the top of a hill in the municipality of Tacotalpa, Tabasco.

in 2500 BC and were Proto-Mixe-Zoque immigrants coming from the greater Isthmian area.

The way of life of the first settlers of the Sierra Tabasqueña was semisedentary. Food was gathered through hunting, fishing, and collecting plants and snails (shotes). The chronology indicates that the first settlements occurred from the end of the third millennium to 1300 BC.

From 1300 BC until the Protoclassic Period, the settlement pattern gradually transformed into a society of sedentary farmers, most likely organized around the political figure of a chief. The evidence, based on some preserved relics, points to a group that took advantage of the abundant natural resources of the area very efficiently.

The aforementioned pottery, in addition to the differences in architecture, the presence of tombs, inlaid dental pieces, ball games, foreign ceramics (pieces of leaded pottery and shards with remnants of chapopote), green stone, sheets of travertino (*tecali*), gray obsidian from El Chayal, and green obsidian from the Sierra de Pachuca, Hidalgo, indicate that the Zoque people settled in the headquarters of economically and politically integrated societies headed by caciques.

In conclusion, the loss of the cultural components that differentiated the Zoque serranos (living in the Sierra or montain) from Tabasco is linked to the deterioration of the geographic environment, relocation and regrouping of the indigenous population, dispossession of land, exploitation of indigenous labor, excessive taxes, emigration, demographic crises, and changes in the form and structure of housing, among others. These components have been evolving from pre-Hispanic times until a few years ago.

References

Agrinier, P. 1984. *The Early Olmec Horizon at Mirador, Chiapas, Mexico*. Papers of the New World Archaeological Foundation No. 48, Provo, UT.

Ball, J. W. 1978. "Archaeological Pottery of the Yucatan-Campeche Coast." In *Studies in the Archaeology of the Coastal Yucatan and Campeche Mexico*. Publication Nr. 46. Middle American Research Institute, Tulane University, New Orleans.

Brown, C. H., S. Wichmann, and D. Beck. 2014. "Chitimacha: A Mesoamerican language in the lower Mississippi Valley 1." *International Journal of American Linguistics* 80, no. 4: 425–474.

Bryant, D., and J. Clark. 1983. "Los primeros mayas precolombinos de la cuenca superior del Río Grijalva [The first pre-Columbian Maya of the upper Grijalva River basin." In *Antropología e historia de los mixe-zoques y mayas: Homenaje a Frans Blom*, edited by L. Ochoa and T. A. Lee, 223–239. UNAM-Brigham Young University. Mexico City.

Clark, J. E., R. D. Hansen, and T. Pérez-Suárez. 2000. "La zona maya en el Preclásico Preclásico [The Mayan zone in the Preclassic]." *Historia Antigua de México,* Vol. I, edited by L. Manzanilla and L. López Luján, 437–510. CONACULTA-INAH-UNAM. Mexico City.

Coe, M. D., and R. A. Diehl. 1980. *The Land of the Olmec: The Archaeology of San Lorenzo Tenochtitlan*. Austin: University of Texas Press.

Cheetham, D. 2006. "The Americas first colony." *Archaeology* 59, no. 1: 40–46.

Cuevas-Reyes, F. 1991. *Proyecto arqueologico Sierras Bajas de Tabasco. Boletin del Consejo de Arqueologia [Archaeological Project, Sierras Bajas of Tabasco. Bulletin of the Council of Archeology],* 60–63. INAH.

García de León, A.. 1971. "El Ayapaneco: una variante del zoqueano en la Chontalpa tabasqueña [The Ayapaneco: a variant of the Zoqueano in the Chontalpa of Tabasco]." Instituto Nacional de Antropología e Historia. Mexico City.

Holman, E.W., C. H. Brown, S. Wichmann, A. Müller, V. Velupillai, H. Hammarström, S. Sauppe, H. Jung, D. Bakker, P. Brown, O. Belyaev, M. Urban, R. Mailhammer, J.-M. List, and D. Egorov. 2011. Automated dating of the world's language families based on lexical similarity. Current Anthropology 52 (6): 841-875.

Ledesma-Gallegos, L. 1992. *La vicaria, de Oxolotan, Tabasco [The vicaria, from Oxolotan, Tabasco]* . INAH, Colección Científica, 257, Serie Arqueología Histórica. Mexico City.

Lee, . A. 2009. "El papel civilizatorio de los olmecas y sus protagonistas, los mixe-zoques en Mesoamérica [The civilizing role of the Olmecs and their protagonists, the Mixe-Zoques in Mesoamerica]." In *Medioambiente, antropología, historia y poder regional en el occidente de Chiapas y el Istmo de Tehuantepec*, edited by T. A. Lee Whiting, V. M. Esponda Jimeno, D. Domenici, and C. Uriel del Carpio Penagos, 67–80. Universidad de Ciencias y Artes de Chiapas. Chiapas, Mexico.

Lee Whiting, Thomas A., and Cheetham. 2008. "Lengua y escritura olmeca [Olmec language and writing]." Olmeca: balance y perspectivas. Memoria de la Primera Mesa Redonda, edited by M. T. Uriarte and R. B. González Lauck, 695–713. UNAM, CONACULTA, NWAF. Mexico City.

Lewis, P. M., ed. 2009. *Ethnologue: Languages of the World*. Dallas: SIL International. www.ethnologue.com/show_family.asp?subid=1865-16

Lowe, G. W. 1983. "Los olmecas, mayas y mixe-zoques [The Olmecs, Mayas and Mixe-Zoques]." In *Antropología e historia de los mixe-zoques y mayas: Homenaje a Frans Blom*, edited by L. Ochoa and T. A. Lee, 125–130. UNAM-Brigham Young University. Mexico City.

Lowe, G. W. 1998. "Los olmecas de San Isidro en Malpaso, Chiapas [The Olmecs of San Isidro in Malpaso, Chiapas]." Víctor Esponda Jimeno, (ed.), México, Instituto Nacional de Antropología e Historia, Centro de Investigaciones Humanísticas de Mesoamérica y El Estado de Chiapas-UNAM (Colección Científica 371, Serie Arqueología). Mexico City.

Ocho. L.. 1997. "Los señoríos prehispánicos. En los límites de la imaginación: la concepción geopolítica de Tabasco al momento del contacto [The pre-Hispanic Lordships. In the limits of the imagination: the geopolitical conception of Tabasco at the time of contact]." In *Tabasco: apuntes de frontera*, edited by Mario Humberto Ruz, 15–40.Coordinación Nacional de Descentralización, Programa de Desarrollo Cultural del Usumacinta, Consejo Nacional para la Cultura y las Artes. Mexico City.

Ochoa, L.. 2011. "Noticias de su historia milenaria: Los pueblos prehispánicos de Tabasco [News of its millenary history: The pre-Hispanic peoples of Tabasco]." In *Tabasco más agua que tierra*, edited by Carlos Martínez Assad, 52–77. Sociedad de Estudios Regionales. Mexico City.

Ochoa, L., and Luis Casasola. 1991. "Tierra Blanca y el Medio Usumacinta. Notas de su cerámica arqueológica [Tierra Blanca and the Middle Usumacinta. Notes of its archaeological pottery]." In *Tierra y Agua: La antropología en Tabasco,* edited by Lorenzo Ochoa, 7–28. Instituto de Cultura de Tabasco, Dirección de Patrimonio Cultural from the Gobierno del Estado de Tabasco,. Tabasco, Mexico.

Ochoa, L., and Olaf J.. 2000. *Un Paseo por el Parque-Museo La Venta* [A Walk through the Park Museum-La Venta]. Tabasco, Mexico: Gobierno de Tabasco, Instituto de Cultura de Tabasco, Consejo Nacional para la Cultura y las Artes.

Pye, M. E., and J. E. Clark. 2006. "Los olmecas son mixe-zoques: contribuciones de Gareth Lowe a la arqueología del Formativo [The Olmecs are Mixe-Zoques: Gareth Lowe's contributions to Formative archeology]." In *Presencia zoque una aproximación multidisciplinaria*, edited by D. A. Calderon, T. A. Lee Whiting, and M. L. Guillen, 207–222. Universidad de Ciencias y Artes de Chiapas. Chiapas, Mexico.

Rands, R.. L. 1967. "Cerámica de la región de Palenque, México [Pottery from the Palenque region, Mexico]." *Estudios de Cultura Maya* VI: 111–147. Mexico City

Ruz, M. H. 2000. *El Magnífico Señor Alonso López, Alcalde Santa María de la Victoria y Aperreador de Indios (Tabasco 1541) [The Magnificent Lord Alonso López, Mayor of Santa María de la Victoria and Aperreador de Indios (Tabasco 1541)].* UNAM, Instituto de Investigaciones Filológicas, Centro de Estudios Mayas, Plaza y Valdés. Mexico City.

Schumann, O.. 1985. "Consideraciones históricas acerca de las lenguas indígenas de Tabasco [Historical considerations about the indigenous languages of Tabasco]." In Olmecas y mayas en Tabasco: cinco acercamientos, edited by Lorenzo Ochoa, 114. , Gobierno del Estado de Tabasco, Instituto de Cultura de Tabasco. Villahermosa, Tabasco, Mexico.

Stirling, M. W.. 1957. "An archeological reconnaissance in Southeastern Mexico." Bureau of American Ethnology Bulletin. 164 (53):213–240. United States.

Thomas, N. D. 1993. "Los Zoques." In *La población indígena de Chiapas*, edited by Víctor M. Esponda, 49–90. Tuxtla Gutiérrez, Gobierno del Estado de Chiapas, Consejo Estatal de Fomento a la Investigación y Difusión de la Cultura, DIF-Chiapas/Instituto Chiapaneco de Cultura, (Serie Nuestros Pueblos). Chiapas, Mexico.

Vargas, E., and Lorenzo O.. 1982. "Navegantes, viajeros y mercaderes: notas para el estudio de la historia de las rutas fluviales y terrestres entre la costa de Tabasco-Campeche y tierra adentro [Navigators, travelers and merchants: notes for the study of the history of the river and land routes between the coast of Tabasco-Campeche and inland]." *Estudios de Cultura Maya* XIV: 59–118. Instituto de Investigaciones Filológicas/Centro de Estudios Mayas, UNAM. Mexico City.

Villa Rojas, A.. 1990. "Configuración cultural de la región zoque de Chiapas [Cultural configuration of the zoque region of Chiapas]." In: *Los zoques de Chiapas*. Alfonso VillaRojas A., José M. Velasco-Toro; F. Báez-Jorge; F. Córdoba Olivares y Norman D. Inoma. Dirección General de Publicaciones del Consejo Nacional para la Cultura y las Artes/Instituto Nacional Indigenista, México, pp. 15–42 (Colección Presencias). Mexico City.

Wichmann S., D. Beliaev, and A. Davletshin. 2008. "Posibles correlaciones lingüísticas y arqueológicas involucrando a los olmecas [Possible linguistic and archaeological correlations involving the Olmecs]." In *Olmeca: balance y perspectivas. Memoria de la Primera Mesa Redonda*, edited by M. T. Uriarte and R. B. González Lauck, 667–684. UNAM, CONACULTA, NWAF. Mexico City.

PART V
DISEASE AND MIGRATION

18

Impact of Human Migration on the Spread of Arboviral Diseases on the United States–Mexico Border

Alvaro Diaz-Badillo and Maria de Lourdes Muñoz-Moreno

Introduction

Migrant healthcare is very complex and is influenced by human migration, resettlement, and the social determinants of critical postmigration health stressors. For regular migrant intakes, recent changes in source-country migration patterns have altered the cultural and linguistic diversity of many host countries, increasing the complexities of delivering health services to various migrant and refugee populations and the impact on health after migration. Migrants face obstacles to accessing essential healthcare services due to factors like language, cultural barriers, financial and legal barriers, and a lack of specific comprehensive health policies. Overcoming obstacles to healthcare access is a priority for the well-being of migrants as well as host communities, for achieving global health goals, and for preventing and eradicating infectious diseases.

On the American continent, most migrants travel to the United States through Mexico; therefore, viral pathogens carried by mosquito vectors may be carried by travelers from south to north and also from north to south. Consequently, mosquito vectors of some viral diseases are expected to increase vector-borne disease in the United States and Mexico (de Lourdes Munoz et al. 2013; Diaz-Badillo et al. 2014; Gorrochotegui-Escalante et al. 2002), which is also influenced by climate variability and extreme weather events increasing mosquito vectors and consequently viral infections.

A local surveillance arrangement to detect emergent and re-emergent infections, awareness on the part of healthcare organizations and laboratories, vector control, and health-protection education programs are needed to stop the spread of arboviral diseases (ADs; Soto 2009.). Surveillance Systems frequently fail due to limited available resources, lack of knowledgeable staff, disorganization, and poor infrastructure for finding and reporting cases. Contrary, stronger public health surveillance systems will allow more accurately deacription and assessment of the state

Alvaro Diaz-Badillo and Maria de Lourdes Muñoz-Moreno, *Impact of Human Migration on the Spread of Arboviral Diseases on the United States–Mexico Border* In: *Human Migration*. Edited by: Maria de Lourdes Muñoz-Moreno and Michael H. Crawford, Oxford University Press. © Oxford University Press 2021. DOI: 10.1093/oso/9780190945961.003.0019 of health. This will improve health promotion programs and be used as a guide by policy makers for allocating resources effectively. Furthermore, understanding the influence of human migration on the spread of ADs is critical to improving control efforts and to preventing future ADs outbreaks..

The main focus of our research is disease caused mainly by arboviruses (e.g., dengue, chikungunya, Zika, and West Nile) and the socioeconomic and environmental factors likely driving transmission of arboviruses. Furthermore, our current efforts are directed at studying the underserved and understudied United States–Mexico border populations to assess differences in immune responses linked to the regulation of numerous genes directly associated with the transcriptional response to infection, which may be associated with ethnic origin. This integrative analysis could potentially identify novel ethnic-specific biomarkers of susceptibility to arboviral infections (AIs) that may aid in disease prognosis and pinpoint novel therapeutic targets for a move toward precision medicine. The innovative features of this study include the use of an exceptional binational border population and state-of-the-art molecular and bioinformatics techniques for understanding the pathogenesis of ADs.

Description of the United States–Mexico Border

The United States–Mexico border area represents a binational geopolitical system with strong socioeconomic, cultural, and environmental connections governed by different policies, customs, and laws. Important dimensions of this binational system include commerce, tourism, sister-city familial ties, and Mexico's assembly plants, or maquiladoras, which import components for processing or assembly by Mexican labor and then export the finished products.

The border between the United States and Mexico stretches for nearly 2,000 miles from the Gulf of Mexico to the Pacific Ocean. The complete United States–Mexico border is limited by 10 states (California, Arizona, New Mexico, and Texas in the United States, and Baja California Norte, Sonora, Chihuahua, Coahuila, Nuevo Leon, and Tamaulipas in Mexico), 25 U.S. counties, 37 Mexican municipalities, and 24 pairs of sister cities (Table 18.1). This which constitutes a total population of a little more than 14 million inhabitants (INEGI 2017; United Nations 2017; Unied States Census Bureau 2017). The border areas of the two countries share environmental, social, economic, cultural, and epidemiological characteristics but operate under different policies, rules, and regulations.

The determinants of health on the Mexican side of the border show more favorable conditions than in the country as a whole. The opposite is observed on the U.S. side of the border, where the determinants of health are usually worse than in the country as a whole. Unknown numbers of the U.S. border population are underinsured; however, this problem is overcome by obtaining healthcare across the border, since medical and dental procedures and medications are cheaper in Mexico than in the United States.

Table 18.1 Sister cities in states along the United States–Mexico border*

Mexican State (N = 6)	Municipality (N = 37)	U.S. State (N = 4)	County (N = 25)	Sister Cities (24 pairs***)
Baja California Norte	Tijuana	California	San Diego	San Diego/Tijuana
(3/5)**	Tecate	(2/58)**		Tecate/Tecate
	Mexicali		Imperial	Calexico/Mexicali
Sonora	San Luis del Rio Colorado	Arizona	Yuma	Yuma/San Luis Rio Colorado
(11/72)**	Puerto Peñasco	(4/15)**		
	Plutarco Elias Calles		Pima	Lukeville/Sonoyta
	Caborca			
	Altar			
	Saric			Sasabe/Sasabe
	Nogales		Santa Cruz	Nogales/Nogales
	Santa Cruz			
	Cananea			
	Naco		Cochise	Naco/Naco
	Agua Prieta			Douglas/Agua Prieta
Chihuahua	Janos	New Mexico	Hidalgo	
(7/67)**	Ascencion	(3/33)**	Luna	Columbus/Puerto Palomas
			Doña Ana	
	Juarez	Texas	El Paso	El Paso/Ciudad Juarez
	Guadalupe I	(16/254)**	Hudspeth	Fabens/Guadalupe
	Praxedis			Fort Hancock/El Porvenir
			Culberson	
			Jeff Davis	
	Ojinaga		Presidio	Presidio/Ojinaga
	Manuel Benavides		Brewster	
Coahuila	Ocampo		Terrell	
(6/38)**	Acuña		Val Verde	Del Rio/Ciudad Acuña
	Jimenez		Kinney	
	Piedras Negras		Maverick	Eagle Pass/Piedras Negras
	Guerrero		Dimmit	
	Hidalgo			
Nuevo Leon	Anahuac			
(1/51)**				
Tamaulipas	Nuevo Laredo		Webb	Laredo/Nuevo Laredo
(9/43)**				
			Zapata	
	Guerrero		Starr	Falcon Heights/Nueva Ciudad Guerrero
	Mier			

Continued

Table 18.1 *Continued*

Mexican State ($N = 6$)	Municipality ($N = 37$)	U.S. State ($N = 4$)	County ($N = 25$)	Sister Cities (24 pairs***)
	Miguel Aleman			Roma/Ciudad Miguel Aleman
	Camargo			Rio Grande City/Ciudad Camargo
	Gustavo Diaz Ordaz			
	Reynosa		Hidalgo	McAllen-Hidalgo/Reynosa
	Rio Bravo			Donna-Weslaco/Rio Bravo
				Progreso/Nuevo Progreso
	Matamoros		Cameron	Brownsville/Matamoros

* The counties and municipalities were included based on the geographical border. ** The number indicates the number of border municipalities/counties of the total municipalities/counties in the states. ***The sister-city pairs were included base on border community feedback in 2017. The table was created with data from the U.S. Environmental Protection Agency and the Texas Commission on Environmental Quality.

Thus, segments of the U.S. border populations optimize their health by using the resources available in both countries (Miller-Thayer 2010). The Texas border makes up about half of the United States–Mexico border, with a population of approximately 3 million residents, and it is one of the busiest international boundaries in the world. Furthermore, air, sea, and land transport networks continue to expand in reach, speed of travel, and volume of passengers and goods carried. This has allowed pathogens and their vectors to move faster and in greater numbers than ever. The Texas Airport Directory contains information on nearly 400 airports that are open to the public (Texas Department of Transportation 2017; Figure 18.1).

Three important consequences of global transport network expansion are infectious disease pandemics, vector invasion events, and vector-borne pathogen importation (Tatem et al. 2006). Furthermore, vector populations are diverse and are capable of transmittingand spreading ADs (TDSHS 2018b).

Human Migration and Infectious Diseases

Since remote times, migrant populations played an essential role in the spread of infectious diseases (Barnett and Walker 2008). Immigrants have ongoing links with populations in their countries of origin that provide a channel through which infectious diseases potentially can be introduced to new areas (Barnett and Walker 2008; Vignier and Bouchaud 2018). Commercial globalization, population movements, and environmental changes are the main factors favoring the international spread of emerging infectious diseases (EIDs), including the viral diseases transmitted by mosquito vectors.

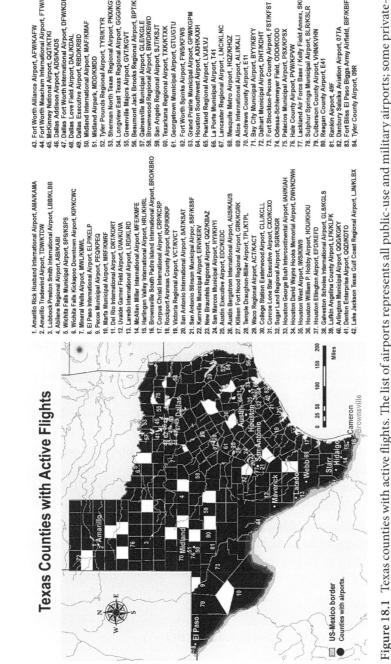

Texas Counties with Active Flights

1. Amarillo Rick Husband International Airport, AMA/KAMA
2. Amarillo Tradewind Airport, TDW/KTDW
3. Lubbock Preston Smith International Airport, LBB/KLBB
4. Abilene Regional Airport, ABI/KABI
5. Wichita Falls Municipal Airport, SPS/KSPS
6. Wichita Falls Kickapoo Downtown Airport, KIP/KCWC
7. Mineral Wells Airport, MWL/KMWL
8. El Paso International Airport, ELP/KELP
9. Pecos Municipal Airport, PEQ/KPEQ
10. Marfa Municipal Airport, MRF/KMRF
11. Del Rio International Airport, DRT/KDRT
12. Uvalde Garner Field Airport, UVA/KUVA
13. Laredo International Airport, LRD/KLRD
14. McAllen Miller International Airport, MFE/KMFE
15. Harlingen Valley International Airport, HRL/KHRL
16. Brownsville South Padre Island International Airport, BRO/KBRO
17. Corpus Christi International Airport, CRP/KCRP
18. Rockport Aransas County Airport, RKP/KRKP
19. Victoria Regional Airport, VCT/KVCT
20. San Antonio International Airport, SAT/KSAT
21. San Antonio Stinson Municipal Airpor, SSF/KSSF
22. Kerrville Municipal Airport, ERV/KERV
23. New Braunfels Regional Airport, HYI/KHYI
24. San Marcos Municipal Airport, HYI/KHYI
25. Austin Executive Airport, EDC/KEDC
26. Austin Bergstrom International Airport, AUS/KAUS
27. Killeen Fort Hood Regional Airport, GRK/KGRK
28. Temple Draughon-Miller Airport, TPL/KTPL
29. Waco Regional Airport, ACT/KACT
30. College Station Easterwood Airport, CLL/KCLL
31. Conroe Lone Star Executive Airport, CXO/KCXO
32. Sugar Land Regional Airport, SGR/KSGR
33. Houston George Bush Intercontinental Airport, IAH/KIAH
34. Houston David Wayne Hooks Memorial Airport, DWH/KDWH
35. Houston West Airport, IWS/KIWS
36. Houston William P. Hobby Airport, HOU/KHOU
37. Houston Ellington Airport, EFD/KEFD
38. Galveston Scholes International Airport, GLS/KGLS
39. Lufkin Angelina County Airport, LFK/KLFK
40. Arlington Municipal Airport, QQG/KGKY
41. Denton Enterprise Airport, QQD/KDTO
42. Lake Jackson Texas Gulf Coast Regional Airport, LJN/KLBX

43. Fort Worth Alliance Airport, AFW/KAFW
44. Fort Worth Meacham International Airport, FTW/KFTW
45. McKinney National Airport, QQT/KTKI
46. Dallas Addison Airport, ADS/KADS
47. Dallas Fort Worth International Airport, DFW/KDFW
48. Dallas Love Field Airport, DAL/KDAL
49. Dallas Executive Airport, RBD/KRBD
50. Midland International Airport, MAF/KMAF
51. Midland Airpark, MDD/KMDD
52. Tyler Pounds Regional Airport, TYR/KTYR
53. Sherman North Texas Regional Airport, PNX/KGYI
54. Longview East Texas Regional Airport, GGG/KGGG
55. Greenville Majors Airport, GVT/KGVT
56. Beaumont Jack Brooks Regional Airport, BPT/KBPT
57. Gainesville Municipal Airport, GLE/KGLE
58. Brownwood Regional Airport, BWD/KBWD
59. San Angelo Regional Airport, SJT/KSJT
60. Texarkana Regional Airport, TXK/KTXK
61. Georgetown Municipal Airport, GTU/GTU
62. Fort Worth Spinks Airport, FWS/KFWS
63. Grand Prairie Municipal Airport, GPM/KGPM
64. Houston Southwest Airport, AXH/KAXH
65. Pearland Regional Airport, LVJ/LVJ
66. La Porte Municipal Airport, T41
67. Lancaster Regional Airport, LNC/KLNC
68. Mesquite Metro Airport, HQZ/KHQZ
69. Alice International Airport, ALI/KALI
70. Andrews County Airport, E11
71. Bay City Municipal Airport, BYY/KBYY
72. Dalhart Municipal Airport, DHT/KDHT
73. Fort Stockton-Pecos County Airport, FST/KFST
74. Odessa-Schlemeyer Field, ODO/KODO
75. Palacios Municipal Airport, PSX/KPSX
76. Hale County Airport, PVW/KPVW
77. Leckland Air Force Base / Kelly Field Annex, SKF/KSKF
78. Sulphur Springs Municipal Airport, SLR/KSLR
79. Culberson County Airport, VHN/KVHN
80. Reagan County Airport, E41
81. Rankin Airport, 49F
82. Danbury Salaika Aviation Airport, 07TA
83. Fort Bliss El Paso Biggs Army Airfield, BIF/KBIF
84. Tyler County Airport, 09R

Figure 18.1 Texas counties with active flights. The list of airports represents all public-use and military airports; some private-use and former airports may be also included, such as airports that were previously for public use, those with commercial enplanements recorded by the Federal Aviation Administration (FAA), or airports assigned an International Air Transport Association (IATA) airport code. (Map created with data from Texas Department of Transportation, Aviation Division, & Airport Directory; Flightradar24.com, AirNav.com.)

The traditional model of infectious disease causation, known as the epidemiologic triad, which is ideal for descibing the relationship between ADs, human migration, and the spread of diseases along the United States–Mexico border (Ansari and Shope 1994). Classic examples of vectors include the *Aedes, Culex,* and *Anopheles* mosquitoes. These mosquitoes can ingest blood from an arbovirus-infected host, after which the arboviruses first replicate in the insect's midgut epithelium, subsequently disseminate through the hemolymph to other organs, and finally infect the salivary glands. At this point, the virus (e.g., dengue, West Nile, chikungunya, or Zika) is stored in the mosquito's saliva and then is injected into a new human host during the insect's next blood meal, causing the viral disease.

AIs are a global health problem accounting for significant morbidity and mortality in human and animal populations. They belong to the families Togaviridae, Flaviviridae, Bunyaviridae, Reoviridae, and Rhabdoviridae, and they are transmitted to humans and domestic/sylvatic animals by the bite of infected arthropods. Rodents and birds are significant vertebrate hosts (reservoirs), while humans are usually not involved in the maintenance and spread of most arboviruses. Ecologic changes and human behavior, including migration, are important in the spread of these infections.

The management of illness and disease in foreign-born populations can be affected by the status of the individuals themselves. Illegal or irregular migrants may not have healthcare services, attend follow-up assessments, or fill prescriptions out of fear of resulting immigration enforcement actions. This behavior can affect compliance in cases of long-term treatment and may delay initial presentation, complicating public health follow-up for communicable diseases.

Dengue, West Nile, chikungunya, Zika, and Mayaro are viruses that evolve rapidly when confronted with new populations or challenges. Due to extreme changes in weather, increasing travel and immigration from tropical and subtropical climates, and increasing numbers of ADs, dengue and Zika infections represent a growing threat at the United States–Mexico border(Gyawali, Bradbury, and Taylor-Robinson 2016; Imperato 2016).

Consequently, good epidemiological surveillance system for rapid detection of emerging diseases will help to prevent spread of arthropod-borne viral infections. Undoubtedly, surveillance systems collaboration in cross-border public health programs between the Unated States and Mexico will enhance the effectiveness of disease prevention strategies and contribute to efforts to prevent arboviral infections.

The Etiological Agents

Previously, ADs were considered to be minor contributors to global mortality and disability, but in the past few decades, the AD epidemics produced by dengue, West Nile, chikungunya, and Zika have significantly increased, as the result of the modern world's urbanization, globalization, and international mobility, together with migration. Common features of these diseases must encourage research in diagnosis, vaccines, biological targets, and host immune responses, as well as environmental determinants and vector-control measures. Therefore, new global alliances are needed to enable the combination of efforts and resources for more effective and timely solutions (Wilder-Smith et al. 2017) .

The major differences in today's global emergence of ADs is that emergence and dispersion are more rapid and geographically extensive, largely due to intensive growth of global transportation systems, arthropod adaptation to increasing urbanization, failure to contain mosquito populations, land perturbation, and massive human migration (Lindahl and Grace 2015). Furthermore, World War II also contributed to the spread of, and increase in, AIs. The pathogenic arboviruses dengue, West Nile, chikungunya, Zika, and Mayaro are good examples of viral emergence and dispersal in the United States–Mexico border area, especially because they each have a mosquito vector, which is an important determinant.

Dengue Virus

Dengue was endemic in most of Latin America before the emergence of chikungunya in 2013 and Zika in the middle of 2015. Dengue has the highest incidence and the highest case-fatality rate, while both chikungunya and Zika viral infections have potentially long-term sequelae. All the viruses co-circulate, and accordingly there are some reports of concurrent infections in Latin America (Furuya-Kanamori et al. 2016; Rodriguez-Morales, Villamil-Gomez, and Franco-Paredes 2016). Dengue infection may be asymptomatic or symptomatic; those with symptoms get ill between 4 to 7 days after the mosquito bite. The infection is characterized by flu-like symptoms, which include: a sudden high fever coming in separate waves; pain behind the eyes; muscle, joint, and bone pain; severe headache; and a skin rash with red spots. Treatment includes supportive care, since there is no antiviral treatment available. Four antigenically similar serotypes of the dengue virus have been identified (DENV-1, 2, 3, and 4). Most dengue virus infections lead to nonsevere dengue (NSD) based on the World Health Organization (WHO) 2009 dengue case classification, and 9% of the patients develop severe dengue (SD), which is characterized by severe plasma leaking, internal bleeding, organ failure, and possible death (WHO 2009). Dengue can be misdiagnosed as other infectious diseases, especially those due to arboviruses like West Nile, chikungunya, Zika, or Japanese encephalitis (JE), because symptoms are very similar. The frequency and magnitude of dengue epidemics have increased dramatically in the past 50 years as the viruses and the mosquito vectors have both expanded geographically in the tropical regions of the world (Cisneros et al. 2006; Cisneros-Solano et al. 2004).

Dengue in Texas

Dengue is a leading cause of febrile illness in travelers returning to the United States from tropical and subtropical regions (Freedman et al. 2006). From the late 1700s until the 1940s, dengue outbreaks occurred regularly in the southern United States (Ehrenkranz et al. 1971). Dengue did not occur after a campaign that rid much of the continental United States of *Aedes (Stegomya) aegypti* mosquitoes, which transmit the four dengue virus serotypes (Soper 1963). After the campaign, mosquito populations and dengue resurged (Hafkin et al. 1982) in southern Texas in 1980 (Bouri et al. 2012). This dengue outbreak and subsequent outbreaks in 1999 and 2005 were associated with epidemics in northern Mexico. During the 1999 outbreak, despite a higher prevalence of mosquito-infested water containers in Texas, dengue virus infection was less frequent among residents of Texas than among those of northern Mexico, partly because of the more prevalent use of air-conditioning, window screens, and other factors that limit human–mosquito contact in Texas. The Texas–Mexico border area is of particular concern, because dengue virus is endemic in some regions of Mexico (Cisneros-Solano et al. 2004; Gardella-Garcia et al. 2008), but not in the United States. Dengue was once common in Texas in 1922, when there were an estimated 500,000 cases. There was a striking contrast in the incidence of dengue in Texas versus three Mexican states that border Texas (43 cases vs. 50,333) in the period from 1980 to 1996 (Figure 18.2). Dengue cases in the states along the United States–Mexico border from 2000 to 2018 were reported by the Centers for Disease Control and Prevention (CDC), Texas Department of State Health Services (TDSHS), SINAVE/Ministry of Health of Mexico, and the Ministries of Health from Chihuahua, Coahuila, Nuevo Leon, and Tamaulipas States.

Chikungunya Virus

In Texas, the first locally acquired chikungunya case reported was in Cameron County in 2015 (Evans and Meires 2016). As of December 11, 2018, a total of 116 chikungunya cases with illness onset in 2018 were reported to ArboNET from 27 U.S. states, including Texas, where seven cases were reported. All reported cases occurred in travelers returning from affected areas. No locally transmitted cases were reported from U.S. states in 2019 (CDC 2018a; TDSHS 2018a).

In Mexico, the first case of chikungunya was reported in 2014 in Jalisco State in a woman who had returned from the Caribbean (Rivera-Ávila 2014). In the same year, 14 locally acquired cases were reported in Chiapas (Secretaria de Salud Gobierno de Mexico [SSMX] 2015). In the border area, 83 and 0 cases were reported by Mexico and the United States, respectively, in 2014; in 2015, there were 11,577 and 92 cases; in 2016, 757 and 114 cases; in 2017, 61 and 3 case in the border area; and in 2018—updated until epidemiological week 48—39, 0 cases in the border area (SSMX 2018) .

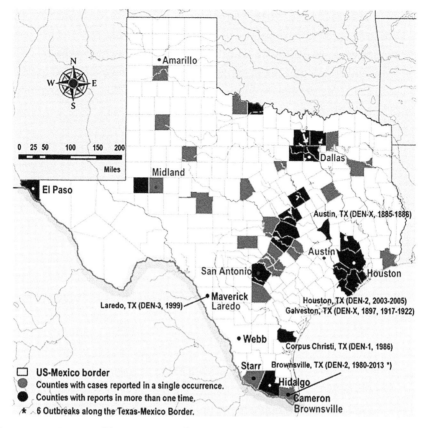

Figure 18.2 Reported human cases of dengue in Texas. Figure was compiled using data reported by CDC and TDSHS, and includes autochthonous cases from Cameron, Hidalgo, Webb, Maverick, Bee, Bexar, Travis, Brazos, and Harris counties (Beaumier et al. 2014; Bouri et al. 2012; Brathwaite-Dick et al. 2012; Hotez 2018; Thomas et al. 2016). All dengue cases in Mexican states were locally acquired.

Zika Virus

There were 5,296 symptomatic Zika virus cases reported to the CDC, 2018c before January 2015. Most of these cases were diagnosed in travelers returning from areas with local transmission, such as the Caribbean, and 224 cases were acquired through presumed local mosquito-borne transmission in two U.S. states—Florida and Texas—reporting local transmission. Confirmed autochthonous vector-borne transmission of Zika virus was reported in the states of Florida (285 cases in 2017) and in Texas (6 cases in 2016). In Texas, three Zika virus cases were confirmed in 2018 (TDSHS 2018c).

In Mexico, the first cases of Zika were reported in November 2015. National health authorities in Mexico notified PAHO/WHO of three cases of Zika virus infection, including two autochthonous cases (residents of Nuevo León and Chiapas) and one imported case, in a person with a history of travel to Colombia (CDC

2018c). Up until December 2016, there were 7,560 confirmed cases of Zika virus infections in Mexico (935 in border states). In 2017, the Ministry of Health reported 3,260 confirmed cases nationally, with 1,010 cases in the states on the U.S. border areas. Until epidemiological week 48 of 2018, 785 cases were reported in México, with one case in Nuevo Leon. In Mexico, no fatal cases of Zika have been reported so far (SSMX 2018).

West Nile Virus

West Nile virus is maintained in nature in a cycle involving transmission between birds and *Culex* mosquitoes as the main vector. Humans, horses, and other mammals can be also infected (WHO 2018). West Nile virus has established itself in North America since its recognition in New York City in 1999. Historically, West Nile virus has been associated with temporally dispersed outbreaks of mild febrile illness, and it is the leading cause of mosquito-borne disease in the continental United States.

In Texas, the first cases of human infection with West Nile virus were reported in 2002, and cases have been reported every year since then (Nolan, Schuermann, and Murray 2013). Historically, 5,539 cases of West Nile disease have been reported since its appearance in 2002 (CDC 2018b). In Mexico, only seven cases were reported in the same period of time, although the vector *Culex pipiens* complex mosquitoes and their hybrids have been reported in Mexico City (Diaz-Badillo 2011).

Mayaro Virus

Mayaro virus was isolated for the first time in Trinidad and Tobago in 1954, although Mayaro virus was identified in sera collected during the construction of channels in Panama and Colombia between 1904 and 1914. Recently, this virus has been spreading into Americas, including the United States–Mexico border area (PAHO/WHO 2019). Mayaro virus symptoms of fever, arthralgia, and maculopapular rash are among the most common symptoms described, being largely indistinguishable from those caused by other arboviruses. However, grave manifestations of the infection, such as chronic polyarthritis, neurological complications, hemorrhage, myocarditis, and death, have been also reported. There are no specific commercial tools for the diagnosis of Mayaro virus.

The Susceptible Host

Human behavior plays an important role in the spread of infectious diseases; consequently, understanding the influence of behavior changes on the spread of these diseases may be a key to improving disease control. EIDs like ADs are increasing and causing losses of both human and animal lives, with large costs to society. Factors like climate change, globalization, and urbanization are contributing to emergence and spread of diseases. Climate-sensitive vector-borne diseases are likely to be emerging due to climate changes and environmental changes, such as temperature or irrigation increases (Funk, Salathe, and Jansen 2010; Lindahl and Grace 2015). Our understanding of EID suggests that humans play a significant role in changing infection patterns not only for themselves but also for domestic animals and wildlife.

Transport and communication development also creates factors that contribute to worldwide dispersion of microorganisms, including chikungunya virus, dengue virus, Zika virus, Mayaro virus, and West Nile virus (Beaumier et al., 2014). This may be also influenced by the lack of some basic living requirements, such as potable water, sewer systems, electricity, and sanitary housing, in Latin-American communities within 50 miles of the United States–Mexico border. Figure 18.3 show the distribution of counties at the border of Texas with Mexico (Mier et al. 2008).

Understanding vector-borne transmission requires a clear understanding of the roles played by wild vertebrate species, domestic pets, or species kept in captivity that may be arbovirus reservoirs and that limit the ability to control disease emergence, and these factors have been neglected by health authorities. Southern Texas, bordering Mexico, is a region where wild animal migration occurs in both directions, which may favor animal infection spread in both directions, making the region a model landscape for AD incursions (Kuno et al. 2017; Liang, Gao, and Gould 2015). With the presence of these reservoirs, periodic outbreaks of viral infections and diseases may be expected among subpopulations in Texas (Figure 18.4).

The Environment

Arboviruses have become important and constant threats in tropical regions, due to rapid climate change, deforestation, population migration, disorderly occupation of urban areas, and precarious sanitary conditions, which favor viral amplification and transmission. These conditions accelerate arbovirus epidemics, directly affecting public health. Abnormally high temperatures, for example, affect populations of insect vectors, as well as ADs, by influencing virus survival, virus replication, vector susceptibility to viruses, vector distribution, extrinsic incubation period of the virus in the insect vector, and seasonality of virus transmission patterns. For example, the vector mosquito *Aedes aegypti* tend to be more active at higher temperatures. In addition, arboviruses are highly spreadable because their vectors can fly long distances, even between countries or continents, which can lead to pandemics (Lorenz et al. 2017).

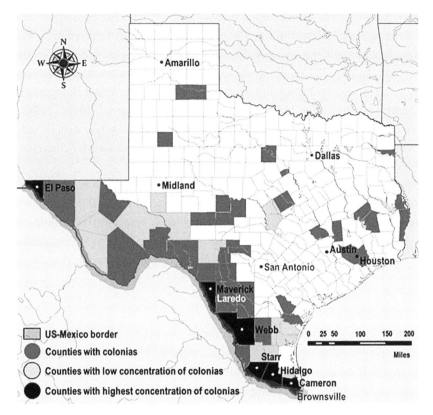

Figure 18.3 Texas counties with colonias. Colonias are economically distressed areas that often lack the most basic necessities, such as drinkable water, septic or sewer systems, electricity, paved roads, or safe, sanitary housing. While colonias are found all across the state of Texas, most of them are located on the border with Mexico. (Map created with data from the US Department of Housing and Urban Development, Texas Commission on Environmental Quality, & USGS.)

In addition, the mosquito *A. aegypti* has effectively adapted to urban habitats, and it has been present throughout the Americas since 1930s. The United States was one of the geographic reservoirs of *A. aegypti* in the Americas. This dramatically increased ADs in America and consequently at the United States–Mexico border (Figure 18.5; Eisen and Moore 2013; Hotez 2018).

The Vector

The dramatic global expansion of *A. albopictus* in the last three decades has increased public health concern because it is a potential vector of numerous arthropod-borne viruses (arboviruses), including the most prevalent arboviral pathogen of humans, dengue virus. *A. aegypti* is considered the primary dengue virus vector and has repeatedly been incriminated as a driving force in dengue's worldwide emergence

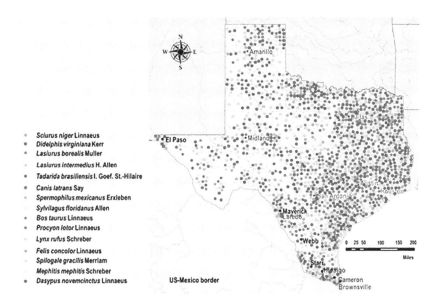

Figure 18.4 Distribution of mammals susceptible to mosquito bites in Texas. Common species representing the distribution of mammals (shown on the left side) were obtained from collection records. Unmarked areas may be the result of limited fieldwork records. (Map created with data from The Mammals of Texas-Online Edition TTU, Texas Park & Wildlife Department, and The Valley Nature Center.)

(Figure 18.5). Lambrechts, Scott, and Gubler (2010) showed that *A. albopictus* plays a relatively minor role in dengue transmission in comparison to *A. aegypti*, in part due to differences in host preferences and reduced vector competence. In addition, places where *A. albopictus* predominates over *A. aegypti* have never experienced a typical dengue epidemic with severe cases of the disease. Meta-analysis of experimental laboratory studies revealed that although *A. albopictus* is overall more susceptible to dengue virus midgut infection, rates of virus dissemination from the midgut to other tissues are significantly lower than in *A. aegypti* (Lambrechts et al. 2010).

The presence of *A. aegypti* and associated ADs on the Gulf Coast of the United States appears to have been unchanged for at least a century. In addition, there have been ecological changes, with the introduction of *A. albopictus* (Asian tiger mosquito), which also transmits dengue, West Nile, chikungunya, and Zika, and which has displaced or co-habited with *A. aegypti* in some of these areas (Figure 18.6). The etiology of ADs and their transmission by mosquitoes were deciphered in the 20th century. Nowadays, arboviruses have spread to more than 100 countries in Asia, the Pacific, the Americas, Africa, and the Caribbean, and about 2.5 billion people (40% of the world's population) live in areas where there is a risk of arboviral transmission.

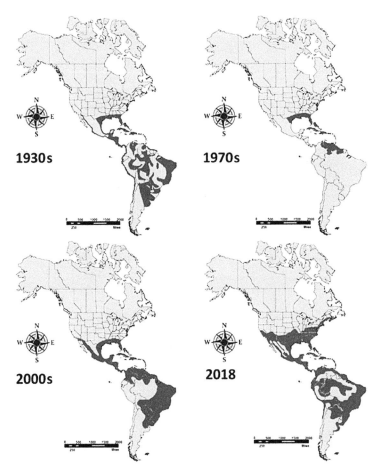

Figure 18.5 Distribution of *A. (Stegomyia) aegypti* in the Americas. The maps show the best knowledge of the geographical re-infestation process of *A. aegypti* in tropical America. The maps represent the risk of arboviral disease dispersion. Unmarked areas indicate areas where the vector has not been reported. (Maps were adapted from Gubler 1998, and were created with data from the CDC ArboNET surveillance system database, VectorMap, Linss et al. 2014; Kraemer et al. 2015; Yakob and Walker 2016; Monaghan et al. 2016; and Chuchuy et al. 2018.)

Globalization of Arboviral Diseases

The eastern Texas–Mexico border region is at risk of arbovirus endemicity due to the presence of the mosquito vectors *A. aegypti* and *A. albopictus* and the circulation of all four dengue virus serotypes, Zika virus, and chikungunya virus. Dengue virus has been detected in residents of five lower Rio Grande Valley counties of Texas, including the first case of dengue hemorrhagic fever (DHF), in Brownsville in 2005. Estimates suggest that up to 40% of the U.S. population in some border communities have been infected by dengue virus and that 110,000 to 200,000 dengue virus infections may occur annually in the southern United States along the Mexican border (Murray, Quam, and Wilder-Smith 2013).

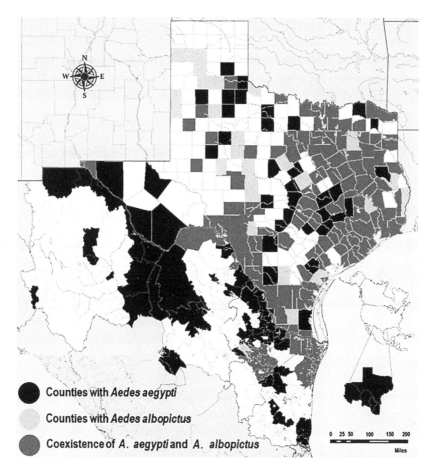

Figure 18.6 Updated reported distribution of *A. (Stegomyia)* spp. on the Texas–Mexico border. The map shows the reported occurrence of *A. aegypti* (black) and *A. albopictus* (titania grey), and their coexistence (sandstone gray) by county, representing the best current distribution of these mosquitoes based on collection records. The map probably indicates dispersion areas of dengue, and unmarked areas indicate areas where the vector has not been reported. (Map created with data from the CDC ArboNET surveillance system database, VectorMap, Texas Department of Health Services, & Texas A&M AgriLife Extension Service.)

Diagnosis

The common detectable problems in arbovirus diagnosis are the limited diagnostic resources in endemic areas and misdiagnoses because symptoms are similar to other tropical diseases. Therefore, other tropical diseases might be also misdiagnosed as dengue infection. Therefore, diagnosing the disease with the appropriate procedure is very important, since procedures should be accurate and specific, as has been described previously (Diaz-Badillo et al. 2014; Iyer et al. 2014).

Contributing to the spread of the disease are also travelers returning to their homes in a nontropical country, who might bring disease from an endemic tropical country. Without proper attention to the concept of travel medicine, dengue can be underdiagnosed, overdiagnosed, misdiagnosed, or receive delayed diagnosis (Ayukekbong et al. 2017; CDC 2001; Mallhi, Khan, and Sarriff 2016).

Human Migration and Its Consequences in Disease

The mass movement of large numbers of people creates new opportunities for the spread and establishment of infectious diseases like dengue, West Nile, chikungunya, and Zika. Furthermore, environmental changes have also been identified as having a huge impact on the emergence of certain infectious diseases, mostly in countries with high biodiversity and serious unresolved environmental, social, and economic issues (Lahiri et al. 2008). An example of a country with biodiversity is Brazil, which encloses a broad spectrum of infectious diseases of present public health importance, such as dengue, West Nile, chikungunya, Zika, Hantavirus pulmonary syndrome, leptospirosis, leishmaniasis, and Chagas. This is supported by extensive studies that revealed a relationship between infectious disease outbreaks and climate change events like El Niño, La Niña, heatwaves, droughts, floods, increased temperature, higher rainfall, and others, or environmental changes like habitat fragmentation, deforestation, urbanization, bushmeat consumption, and others. To avoid or control outbreaks, integrated surveillance systems and effective outreach programs are essential (Nava, Shimabukuro, and Chmura 2017).

Africa is also a continent with a high biodiversity, and the main source of most of the mosquito-borne viruses of medical importance that currently constitute serious global public health threats. Also, several other viruses have the potential for international challenge due to increased human population growth coupled with increased international travel and trade that likely may sustain and increase the threat of further geographical spread of current and new arboviral disease. Kim, Tridane, and Chang (2016) developed a mathematical model of human mobility screening in and out of an endemic country (Nava et al. 2017).

Conclusion

It is critical that the world community understands and responds in more coordinated ways to migration and the associated infectious disease risk associated with globalization. Multidisciplinary research from anthropology, demography, sociology, law, political science, psychology, policy analysis, public health, climate change, and epidemiology is required to identify the complex relationships between migration and health, since human mobility continues to increase, mostly from South to North America. This is also very important since, transport

mosquito vectors that may carry different virus and parasites that are very important in human infection diseases like dengue, West Nile, chikungunya, or Zika.
International globalization from urban communities, travel, immigration, and
medical mission trips are aspects of the burgeoning number of mosquito-borne
illnesses, which was predicted 15 years ago. As the spread of mosquito-borne illness
continues, clinicians will play a critical role in prevention, early diagnosis, treatment, and health education about AIs. The pivotal point in prevention is the ability
to recognize the symptoms in initial assessment; knowledge of management of the
disease phases will result in an appropriate treatment plan.

References

Ansari, M. Z., and R. E. Shope. 1994. "Epidemiology of arboviral infections." *Public Health Reviews* 22: 1–26.

Ayukekbong, J. A., O. G. Oyero, S. E. Nnukwu, et al. 2017. "Value of routine dengue diagnosis in endemic countries." *World Journal of Virology* 6: 9–16.

Barnett, E. D., and P. F. Walker. 2008. "Role of immigrants and migrants in emerging infectious diseases." *Medical Clinics of North America* 92: 1447–1458, xi–xii.

Beaumier, C., M. N. Garcia, and K. O. Murray. 2014. "The history of dengue in the United States and its recent emergence." *Current Tropical Medicine Reports* 1: 32–35.

Bouri, N., T. K. Sell, C. Franco, et al. 2012. "Return of epidemic dengue in the United States: Implications for the public health practitioner." *Public Health Reports* 127: 259–266.

Brathwaite-Dick, O., J. L. San Martin, R. H. Montoya, et al. 2012. "The history of dengue outbreaks in the Americas." *American Journal of Tropical Medicine and Hygiene* 87: 584–593.

Centers for Disease Control and Prevention (CDC). 2001. "Underdiagnosis of dengue--Laredo, Texas, 1999." *Morbidity and Mortality Weekly Report* 50: 57–59.

Centers for Disease Control and Prevention (CDC). 2018a. *Chikungunya Virus in the United States.* https://www.cdc.gov/chikungunya/geo/united-states.html

Centers for Disease Control and Prevention (CDC). 2018b. West Nile Virus in the United States. https://www.cdc.gov/westnile/index.html

Centers for Disease Control and Prevention (CDC). 2018c. *Zika Virus in the United States.* https://www.cdc.gov/zika/reporting/2016-case-counts.html.

Cisneros, A., A. Diaz-Badillo, G. Cruz-Martinez, et al. 2006. "Dengue 2 genotypes in the state of Oaxaca, Mexico." *Archives of Virology* 151, no. 1: 113–125.

Cisneros-Solano, A., M. M. B. Moreno-Altamirano, U. Martínez-Soriano, F. et al. 2004. "Sero-epidemiological and virological investigation of dengue infection in Oaxaca, Mexico, during 2000–2001." *WHO/SEARO Dengue Bulletin*, 28: 28–34.

de Lourdes Munoz, M., R. F. Mercado-Curiel, A. Diaz-Badillo, et al. 2013. "Gene flow pattern among *Aedes aegypti* populations in Mexico." *Journal of the American Mosquito Control Association* 29, no. 1: 1–18.

Diaz-Badillo, A., B. G. Bolling, G. Perez-Ramirez, et al. 2011. "The distribution of potential West Nile virus vectors, *Culex pipiens pipiens* and *Culex pipiens quinquefasciatus* (Diptera: Culicidae), in Mexico City." *Parasites & Vectors* 4, 70. doi:10.1186/1756-3305-4-70

Diaz-Badillo, A., M. de Lourdes Munoz, G. Perez-Ramirez, et al. 2014. "A DNA microarray-based assay to detect dual infection with two dengue virus serotypes. Sensors (Basel). 2014, 14(5): 7580–7601.

Ehrenkranz, N. J., A. K. Ventura, R. R. Cuadrado, et al. 1971. "Pandemic dengue in Caribbean countries and the southern United States--Past, present and potential problems." *New England Journal of Medicine* 285: 1460–1469.

Eisen, L., and C. G. Moore. 2013. "*Aedes (Stegomyia) aegypti* in the continental United States: A vector at the cool margin of its geographic range." *Journal of Medical Entomology* 50: 467–478.

Evans, D., and J. L. Meires. 2016. "Chikungunya virus: A rising health risk in the United States and how nurse practitioners can help address and reduce the risk." *The Journal for Nurse Practitioners* 12: 289–298.

Freedman, D. O., L. H. Weld, P. E. Kozarsky, et al. 2006. "Spectrum of disease and relation to place of exposure among ill returned travelers." *New England Journal of Medicine* 354: 119–130.

Funk, S., M. Salathe, and V. Jansen. 2010. "Modelling the influence of human behaviour on the spread of infectious diseases: A review." *Journal of the Royal Society Interface* 7: 1247–1256.

Furuya-Kanamori, L., S. Liang, G. Milinovich, et al. 2016. "Co-distribution and co-infection of Chikungunya and dengue viruses." *BMC Infectious Diseases* 16: 84.

Gardella-Garcia, C. E., G. Perez-Ramirez, J. Navarrete-Espinosa, et al. 2008. "Specific genetic markers for detecting subtypes of dengue virus serotype-2 in isolates from the states of Oaxaca and Veracruz, Mexico." *BMC Microbiology* 8: 117.

Gorrochotegui-Escalante, N., M. L. Munoz, I. Fernandez-Salas, et al. 2000. "Genetic isolation by distance among *Aedes aegypti* populations along the northeastern coast of Mexico." *American Journal of Tropical Medicine and Hygiene* 62, no. 2: 200–209.

Gubler, D. J. 1998. "Resurgent vector-borne diseases as a global health problem." *Emerging Infectious Diseases* 4: 442–450.

Gyawali, N., R. S. Bradbury, and A. W. Taylor-Robinson. 2016. "The global spread of Zika virus: Is public and media concern justified in regions currently unaffected?" *Infectious Diseases of Poverty* 5: 37.

Hafkin, B., J. E. Kaplan, C. Reed, et al. 1982. "Reintroduction of dengue fever into the continental United States. I. Dengue surveillance in Texas, 1980." *American Journal of Tropical Medicine and Hygiene* 31: 1222–1228.

Hotez, P. J. 2018. "The rise of neglected tropical diseases in the 'new Texas.'" *PLOS Neglected Tropical Diseases* 12: e0005581.

Hotez, P. J., and K. O. Murray. 2017. "Dengue, West Nile virus, Chikungunya, Zika—and now Mayaro?" *PLOS Neglected Tropical Diseases* 11: e0005462.

Imperato, P. J. 2016. "The convergence of a virus, mosquitoes, and human travel in globalizing the Zika epidemic." *Journal of Community Health* 41: 674–679.

Instituto Nacional de Estadistica y Geografia (INEGI) [National Institute of Statistic and Geography (Institute of the Mexican Federal Government)]. 2017. Panorama Sociodemográfico de las Entidades Federativas de Mexico. Principales resultados de la Encuesta Intercensal [Sociodemographic Panorama of the Federative Entities of Mexico. Main results of the Intercensal Survey.]. http://www.beta.inegi.org.mx/

Iyer, M. A., G. Oza, S. Velumani, et al. 2014. "Scanning fluorescence-based ultrasensitive detection of dengue viral DNA on ZnO thin films." *Sensors and Actuators B: Chemical* 202: 1338–1348.

Kim, S., A. Tridane, and D. E. Chang. 2016. "Human migrations and mosquito-borne diseases in Africa." *Mathematical Population Studies* 23: 123–146.

Kraemer, M. U., M. E. Sinka, K. A. Duda, et al. 2015. "The global distribution of the arbovirus vectors *Aedes aegypti* and *Ae. albopictus.*" *Elife* 4: e08347.

Kuno, G., J. S. Mackenzie, S. Junglen, et al. 2017. "Vertebrate reservoirs of arboviruses: Myth, synonym of amplifier, or reality?" *Viruses* 9.

Lahiri, M., D. Fisher, and P. A. Tambyah. 2008. "Dengue mortality: Reassessing the risks in transition countries." *Transactions of the Royal Society of Tropical Medicine and Hygiene* 102: 1011–1016.

Lambrechts, L., T. W. Scott, and D. J. Gubler. 2010. "Consequences of the expanding global distribution of *Aedes albopictus* for dengue virus transmission." *PLOS Neglected Tropical Diseases* 4: e646.

Liang, G., X. Gao, and E. A. Gould. 2015. "Factors responsible for the emergence of arboviruses; Strategies, challenges and limitations for their control." *Emerging Microbes & Infections* 4: e18.

Lindahl, J. F., and D. Grace. 2015. "The consequences of human actions on risks for infectious diseases: A review." *Infection Ecology & Epidemiology* 5: 30048.

Linss, J. G., L. P. Brito, G. A. Garcia, et al. 2014. "Distribution and dissemination of the Val1016Ile and Phe1534Cys Kdr mutations in *Aedes aegypti* Brazilian natural populations." *Parasites & Vectors* 7: 25.

Lorenz, C., T. S. Azevedo, F. Virginio, et al. 2017. "Impact of environmental factors on neglected emerging arboviral diseases." *PLOS Neglected Tropical Diseases* 11: e0005959.

Mallhi, T. H., A. H. Khan, and A. Sarriff. 2016. "Patient-related diagnostic delay in dengue: An important cause of morbidity and mortality." *Clinical Epidemiology and Global Health* 4: 200–201.

Mier, N., M. G. Ory, D. Zhan, et al. 2008. "Health-related quality of life among Mexican Americans living in colonias at the Texas–Mexico border." *Social Science & Medicine* 66: 1760–1771.

Miller-Thayer, J. 2010. "Health migration: Crossing borders for affordable health care." *Field Actions Science Reports* [Online] Special Issue 2.

Monaghan, A. J., C. W. Morin, D. F. Steinhoff, et al. 2016. "On the seasonal occurrence and abundance of the Zika virus vector mosquito *Aedes aegypti* in the contiguous United States." *PLOS Currents* 8. Doi: 10.1371/currents.outbreaks.50dfc7f46798675fc63e7d7da563da76

Murray, N. E., M. B. Quam, and A. Wilder-Smith. 2013. "Epidemiology of dengue: Past, present and future prospects." *Clinical Epidemiology* 5: 299–309.

Nava, A., J. S. Shimabukuro, and A. A. Chmura. 2017. "The impact of global environmental changes on infectious disease emergence with a focus on risks for Brazil." *ILAR Journal* 58: 393–400.

Nolan, M. S., J. Schuermann, and K. O. Murray. 2013. "West Nile virus infection among humans, Texas, USA, 2002–2011." *Emerging Infectious Diseases* 19: 137–139.

Pan American Health Organization (PAHO)/World Health Organization (WHO). 2019. *Epidemiological Alert: Mayaro Fever.* https://www.paho.org/hq/index.php?option=com_docman&view=download&category_slug=mayaro-fever-2323&alias=48374-1-may-2019-mayaro-fever-epidemiological-alert&Itemid=270&lang=en

Rivera-Ávila, R. C. 2014. "Fiebre chikungunya en México: caso confirmado y apuntes para la respuesta epidemiológica [Chikungunya fever in Mexico: confirmed case and notes for the epidemiological response]." *Salud Publica de Mexico* 56: 402–404.

Rodriguez-Morales, A. J., W. E. Villamil-Gomez, and C. Franco-Paredes. 2016. "The arboviral burden of disease caused by co-circulation and co-infection of dengue, Chikungunya and Zika in the Americas." *Travel Medicine and Infectious Disease* 14: 177–179.

Soper, F. L. 1963. "The elimination of urban yellow fever in the Americas through the eradication of *Aedes aegypti*." *American Journal of Public Health and the Nation's Health* 53: 7–16.

Soto, S. M. 2009. "Human migration and infectious diseases." *Clinical Microbiology and Infection* 15, Suppl 1: 26–28.

Secretaria de Salud Gobierno de Mexico (SSMX). 2015. Unidad de inteligencia Epidemiologica y Sanitaria. Suplemento Chikungunya [Chikungunya Supplement]. https://www.gob.mx/cms/uploads/attachment/file/25675/Suplemento_chikungunya.pdf

Secretaria de Salud Gobierno de Mexico (SSMX). 2018. Boletín Epidemiológico. Sistema Nacional de Vigilancia Epidemiológica. Sistema Único de Información. Dirección General de Epidemiología [Secretary of Health. Mexican government (SSMX). 2018. Epidemiological Bulletin. National Epidemiological Surveillance System. Unique Information System. General Direction of Epidemiology]. Retrieved from https://www.gob.mx/salud/acciones-y-programas/direccion-general-de-epidemiologia-boletin-epidemiologico

Tatem, A. J., D. J. Rogers, and S. I. Hay. 2006. "Global transport networks and infectious disease spread." *Advances in Parasitology* 62: 293–343.

Texas Department of State Health Services (TDSHS). 2016. TexasZika.org. http://texaszika.org/historicaldata.htm

Texas Department of State Health Services (TDSHSa). 2018a. *Chikungunya Updates.* https://www.dshs.texas.gov/news/releases/2016/20160531.aspx

Texas Department of State Health Services (TDSHS). 2018b. *US-Mexico Border.* Sister Cities and Population. https://www.dshs.texas.gov/borderhealth/

Texas Department of State Health Services (TDSHS). 2018c. *Zika Updates.* https://dshs.texas.gov/news/releases/2016/20161209.aspx

Texas Department of Transportation (TXDOT). 2017. *Texas Airport Directory*. Retrieved from https://www.txdot.gov/inside-txdot/division/aviation/airport-directory-list.html

Thomas, D. L., G. A. Santiago, R. Abeyta, et al. 2016. "Reemergence of dengue in southern Texas, 2013." *Emerging Infectious Diseases* 22: 1002–1007.

United Nations (UN) Department of Economic and Social Affairs, Population Division. 2017. *International Migration Report 2017*. United Nations (ST/ESA/SER.A/403).

United States Census Bureau (USCB). 2020. *United States Population Growth by Region*. Annual Population Estimates. https://www.census.gov/

Vignier, N., and O. Bouchaud. 2018. "Travel, migration and emerging infectious diseases." *Electronic Journal of the International Federation of Clinical Chemistry* 29: 175–179.

World Health Organization (WHO). 2009. *Dengue: Guidelines for Diagnosis, Treatment, Prevention and Control,* new ed., 11–12. http://www.who.int/tdr/publications/documents/dengue-diagnosis.pdf?ua=1

World Health Organization (WHO). 2018. "West Nile virus." *Fact sheets* https://www.who.int/news-room/fact-sheets/detail/west-nile-virus

Wilder-Smith A, Gubler DJ, Weaver SC, Monath TP, Heymann DL, Scott TW. Epidemic arboviral diseases: priorities for research and public health. Lancet Infect Dis. 2017 Mar;17(3):e101-e106. doi: 10.1016/S1473-3099(16)30518-7

Yakob L, Walker T. Zika virus outbreak in the Americas: the need for novel mosquito control methods. Lancet Glob Health. 2016 Mar;4(3):e148-149.

19

Major Impact of Massive Migration on Spread of *Mycobacterium tuberculosis* Strains

Igor Mokrousov

Introduction

Mycobacterium tuberculosis (*M. tuberculosis*) *sensu stricto* is an exclusively human pathogen (*M. bovis, M. pinnipedii,* and other animal ecotypes of the *M. tuberculosis* complex are beyond the scope of this chapter). The tubercle bacillus can persist in the human host as a long-term asymptomatic infection, and it is estimated that one third of the world's population has a latent tuberculosis (TB) infection. Due to lack of horizontal gene transfer, the species *M. tuberculosis* is strictly clonal; consequently, the use of evolutionarily robust markers permits the building of congruent phylogenies. Further, the use of markers with different discriminatory power permits the delineation of large-scale phylogenetic lineages, genotype families, and compact clonal clusters. It has been recognized that different phylogenetic/epidemic groups may have followed different evolutionary pathways and differ in their pathogenic capacities.

TB is an ancient human disease that most likely has plagued humans since prehistory. However, the exact dating of its first appearance remains unclear. According to long-lasting dogma, humans acquired TB from animals during cattle domestication in the Neolithic period. Evolutionary genetics reconstruction demonstrated that this was not the case, and bovine *M. tuberculosis* complex ecotypes emerged later (Brosch et al. 2002). A recent paleopathological study in the Middle Euphrates Valley from a site dated 9310–8290 cal. BC gave evidence of the most ancient paleopathological cases of human TB, predating or accompanying the emergence of animal domestication (Baker et al. 2017).

The Industrial Revolution, overcrowding, and poor social conditions led to the high TB mortality levels of 400 to 500/100,000 in Europe in the 17th to 18th centuries and in Russia in the 19th century. The list of famous persons who died from TB, from poets to kings, is long and impressive (https://www.thefamouspeople.com/tuberculosis.php).

The local population structure of *M. tuberculosis* is affected by multiple factors (strain pathobiology, host genetics, host migration, and environment, including

Igor Mokrousov, *Major Impact of Massive Migration on Spread of* Mycobacterium tuberculosis *Strains* In: *Human Migration.*
Edited by: Maria de Lourdes Muñoz-Moreno and Michael H. Crawford, Oxford University Press. © Oxford University Press 2021.
DOI: 10.1093/oso/9780190945961.003.0020

climate and social conditions) whose weights are different and remain to be clarified. The factors directly related to the host are human migration and human genetics.

Mass migration has dramatically changed human population structures in the past. Different hypotheses have been offered about the origin and dispersal of *M. tuberculosis* strains in different parts of the world. For example, the introduction of the Beijing genotype to the Cape area in South Africa resulted from the activities of the Dutch East India Company, which had actively brought slaves from East Asia since the 16th century (van Helden et al. 2002). The Mongol invasion in the 13th century and subsequent close and long interaction between Rus and Orda were speculatively a primary vehicle that brought *M. tuberculosis* Beijing genotype strains to northern Eurasia (Mokrousov et al. 2005). Analysis of *M. tuberculosis* genotypes in Madagascar demonstrated their correlation with the history of human migration across the Indian Ocean (Ferdinand et al. 2005).

However, the correlation between human migration and the spread of tuberculosis has never been straightforward. A study of TB sites across the Roman Empire suggested that TB was spread from the city of Rome not via demic diffusion but rather through the transmission of lifestyle, leading to increased urbanization and the subsequent increase of TB cases seen at archeological sites in locations with a low prevalence of TB (Eddy 2015).

In my previous work, I correlated evolutionary trajectories of *M. tuberculosis* genotypes with distant human history. In this chapter, I analyze a recent impact of human migration on the spread of epidemiologically successful *M. tuberculosis* clones that started to emerge in recent decades and even years and that represent the most widespread East-Asian and Euro-American lineages: the Beijing genotype, which is more and more globally disseminated; the Latin American Mediterranean (LAM) genotype, which is widespread in many world regions; and the underestimated Ural and NEW-1 families. I demonstrate the importance of a more critical view of some clichés regarding transmission patterns of *M. tuberculosis*.

Analysis

M. tuberculosis Genotypes Included in the Analysis

Among eight phylogenetic lineages of *M. tuberculosis*, Lineage 2 (East-Asian) and Lineage 4 (Euro-American) are most globally spread, although their distribution patterns differ.

The most well-known component of the East Asian lineage is the Beijing genotype. It was first detected at 90% among isolates in the Beijing area in North China, hence the name (van Soolingen et al. 1995). The use of high-resolution typing identified interesting clusters within the Beijing family, such as the W strain that caused serious multidrug-resistant (MDR) TB epidemics in New York City in the mid-1990s (Bifani et al. 2002), and the quite notorious Russian strain Beijing B0/

W148, the first appearance of which was noted in articles published 20 years ago (reviewed in Mokrousov 2013). B0/W148 constitutes one fourth of Beijing genotype isolates in Russia and is circulating in immigrant communities in western Europe and the United States. Based on phylogenetic and meta-analyses, it was concluded that Beijing B0/W148 was recently spread throughout the former Soviet Union (FSU) due to its strong association with drug resistance (its capacity to rapidly acquire drug-resistant mutations) and presents a successful clone of *M. tuberculosis* (Mokrousov 2013)

The Euro-American lineage encompasses several quite distant genetic families. The LAM family was first suggested in 2001 by analysis of CRISPR-spoligotyping data (Sola et al. 2001). LAM is prevalent in the Americas (up to 50% in South America), some parts of Europe, and Africa. It is the second important *M. tuberculosis* family in Russia (up to 30% in central Russia). The LAM family is subdivided into large sublineages based on unique genomic deletions. One of them, the RD-Rio sublineage, was first found in Rio de Janeiro, Brazil, and attracted special interest due to the unusually large size of its name-giving genomic deletion (> 23 kb), its MDR properties, and its purported global spread due to special pathogenic properties (Gibson et al. 2008).

The next two families in Lineage 4 have been an issue of controversy and persisting dogmatism. In spite of similar spoligotype profiles (see below), Ural and NEW-1 are different genetic families, as is evident from the use of evolutionarily robust markers (reviewed in Mokrousov et al. 2017). The Ural family is found at a moderate prevalence rate (7%–15%) and has a limited circulation, mainly in northern Eurasia. Until recently, it was regarded as a low-virulence genotype. Previously, I placed its origin in the North Pontic area (Mokrousov 2015). The NEW-1 family is circulating mainly in Iran; its distribution pattern was speculatively correlated with the Iranian language continuum, and to some extent, with interaction within the Islamic world (Mokrousov et al. 2017).

The Volume of Migration and Spread of *M. tuberculosis* Strains

Beijing Genotype and Impact of Migration

Indeed, the Beijing genotype is a globally distributed family, but its prevalence rates range from 60% to 90% in China to 40% to 60% in Russia and 70% in Kazakhstan (Figure 19.1). Beyond Eurasia, the Beijing genotype has a relatively higher prevalence of 20% in South Africa (van Helden et al. 2002) and 10% in Peru (Grandjean et al. 2012). Unlike in China, where the Beijing strains are not MDR and exhibit a certain level of diversity, in the FSU the Beijing genotype is MDR/XDR associated, sufficiently highly prevalent, and shows low diversity. Under principal component analysis (PCA), the Beijing populations from geographically distant Russian regions nonetheless formed a compact cluster (and the same situation is observed for

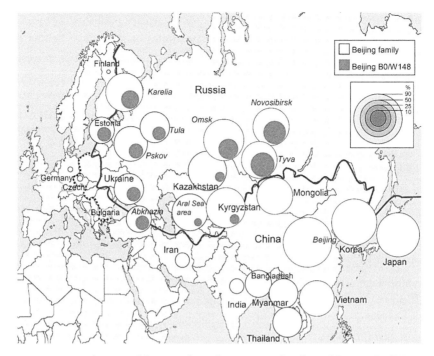

Figure 19.1 Distribution of the *M. tuberculosis* Beijing family and Beijing B0/W148 cluster in Eurasia. The solid line shows the border of the Soviet Union (USSR) in 1991. The dotted line depicts the Western border of the Warsaw Pact countries in 1989 (USSR and its allies in Europe). Free map: http://www.amaps.com/mapstoprint/ OUTLINE%20MAPS/free_map_of_world.htm

the LAM populations in the FSU; Mokrousov 2015; Mokrousov et al. 2016). This situation may be due to the rapid spread of *M. tuberculosis* strains across Russia only recently, during the 20th century. I speculate that this can be explained by mass and partly forced population mixing, especially via GULAG [https://en.wikipedia. org/wiki/Gulag] concentration camps. Historical estimates of the GULAG population size suggest that in 1928–1953, about 14 million prisoners passed through the system of GULAG labor camps and 4 to 5 million passed through the labor colonies (https://en.wikipedia.org/wiki/Gulag); that is, epidemiologically speaking, they lived under adverse conditions that facilitated the transmission of *M. tuberculosis*. A meta-analysis demonstrated an increased prevalence of the Beijing genotype in Russian prison settings compared to the general population ($P < 0.0001$; OR 2.35 [1.59–3.47]; Mokrousov 2015).

It has long been an issue of surprise that the Beijing genotype was not found in East Europe in local populations, in spite of close links with the Russian Empire/ Soviet Union in recent and historical past. For example, Beijing family strains were not described in different studies in Bulgaria (Valcheva et al. 2008, and references therein). In contrast, the Beijing genotype was found in up to 30% to 40% in Estonia (Krüüner et al. 2001; Vyazovaya et al. 2018). A closer look at migration patterns and demographics can inform us about the reasons underlying this difference.

Due to cultural reasons, Bulgaria had long-lasting links and human exchange with Russia (https://en.wikipedia.org/wiki/Russians_in_Bulgaria). However, the absolute number of Russians permanently residing in Bulgaria never exceeded 20,000 (i.e., 0.4% of the total population of the country). Reciprocally, Bulgarians represent 0.02% of the total permanent population of the Russian Federation. Two main groups of Bulgarians in the Soviet Union were workers and students whose maximum numbers reached 32100 in 1975 and 10900 in 1985, respectively (i.e., ~ 0.1% of the total population of the USSR; Bolgarskie rabochie v SSSR; Goskomstat SSSR). Regarding tourist influx, 580,000 and 695,000 tourists from Russia visited Bulgaria in 2013 and 2016, respectively (http://www.atorus.ru/news/press-centre/new/40519.html), which is a visible proportion of the 7.1-million population of Bulgaria.

In contrast to Bulgaria, the demographic situation in Estonia in the 20th century was heavily and dramatically influenced by interaction with Soviet Union (Figure 19.2). Following Soviet occupation in 1940 and 1945, the pre-war population structure has changed significantly and was shaped by two mutually enhancing events. Firstly, Estonia lost approximately 17.5% of its population due to deportations [http://estonianworld.com/life/estonia-remembers-the-soviet-deportations]). According to other estimates, 25% of the population were lost through deaths, deportations, and evacuations in World War II (Kangilaski and Salo 2005; https://en.wikipedia.org/wiki/Soviet_deportations_from_Estonia). Secondly, the proportion of the population of foreign origin increased, mostly due to influx from the Soviet Union, from 1% in 1934 to 34% in 1989 (Katus et al. 2000). Thus, a high 30%

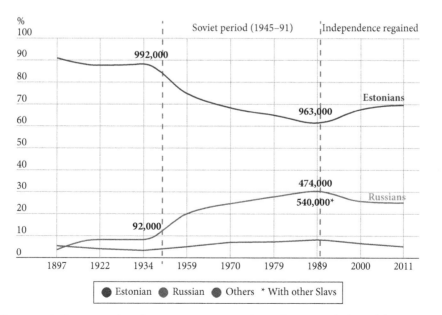

Figure 19.2 Demographic changes in Estonia in the 20th century (adapted from https://commons.wikimedia.org/wiki/File:Estonia_ethnic_makeup.png).

prevalence of the Beijing genotype in the mid-1990s (and consequently, high MDR-TB rate) in Estonia can be linked to massive immigration from Russia in 1945 to 1990. In neighboring Finland, a country that remained independent since 1918 and without any significant immigration, the Beijing strains are found in only 1% of Finnish-born (Smit et al. 2013).

The role of the migration volume and direction can also be seen in the example of the Russian successful strain Beijing B0/W148. Its geographic distribution pattern across FSU is quite peculiar from the highest prevalence in Siberia (~ 20%) to somewhat smaller circulation in the European part of FSU and negligible presence in former Soviet Central Asia (2%–3%; Figure 19.1). Based on the phylogenetic and phylogeographic analysis, I have previously proposed that West Siberia could be a hypothetical area of origin of B0/W148 whereas its primary dispersal started only after the 1950s (otherwise, it would have been found similarly prevalent across all FSU). A remarkable population outflow from Siberia, especially West Siberia, due to social and economic reasons took place in the 1960s to 1980s. It was directed to the European (but not Asian) part of the former Soviet Union and 800,000 migrated from West Siberia to European Russia in 1961 to 1970 (Ball and Demko 1978; Perevedentsev 1965, 1989). This may explain why the B0/W148 strain remains low-prevalent in Central Asia that never received any significant human influx from Siberia nor the European part of Russia (after the mid-1960s).

The above-hypothesized migration routes related to the dispersal of the Beijing genotype and B0/W148-cluster are presented in Figure 19.3 (along with prototype spoligotyping profiles of the four *M. tuberculosis* genotypes discussed in this chapter).

Latin American Mediterranean (LAM) Genotype and Impact of Migration

LAM strains present the main component of the local *M. tuberculosis* populations in South America and many parts of Europe, including the European part of Russia. LAM prevalence is decreasing toward Siberia and is negligible in China (Figure 19.4).

Although the location of origin of the LAM RD-Rio sublineage is unclear (Perdigão et al. 2018), its primary dispersal took place in Brazil where LAM on a whole constitutes 50% to 60% of the total *M. tuberculosis* population and RD-Rio strains make up to 50% of the LAM strains. A multicenter study identified these isolates at high prevalence in South America, in different parts of the world and postulated a global spread of RD-Rio due to its special pathogenic properties (Gibson et al. 2008). In reality, the situation is more complex. A comparison of the distant settings secondary with regard to the primary focus of RD-Rio (Brazil) permits to assess the role of migration in the spread of the LAM RD-Rio strains (Figure 19.3 and Figure 19.4). In the USA, and in particular, in New York City,

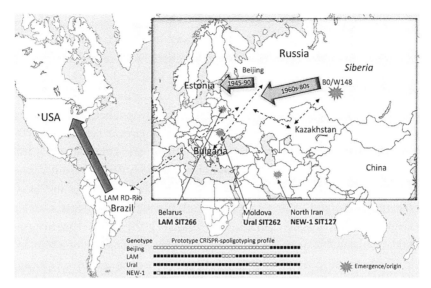

Figure 19.3 Major and minor migration routes discussed in this chapter in the context of the phylogeography of particular *M. tuberculosis* strains and lineages.

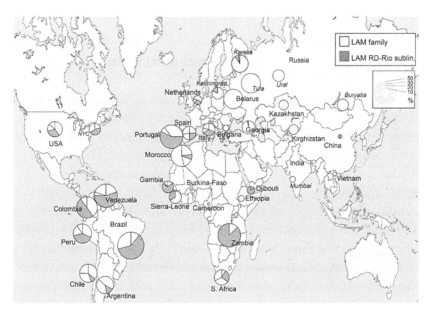

Figure 19.4 Global distribution of the *M. tuberculosis* LAM family and its RD-Rio sublineage. Free map: http://www.amaps.com/mapstoprint/OUTLINE%20MAPS/free_map_of_world.htm

RD-Rio strains constitute one-third of the LAM population (which in its turn makes 15-20% of all *M. tuberculosis* population). In Russia, LAM on a whole is the second most prevalent *M. tuberculosis* family (35% in central Russia, 10% in Siberia) but RD-Rio isolates were described only in sporadic cases, and, interestingly, mainly in the northwestern Russia areas bordering the European Union.

As for the human exchange and demographic situation, below are numbers of migrants from South American countries with a high prevalence of LAM RD-Rio in Russia and the United States (https://www.iom.int/world-migration). A total of 433 migrants from Brazil, 206 from Colombia, 277 from Peru, 98 from Venezuela live in Russia. In contrast, 348,000 migrants from Brazil, 688,000 from Colombia, 442,000 from Peru, and 197,000 from Venezuela live in the United States.

Therefore, it seems that the claimed global spread of LAM RD-Rio (due to its special properties) is defined by mass migration rather than the pathobiology of strains. In case of a low level of immigration, the RD-Rio isolates are rare encounters in such locations.

Summing up the above analysis, the following conclusion can be made. Ordinary human exchange/travel is not enough to bring and settle down a new *M. tuberculosis* strain in an indigenous human population. In contrast, a *massive* influx of migrants will change dramatically the population structure (both human and the pathogen's).

Emerging Strain: A Conditional Biohazard

Russian MDR-TB epidemics is a recognized problem for Russia and all FSU countries with postulated adverse global impact, especially on the countries receiving their immigrants. In reality, the situation is more complex as I show below on the examples of Spain and Greece.

A valuable study by Pérez-Lago et al. (2016) demonstrated the importance and capacity of development and implementation of rapid molecular tests to trace clonal transmission of isolates resulting from an index case. However, that study had an additional and unnoticed insight. In brief, in September 2015, in Almería, Southeast Spain, two cases of XDR-TB (extensively drug-resistant tuberculosis) were diagnosed. Both cases were Russian immigrants, female commercial sex workers, who had lived in Almería for 2 and 3 years. Both strains were Beijing genotype and the epidemiological alarm of potential extremely high risk of transmission was activated. WGS (whole genome sequencing) and bioinformatics analysis assigned the strains to the two major Beijing groups: Russian successful B0/W148 and Central Asian/Russian 94-32 clusters. Strain-specific SNPs were identified and PCR assays for their screening were developed. A prospective survey was performed on all new cases recovered in Almeria since November 2015 but no secondary cases resulting from exposure to those imported XDR cases had been found as of the end of 2017 (Dario Garcia de Viedma, pers. comm.). In other words, no traces of active transmission of these two strains to locals were found.

The second example is Greece, where the immigrant population (including Greeks repatriated from the FSU) in 2015 was 11.34% of the total resident population (https://www.iom.int/world-migration). A comprehensive study of all MDR-TB cases in Greece demonstrated that MDR-TB in Greece is an "imported" problem, mainly from the FSU, and due to Beijing and LAM strains (neither genotype is endemic in the Eastern Mediterranean; Ioannidis et al. 2017). In particular, the Beijing genotype represented 40% of the MDR/XDR strains, mostly from patients from the FSU. The major variable number tandem repeat (VNTR) clusters were Beijing B0/W148 and 94-32, and they included only single isolates from Greek patients, which implies only a limited transmission from immigrants to the autochthonous population.

Another example is the emerging LAM drug-resistant spoligotype SIT266 (within the LAM-RUS branch). It is genetically homogeneous (based on WGS and VNTR analysis) and geographically confined. SIT266 mainly circulates in Belarus and is associated not only with MDR-TB but also with XDR-TB. Its parental genotype SIT264 is found only in low prevalence across the larger areas from Georgia to northwestern Russia. In spite of active human exchange (but not resettlement) between Belarus and different Russian regions in the last 30 years, those MDR/XDR SIT266 isolates have been described only sporadically beyond Belarus.

Summing up the above analysis, the following conclusion can be drawn. A new emerging strain becomes *emerging* in its area of origin, where its parental strain was circulating, but it will not necessarily become immediately successful elsewhere, in a genetically distinct human population.

Imported Strain: High Prevalence at Origin Has a Visible Impact at Destination (and Its Lack Thereof)

The previous sections discuss the role of the volume of incoming human migration and, indirectly, the interaction between imported pathogen and receiving host. The examples concerned Beijing genotype (~ 50% in Russia) and LAM RD-Rio (> 50% in Brazil). As a result, the massive human influx from Russia to Estonia and from Brazil to the United States have led to a significant shift in the local *M. tuberculosis* population structures and an increasing proportion of the Beijing genotype in Estonia and LAM RD-Rio sublineage in the United States.

In this light, one extra point should be taken into consideration: how prevalent the strain is in its location of origin. The Ural genotype, as an opposite case to the Beijing and LAM families, is used illustrate this point. The Ural genotype is circulating in northern Eurasia (which loosely correlates with the FSU) and it is not especially prevalent anywhere. Initially, the family was named for the Ural region where these strains were first identified as a separate cluster (Kovalev et al. 2005). Subsequently, it became evident that the strains were more prevalent not in the Ural area in Russia, but in the areas to the north and east of the Black Sea (Mokrousov

2015). However, the percentages are low, reaching at most 15% to 19% in southern Ukraine and Georgia. As far as virulence, transmissibility, and drug resistance are concerned, the Ural strains in the studies published before 2014 were not associated with these pathobiologically relevant properties. Indeed, a particular drug-resistant clone is emerging within the Ural family in eastern Europe (Moldova, Lithuania, northwestern Russia) but only a single Ural MDR strain from Moldova was described in the above study in Greece.

Taken together, it seems that a modest initial prevalence of the Ural strains, even in the areas of their endemic circulation, is one of the key features determining its negligible presence in other world regions, even in not so distant central and Eastern Europe.

Summing up, the following auxiliary conclusion can be drawn. To be efficiently imported to a new location, a strain, even if emerging and drug-resistant, should be *sufficiently prevalent* in its country of origin.

Concluding Considerations

In summary, three interconnected conclusions about the impact of human migration and demographics on the current spread of emerging and epidemic strains of *M. tuberculosis* can be drawn.

First, ordinary human exchange/travel is not sufficient to bring and settle down a new *M. tuberculosis* strain in an indigenous human population. In contrast, a massive influx of migrants may change dramatically the population structure (both human and the pathogen's). Second, the new emerging strain can become emerging/epidemic in its area of origin, but it may remain a low-transmission pathogen in other distant, countries, perhaps due to differences in the genetic/ethnic background of the human host populations. Third, to be efficiently imported to a new location, a strain should be sufficiently prevalent in its country of origin.

In other words, and opposite to the classical view, the contagiosity of an *M. tuberculosis* strain is a conditional, not an absolute, feature. Speculatively, a kind of human resistance is developed in a local population through the population's co-existence with historical local clones and its defenses against imported clones. This emphasizes the role of host genetics, which, although inseparable from host–pathogen interactions, is beyond the scope of this chapter.

Furthermore, it will be interesting in the coming years to follow the situation with one particular strain within the Euro-American lineage that has recently been designated an "emerging resistant strain in West Asia" (Mokrousov 2016). This strain belongs to the still awkwardly named NEW-1 family and its prototype spoligotype is SIT127 (Figure 19.3). Some 20 years ago, this strain was highly prevalent only in the south of Iran, was absent in its north, and had low prevalence elsewhere, with a decreasing gradient toward northeast Iran and central Asia. Recent studies have shown that SIT127 isolates are increasingly prevalent in Iran and its eastward neighbors (Afghanistan and Pakistan) and are significantly associated with MDR-TB. According to the International Organisation of Migration, 4.8 million Afghans live abroad, and the largest host countries are Iran (2.35 million) and Pakistan (1.62 million). Current migration flows in this region in Asia, especially ongoing immigration from Afghanistan, highlight a possibility of the wider dissemination of the MDR subtype of the NEW-1 family. Soon, we will be able to observe how this migration will (or will not) influence the spread of this emerging resistant *M. tuberculosis* strain in West Asia.

References

Baker, Oussama, Berenice Chamel, Éric Coqueugniot, et al. 2017. "Prehistory of human tuberculosis: Earliest evidence from the onset of animal husbandry in the Near East." *Paléorient* 43, no. 2: 35–51.

Ball, Blaine, and George J. Demko. 1978. "Internal migration in the Soviet Union." *Economic Geography* 54, no. 2: 95–114.

Bifani, Pablo J., Barun Mathema, Natalia E. Kurepina, and Barry N. Kreiswirth. 2002. "Global dissemination of the *Mycobacterium tuberculosis* W-Beijing family strains." *Trends in Microbiology* 10, no. 1: 45–52. doi:10.1016/s0966-842x(01)02277-6

Bolgarskie rabochie v SSSR [Bulgarian workers in USSR]. http://www.maxi4.narod.ru/03_mezen/02.htm (In Russian)

Brosch, R., S. V. Gordon, M. Marmiesse, et al. 2002. "A new evolutionary scenario for the *Mycobacterium tuberculosis* complex." *PNAS* 99, no. 6: 3684–3689. doi:10.1073/pnas.052548299

Eddy, Jared J. 2015. "The ancient city of Rome, its empire, and the spread of tuberculosis in Europe." *Tuberculosis (Edinburgh)* 95, Suppl 1: S23–S28. doi:10.1016/j.tube.2015.02.005

Ferdinand, Séverine, Christophe Sola, Suzanne Chanteau, Herimanana Ramarokoto, Tiana Rasolonavalona, Voahangy Rasolofo-Razanamparany, and Nalin Rastogi. 2005. "A study of spoligotyping-defined *Mycobacterium tuberculosis* clades in relation to the origin of peopling and the demographic history in Madagascar." *Infection Genetics and Evolution* 5, no. 4: 340–348. doi:10.1016/j.meegid.2004.10.002

Gibson, A. L., R. C. Huard, N. C. Gey van Pittius, et al. 2008. "Application of sensitive and specific molecular methods to uncover global dissemination of the major RDRio sublineage of the Latin American-Mediterranean *Mycobacterium tuberculosis* spoligotype family." *Journal of Clinical Microbiology* 46, no. 4: 1259–1267. doi:10.1128/JCM.02231-07

Goskomstat SSSR. http://www.gks.ru/free_doc/new_site/perepis2010/croc/Documents/Vol4/pub-04-01.pdf (In Russian)

Grandjean, Louis, Tomotada Iwamoto, Anna Lithgow, et al. 2015. "The association between *Mycobacterium tuberculosis* genotype and drug resistance in Peru." *PLOS One* 10, no. 5: e0126271. doi:10.1371/journal.pone.0126271

Ioannidis, P., D. van Soolingen, I. Mokrousov, et al. 2017. "Multidrug-resistant/extensively drug-resistant tuberculosis in Greece: Predominance of *Mycobacterium tuberculosis* genotypes endemic in the former Soviet Union countries." *Clinical Microbiology and Infection* 23, no. 12: 1002–1004. doi:10.1016/j.cmi.2017.07.002

Kangilaski, Jaak, and Vello Salo. 2005. *The White Book: Losses Inflicted on the Estonian Nation by Occupation Regimes, 1940–1991*, 37. Tallinn: Estonian Encyclopaedia Publishers.

Katus, Kalev, Allan Puur, and Luule Sakkeus. 2000. "Immigrant Population of Estonia." In *The Demographic Characteristics of Immigrant Populations,* edited by Werner Haug, Paul Compton, and Youssef Courbage, 131–191. Strasbourg: Council of Europe Publishing.

Kovalev, S. Y., E. Y. Kamaev, M. A. Kravchenko, N. E. Kurepina, and S. N. Skorniakov. 2005. "Genetic analysis of *Mycobacterium tuberculosis* strains isolated in Ural region, Russian

Federation, by MIRU-VNTR genotyping." *International Journal of Tuberculosis and Lung Disease* 9, no. 7: 746–752.

Krüüner, Annika, Sven E. Hoffner, Heinart Sillastu, Manfred Danilovits, Klavdia Levina, Stefan B. Svenson, Solomon Ghebremichael, Tuija Koivula, and Gunilla Källenius. 2001. "Spread of drug-resistant pulmonary tuberculosis in Estonia." *Journal of Clinical Microbiology* 39, no. 9: 3339–3345. doi:10.1128/jcm.39.9.3339-3345.2001

Mokrousov, Igor. 2013. "Insights into the origin, emergence, and current spread of a successful Russian clone of *Mycobacterium tuberculosis.*" *Clinical Microbiology Review* 26, no. 2: 342–360. doi:10.1128/CMR.00087-12

Mokrousov, Igor. 2015. "*Mycobacterium tuberculosis* phylogeography in the context of human migration and pathogen's pathobiology: Insights from Beijing and Ural families." *Tuberculosis (Edinburgh)* 95, Suppl 1: S167–S176. doi:10.1016/j.tube.2015.02.031

Mokrousov, Igor. 2016. "Emerging resistant clone of *Mycobacterium tuberculosis* in West Asia." *Lancet Infectious Diseases* 16, no. 12: 1326–1327. doi:10.1016/S1473-3099(16)30460-1

Mokrousov, Igor, Ho Minh Ly, Tatiana Otten, Nguyen Ngoc Lan, Boris Vyshnevskyi, Sven Hoffner, and Olga Narvskaya. 2005. "Origin and primary dispersal of the *Mycobacterium tuberculosis* Beijing genotype: Clues from human phylogeography." *Genome Research* 15, no. 10: 1357–1364. doi:10.1101/gr.3840605

Mokrousov, Igor, Egor Shitikov, Yuriy Skiba, Sergey Kolchenko, Ekaterina Chernyaeva, and Anna Vyazovaya. 2017. "Emerging peak on the phylogeographic landscape of *Mycobacterium tuberculosis* in West Asia: Definitely smoke, likely fire." *Molecular Phylogenetics and Evolution* 116: 202–212. doi:10.1016/j.ympev.2017.09.002

Mokrousov, Igor, Anna Vyazovaya, Tomotada Iwamoto, et al. 2016. "Latin-American-Mediterranean lineage of *Mycobacterium tuberculosis*: Human traces across pathogen's phylogeography." *Molecular Phylogenetics and Evolution* 99: 133–143. doi:10.1016/j.ympev.2016.03.020

Perdigão, João, Carla Silva, Jaciara Diniz, et al. 2019. "Clonal expansion across the seas as seen through CPLP-TB database: A joint effort in cataloguing *Mycobacterium tuberculosis* genetic diversity in Portuguese-speaking countries." *Infection Genetics and Evolution* 72: 44–58. doi:10.1016/j.meegid.2018.03.011

Perevedentsev, Vladimir I. 1965. *Modern Migration of Population of West Siberia.* Novosibirsk: West-Siberian Book Publishing. (In Russian)

Perevedentsev, Vladimir I. 1989. *Who are we? How many are we?* Moscow: Mysl. (In Russian)

Pérez-Lago, Laura, Laura Pérez-Lago, Miguel Martínez-Lirola, Sergio García, Marta Herranz, Igor Mokrousov, Iñaki Comas, Llúcia Martínez-Priego, Emilio Bouza, and Darío García-de-Viedma. 2016. "Urgent implementation in a hospital setting of a strategy to rule out secondary cases caused by imported extensively drug-resistant *Mycobacterium tuberculosis* strains at diagnosis." *Journal of Clinical Microbiology* 54, no. 12: 2969–2974. doi:10.1128/JCM.01718-16

Smit, Pieter W., Marjo Haanperä, Pirre Rantala, David Couvin, Outi Lyytikäinen, Nalin Rastogi, Petri Ruutu, and Hanna Soini. 2013. "Molecular epidemiology of tuberculosis in Finland, 2008–2011." *PLOS One* 8, no. 12: e85027. doi:10.1371/journal.pone.0085027

Sola, Christophe, Ingrid Filliol, Eric Legrand, Igor Mokrousov, and Nalin Rastogi. 2001. "*Mycobacterium tuberculosis* phylogeny reconstruction based on combined numerical analysis with IS1081, IS6110, VNTR, and DR-based spoligotyping suggests the existence of two new phylogeographical clades." *Journal of Molecular Evolution* 53, no. 6: 680–689. doi:10.1007/s002390010255

V'yezdnoy turizm rastet medlenno [Inbound tourism is growing slowly]. http://www.atorus.ru/news/press-centre/new/40519.html

Valcheva, Violeta, Igor Mokrousov, Olga Narvskaya, Nalin Rastogi, and Nadya Markova. 2008. "Molecular snapshot of drug-resistant and drug-susceptible *Mycobacterium tuberculosis* strains circulating in Bulgaria." *Infection Genetics and Evolution* 8, no. 5: 657–663. doi:10.1016/j.meegid.2008.06.006

van Helden, Paul D., Robin M. Warren, Thomas C. Victor, Gian van der Spuy, Madalene Richardson, and Eileen Hoal-van Helden. 2002. "Strain families of *M. tuberculosis*." *Trends in Microbiology* 10, no. 4: 167–168. doi:10.1016/s0966-842x(02)02317-x

van Soolingen, D., Qian, L., de Haas, P.E.W., Douglas, J.T., Traore, H., Portaels, F., Quing, H.Z., Enkhasaikan, D., Nymadawa, P., and van Embden, J.D. 1995. Predominance of a single genotype of Mycobacterium tuberculosis in countries of east Asia. Journal of Clinical Microbiology 33: 3234-3238.

Vyazovaya, Anna, Klavdia Levina, Viacheslav Zhuravlev, Piret Viiklepp, Marge Kütt, and Igor Mokrousov. 2018. "Emerging resistant clones of *Mycobacterium tuberculosis* in a spatiotemporal context." *Journal of Antimicrobial Chemotherapy* 73, no. 2: 325–331. doi:10.1093/jac/dkx372

Conclusion

Michael H. Crawford and Maria de Lourdes Muñoz-Moreno

This volume on human migration has a biocultural and environmental emphasis. The chapters are arranged into five topical sections, including a theoretical overview of migration from recent African origins to Asia, Siberia, and the Americas; reconstruction of migration patterns based on ancient DNA; variation in regional movements of peoples from Africa, Siberia, the Americas, and the Caribbean; insights into patterns of migration as revealed by sociocultural observations and the use of qualitative and quantitative approaches in North Africa and Central and South America; and the spread of disease accompanying migrations. The diseases discussed are tuberculosis in Euro-Asia and arboviral infections at the United States–Mexican border.

In this volume, the following methods have been applied to estimate/measure migration and its possible consequences.

Ethnographic/Historical Observations

Chapter 15 by Brent E. Metz applies historical and ethnographic observations to the causes and consequences of migration of the Ch'orti' Maya in Guatemala, Honduras, and El Salvador.

In Chapter 13, Abdelmajid Hannoum examines the historical and social impacts of the migration of Moroccan youth to Europe. He interviewed migrants from Tangier, Larache, Rabat, and Meknes and reconstructed causes and consequences of migration. Despite the great dangers and high risk of migration to Europe, many of the youthful migrants want to experience the world fully and to feel the freedom of leaving home. Hannoum explores how poverty causes migration, despite the high cost of migration that families have to subsidize.

Public and Church Records

Demographic measures of migration in the literature are often based on public or church records. Community or national censuses provide insights into population movements.

Michael H. Crawford and María de LourdesMaria de Lourdes Muñoz-Moreno, *Conclusion* In: *Human Migration*. Edited by: Maria de Lourdes Muñoz-Moreno and Michael H. Crawford, Oxford University Press. © Oxford University Press 2021. DOI: 10.1093/oso/9780190945961.003.0021

In Chapter 9, Larissa Tarskaia and colleagues utilized census records to reconstruct migration patterns for the Yakut people of the Sakha Republic.

Archaeological/Osteological Data

Archaeological excavations and generated chronologies have been used to reconstruct regional patterns of migration in Southern Mexico and Central America. Enrique Alcala-Castaneda, in Chapter 16, skillfully applies these methodologies to the Puyil cave in Tabasco, Mexico. The settlements of Zoque populations in the Sierra Tabasquena, between pre-Hispanic times and the first half of the last century, provides insight to ancient migrations (see chapter 17).

Linguistic/Genetic Data

In Chapter 8, Franz Manni and John Nerbonne reconstruct the spread of Bantu languages from Gabon to most of sub-Saharan Africa. They utilized computational linguistic methods (i.e., Levenshtein distance) to classify major linguistic groups and used paleoclimatic, archaeological, and genetic evidence to test hypotheses surrounding the rapid peopling of early Bantu farmers.

DNA in Contemporary Populations

In Chapter 7, Tatiana Karafet and her research collaborators used SNPs and Y-chromosome markers from four Samoyedic-speaking populations (Nganasan, Nentsi, and North Selkups) to examine their origins and possible migratory patterns. Selkups differ genetically from the Ngansans and Forest and Tundra Nentsi, but they share genetic similarity with the Kets.

In Chapter 11, Gómez and colleagues examine Y-chromosome diversity in Mexico and consider its implications for the history of Central America. The paternal contributions of the Europeans to the mestizo populations of Mexico are considered. The impact of the genetic makeup of the conquering Spanish army, which included Sephardic Jews, Basques, and North Africans, added to the genetic heterogeneity of the mestizos of Mexico.

Based on the sampling of 11 contemporary Aleut island populations, Crawford et al., in Chapter 2, trace the migration of Native populations from Siberia to the Americas and the islands of the Aleutian archipelago. They demonstrate the loss of genetic diversity due to population fission and stochastic processes. Utilizing 750,000 SNPs, they calculated Russian gene flow and admixture.

Ancient DNA

Dennis H. O'Rourke, Justin Tackney, and Lauren Norman (Chapter 4) examine the discordance in timing of population migration from Siberia to the Americas. N.E. Siberia in the Yana-Indigurka plain was occupied 32 kya. They hypothesize a potential bias in molecular dating was due to undersampling of North American populations and the use of nonstable population histories. The latest archaeological and paleoecological data from the Last Glacial Maximum (LGM) brings closer concordance between the molecular and archaeological data.

In Chapter 1, Mark Stoneking reviews genomic evidence to address the question of single versus multiple dispersal of modern humans out of Africa. At this time, the genomic data supports a single dispersal, but multiple dispersals cannot be ruled out.

Constanza de la Fuente, J. Víctor Moreno-Mayar, and Maanasa Raghavan, in Chapter 3, cast their nets widely, using populations from North to South America to reconstruct the patterns of the peopling of the Americas based on ancient genomic evidence.

In Chapter 5, Maria de Lourdes Muñoz-Moreno and her research collaborators focus their reconstruction of Maya migration on the integration of effects of maize cultivation, Maya language, climatic factors, and ancient DNA. This integrative chapter compares the historic, ecological, and linguistic data with the mtDNA hypervariable sequences.

María Teresa Navarro-Romero et al. (Chapter 6) used mtDNA sequences to analyze patterns of regional ancient migration, based on human skeletal remains from Puxcatán, Tacotalpa, Tabasco, Mexico.

Disease and Migration

In Chapter 18, Alvaro Diaz-Badillo and Maria de Lourdes Muñoz-Moreno examine the effects of migration on arboviral diseases. The arboviral diseases, such as dengue, chikungunya, Zika, and West Nile fever, are mainly transmitted by vectors and are influenced by human migration and socioeconomic and environmental factors. Therefore, the chapter analyzes human migrations to assess the effects of ethnic origin and mosquito-borne spread of these diseases. This kind of analysis can potentially aid understanding of the pathogenesis and dispersion of arboviral diseases.

Igor Mokrousov (Chapter 19) focuses on the impact of human migration and demographics on the spread of tuberculosis. He observes the importance of human migration with regard to the virulence of TB. Human travel by itself is usually not sufficient to introduce a new *M. tuberculosis* strain into an indigenous human population. However, massive migrations may alter the structure of both human and pathogen populations. New emerging strains may become epidemic in their area of origin. However, although such strains can be epidemic in their country of origin, they may have low transmission rates in other countries due to differences in the genetic background of the human host populations. In order for migrants to introduce TB strains into other populations, this disease has to be of high frequency in the parental population..

Integrative Methods

Most of the chapters in this volume consider the interactions of historical, social, and biological causes and consequences of migration. However, the chapters tend to emphasize specific approaches based on the expertise of the researchers. In several chapters, written by several collaborators, diverse disciplines are used. For example, in Chapter 10, Emily T. Norris and her colleagues examine the roles of admixture and introgression in the evolution of populations. They focus on how introgression facilitates rapid evolutionary change. The introduction of beneficial genes from ancestral populations allowed colonization of new environments. Similarly, in Chapter 12, Crawford and colleagues document the evolutionary success of the Garifuna residing in malaria-infested environments due to the introduction of hemoglobinopathies from African populations to Native American populations.

Randy David and Bartholomew Dean, in Chapter 14, provide an excellent example of research and analytical collaboration between an anthropological geneticist and a social anthropologist using historical, ethnographic, and molecular genetic data. They examine the genetic effects of rural migration and urbanization in Yurimaguas, a town of Amazonian Peru.

There is some biological evidence, such as dopamine receptor genes, novelty seeking and personalities of migrants to suggest that humans are hard-wired to migrate. Migration in most animal species is genetically based. Humans share similar neurotransmitters, novelty seeking behavior. We can extrapolate behavioral and genetic bases for migration throughout the animal kingdom. For example, Monarch butterflies migrate on a yearly basis from the United States 4,000 km south to Mexico. Butterfly orientation is linked to molecular genetic evidence associated with 40JH genes that provide internal "sun compasses" (Zhu et al. 2008).

Why did early hominins migrate? Was this migration driven by population size increases, necessitating the search for additional sources of food? Chapter 2 on the Aleuts (Unangan) suggests optimal foraging theory applies in the colonization of the Aleutian Islands (Winterhalder and Smith 1981). Modern humans are extremely curious about their surrounding environments and share genetic variation of dopamine receptor genes (DRD4/7R+) associated with novelty-seeking (Ebstein et al. 1996).

Why do contemporary human populations migrate? To date, several sociocultural factors have been implicated in decision-making about migration:

1. Push-pull phenomena—individuals often migrate seeking economic opportunities. In Chapter 13, Abelmajid Hannoum documents the movement of North African youths from Morocco to Europe in search of better economic opportunities.
2. Chain migration is the movement of related individuals in a population following family members who migrated at an earlier time. Chapter 2, on the migration of Aleuts, documents the molecular genetic evidence for kin migration and its evolutionary consequences.
3. Rational choices can motivate individuals or populations to move—for example, due to religious or social discrimination. Anabaptist pacifist groups, such as Mennonites, Amish, and Hutterites, migrated from Central Europe to the Crimea, Russia, and the United States to avoid persecution and military service (Crawford 2000).
4. Life-course factors have peak influence in young adulthood and are associated with family formation. For example, in villages of Alpine Italy, young couples move from the horticultural villages to the surrounding industrial cities, Torino and Milano, in search of work (Crawford 1981).

Environmental factors are also involved in making decisions whether to migrate. The pre-historic Aleut (Unangan) populations, discussed in Chapter 2, were affected by volcanic eruptions and climate changes, such as the cooling patterns during the Holocene that caused calmer seas and facilitated migration of Aleuts between distant islands in relatively flimsy boats.

Who migrates from within a heterogeneous population?

Psychological studies suggest that certain personality traits increase the predisposition to migrate. For example, Jokela (2008 demonstrated that high openness increases migration in the United States within and between the states. High agreeableness decreases the likelihood of migration—that is, people who get along in the society are less likely to migrate. Silventoinen et al. (2008), in a study of migration from Finland to Sweden, showed that higher extraversion and neuroticism (tendency to experience negative emotional states) increase likelihood of migration. These studies suggest that personality characteristics (with genetic roots) play a major role in decisions to migrate.

Genetic roots have been linked to the incidence and patterns of migration. In the 1960s, novelty-seeking and movement of individual organisms were noted in *Drosophila*. Similarly, the dopamine reward receptor DRD4/7R+ has been found to be associated with novelty-seeking in human populations (Ebstein et al. 1996). DRD4/7R+ mutation arose approximately 45,000 years ago (Wang et al. 2004). This mutation coincides with the increased migration of humans out of Africa, and its incidence varies across populations in association with the geographic distance the group has migrated, $r = 0.85$; $n = 39$ (Chen et al. 1999). Ding et al. (2002) described evidence of positive selection acting at the human dopamine receptor D4 gene locus.

This volume demonstrates the complexity of human migration, and its study requires input and integration from several disciplines and an assortment of research specialists. Focusing only on genetics, on history, or strictly on social anthropology provides a partial picture of the causes and consequences of migration.

References

Chen, C. S., M. Burton, E. Greenberger, et al. 1999. "Population migration and the variation of dopamine D4 receptor (DRD4) allelic frequencies allelic frequencies around the globe." *Evolution and Human Behavior* 20, no. 5: 309–324.

Crawford, M. H. 1980. "The Breakdown of Reproductive Isolation in an Alpine Genetic Isolate: Acceglio, Italy." In *Population Structure and Genetic Diseases,* edited by A.W. Ericksson, H.R. Forsius, H.R. Nevanlinna, R.K. Norio, and P.L. Workman, 57–71. Academic Press. London, U.K.

Crawford, M. H., ed. 2000. *Different Seasons: Biological Aging of Mennonites of the Midwest.* Anthropology Series 21, University of Kansas Press, Lawrence, Kansas.

Ding, Y. C., H. Chung Chi, D. L. Grady, et al. 2002. "Evidence of positive selection acting at the human dopamine receptor D4 gene locus." *PNAS* 99, no. 1: 309–314.

Ebstein, R. P., O. Novick, R. Umansky, et al. 1996. "Dopamine D4 receptor (D4DR) exon III polymorphism associated with the human personality trait of novelty seeking." *Nature Genetics* 12: 78–82.

Jokela, M. 2009. "Personality predicts migration within U.S. states." *Journal of Research in Personality.* 43(1): 79-83. doi:10.1016

Silventoinen, K., N. Hammar, E. Hedlund, et al. 2008. "Selective international migration by social position, health behavior and personality." *European Journal of Public Health* 18, no. 2: 150–155.

Wang, Y. C., Y. C. Ding, P. Floodman, et al. 2004. "The genetic architecture of selection at the human dopamine receptor D4 (*DRD4*) gene locus." *American Journal of Human Genetics* 74, no. 5: 931–944.

Winterhalder, B., and E. A. Smith, eds. 1981. *Hunter and Gatherer Foraging Strategies: Ethnographic and Archaeological Analyses.* Chicago: University of Chicago Press.

Zhu, H., A. Casselman, and S. M. Rapport. 2008. "Chasing migration genes: A brain expressed sequence tag resource for summer and migratory monarch butterflies (*Danaus plexippus*)." *PLOS One* 3, no. 1: e1345.

Index